24th International Laser Physics Workshop (LPHYS'15)

AA005908

Journal of Physics: Conference Series Volume 691

Shanghai, China
21 - 25 August 2015

ISBN: 978-1-5108-2184-2
ISSN: 1742-6588

Printed from e-media with permission by:

Curran Associates, Inc.
57 Morehouse Lane
Red Hook, NY 12571

CURRAN ASSOCIATES INC. proceedings.com

Some format issues inherent in the e-media version may also appear in this print version.

Printed by Curran Associates, Inc. (2016)

For permission requests, please contact the Institute of Physics
at the address below.

Institute of Physics
Dirac House, Temple Back
Bristol BS1 6BE UK

Phone: 44 1 17 929 7481
Fax: 44 1 17 920 0979

techtracking@iop.org

Additional copies of this publication are available from:

Curran Associates, Inc.
57 Morehouse Lane
Red Hook, NY 12571 USA
Phone: 845-758-0400
Fax: 845-758-2633
Email: curran@proceedings.com
Web: www.proceedings.com

Table of contents

Volume 691

2016

24th International Laser Physics Workshop (LPHYS'15)
21–25 August 2015, Shanghai, China

Accepted papers received: 2 February 2016
Published online: 7 March 2016
Preface

Papers

Modern Trends in Laser Physics

Strong Field & Attosecond Physics

Biophotonics

Physics of Cold Trapped Atoms

Journal of Physics: Conference Series **691** (2016) 011001 doi:10.1088/1742-6596/691/1/011001

Dear Readers,

The 24th annual International Laser Physics Workshop, LPHYS'15, took place in the City of Shanghai, China. 357 participants from 29 countries attended the conference. It was hosted by the Shanghai Institute of Optics and Fine Mechanics, Chinese Academy of Sciences.

The LPHYS'15 Steering Committee and the Advisory & Program Committee would like to extend their sincere gratitude to Professor Ruxin Li (Co-Chair), Professor Heping Zeng, Professor Zhenrong Sun (Co-Chairs of the Local Organizing Committee) and to their team for the outstanding job performed in organizing, arranging, managing and putting in order the conference. Their combined efforts lead to a successful result.

In this volume of the IOP *Journal of Physics: Conference Series* you will find selected proceedings of the Workshop in Shanghai.

Please make a note that the 25rd annual International Laser Physics Workshop (LPHYS'16) will take place from July 11 to July 15, 2016 in the city of Yerevan, Republic of Armenia hosted by Yerevan Physics Institute, A Alikhanyan National Laboratory.

With kind regards,

Steering and Advisory & Program committees

LPHYS'15

Journal of Physics: Conference Series **691** (2016) 011002 doi:10.1088/1742-6596/691/1/011002

Peer review statement

All papers published in this volume of *Journal of Physics: Conference Series* have been peer reviewed through processes administered by the proceedings Editors. Reviews were conducted by expert referees to the professional and scientific standards expected of a proceedings journal published by IOP Publishing.

Raman Spectroscopy Evaluation of Polyvinylchloride Structure

K A Prokhorov[1], D A Aleksandrova[1], E A Sagitova[1], G Yu Nikolaeva[1], T V Vlasova[1], P P Pashinin[1], C A Jones[2] and S J Shilton[2]

[1]Prokhorov General Physics Institute, Russian Academy of Sciences, Vavilov St. 38, 119991 Moscow, Russia
[2]Department of Chemical and Process Engineering, University of Strathclyde, 16 Richmond Street, Glasgow G1 1XQ. Scotland, UK

E-mail: nikolaeva@kapella.gpi.ru

Abstract. In this work we investigated Raman spectra of a number of industrial grades of polyvinylchloride powder and films, prepared from solutions in tetrahydrofuran and acetophenone. The number and spectral characteristics of the Raman lines in the spectral regions of the C-Cl stretching vibrations and in the region of the C-H and CH_2 stretching vibrations were evaluated.

1. Introduction

Polyvinylchloride (PVC) is one of the most widely used commercial polymers, and the field of its application continues to expand rapidly. Effective development of state-of-art functional materials, such as PVC-based nanocomposites, copolymers, blends and fabrics, subjected to special kinds of treatment, requires elaboration of new, more informative approaches in PVC structure evaluation.

It is well known that PVC macromolecules can be found in two ordered configurations – syndiotactic and isotactic. Macromolecules of syndiotactic PVC in *trans*-conformation can form the orthorhombic crystalline lattice with the unit cell, containing two monomer units from each of two macromolecules, passing through the unit cell. The amorphous phase of PVC consists of macromolecules in atactic configuration and macromolecules with segments in syndiotactic and isotactic configurations in various conformational states.

Industrial PVC is characterized by variety of configurational and conformational macromolecular states and defects and, as a consequence, by very low degree of crystallinity. Thus, the best way to evaluate PVC structure is to apply vibrational spectroscopy, the unique feature of which is very high informativity relative to different configurational and conformational states of macromolecules.

Raman spectroscopy is a powerful method to analyze polymer structure in terms of chemical and phase compositions, contents of different configurational and conformational states, and orientational order of macromolecules. Most of Raman spectroscopic methods of PVC structure analysis in terms of the phase composition, configurational and conformational order [1 - 13], macromolecular orientation [14 - 18] and the degree of degradation/dehydrochlorination [19 - 26] were published long time ago and were based on the spectral analysis using Lorentz shape of the lines. Therefore, re-examination of the already existing methods and development of modern computer-based approaches with emphasis doi:10.1088/1742-6596/691/1/012001

on exact evaluation of number and spectral characteristics of PVC Raman-active lines are extremely important.

Furthermore, most of the published investigations deal with the region of the C-Cl stretching vibrations (500 – 800 cm^{-1}) and do not consider other PVC lines, which can be also useful for structural analysis. For example, the most intense PVC lines are observed in the region of the C-H and CH_2 stretching vibrations (2750 – 3100 cm^{-1}), and, therefore, this spectral range is very attractive for study of PVC samples with highly luminescent additives, such as fillers, impurities or residues of solvent or catalyst or for study of small amounts of PVC in polymer blends.

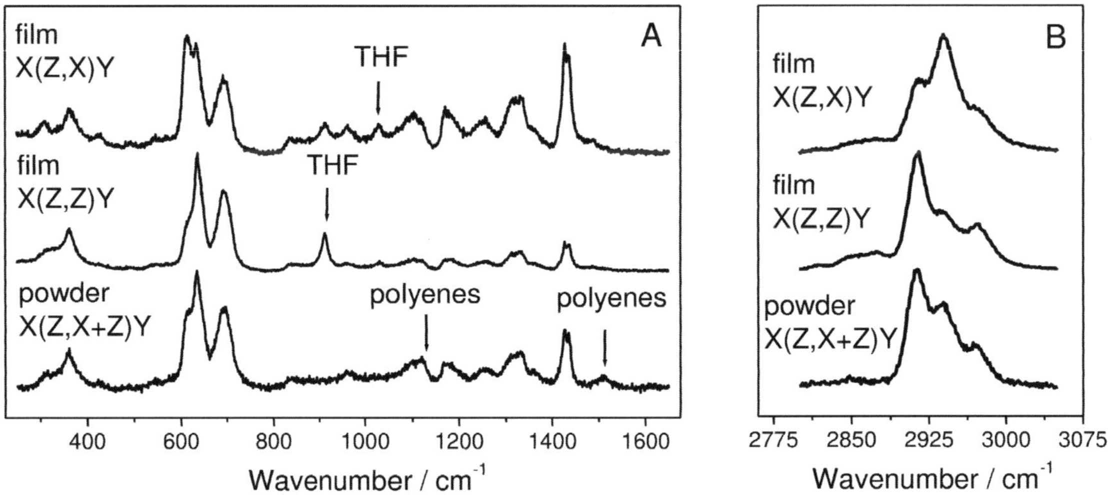

Figure 1. Raman spectra of the PVC powder (Aldrich) and the film, prepared from this powder using THF: A – spectral region 250 – 1650 cm^{-1}, B - spectral region 2800 – 3050 cm^{-1}.

The selection rules and symmetry of PVC Raman- and IR-active vibrations were calculated for an isolated molecule [3, 4] and for the orthorhombic crystalline phase [1, 2]. The A_1, A_2, B_1, and B_2 symmetry vibrations should appear in the Raman spectrum of syndiotactic PVC macromolecule in *trans*-conformation. Due to intermolecular interaction in the orthorhombic unit cell, each Raman line of the two syndiotactic molecules, passing through the unit cell, splits into two lines. Therefore, Raman spectrum of the PVC orthorhombic crystalline phase composes of A_g, B_{1g}, B_{2g}, and B_{3g} symmetry vibrations.

In this work we present an extensive study of Raman spectra of a number of industrial grades of PVC powder and investigation of polarized Raman spectra of films, prepared from solutions in tetrahydrofuran (THF) and acetophenone. Investigation of polarized Raman spectra allows determining depolarization ratios and, thus, the symmetry of Raman-active vibrations. Besides, analysis of polarized spectra is useful for determination of the number and the spectral characteristics of Raman lines, because of significant simplification of the spectra due to damping of the lines with particular symmetry at specified directions of polarization of the exciting and scattered radiation.

2. Experimental

We studied Raman spectra of three industrial grades of PVC powder - 389293 Aldrich, E-62 (emulsion polymerization), C-7059 (suspension polymerization), and spectra of thick fully-isotropic and optically-transparent films, prepared from PVC solution in THF and acetophenone by evaporation of solvent at room temperature. The film thickness (exceeding 200 μ) was enough to exclude molecular orientation, which can form in thin films due to boundary effects. In accordance with data of X-ray analysis and differential scanning calorimetry techniques, the degree of crystallinity of all the samples under investigation was less, than 5 %.

Raman setup included two lasers with 488 and 532 nm excitation lines, double monochromator, and water-cooled photomultiplier. Raman spectra were recorded in the 90°-scattering geometry with spectral resolution of 5 cm^{-1}. Spectra of the films were recorded with parallel (X(Z,Z)Y geometry) and crossed (X(Z,X)Y geometry) directions of polarization of the laser and scattered radiation. Raman spectra of powders were recorded without polarizer - X(Z,X+Z)Y geometry. The reason was strong depolarization of the laser and scattered light due to multiple reflections and refractions during scattering in the powder. Because of the optical properties of the diffraction gratings, used in the monochromator, in the absence of depolarization effects, the non-polarized spectra were similar to the spectra recorded in the X(Z,Z)Y geometry.

Spectral characteristics of the Raman lines were calculated based on the deconvolution analysis of the experimental spectra, using in-house software. The shape of each line was described by the weighted sum of Lorentz and Gauss functions:

$$y = A_0\left[\mu\frac{2}{\pi}\frac{\gamma}{4(\nu-\nu_0)^2+\gamma^2} + (1-\mu)\frac{2\sqrt{\ln 2}}{\sqrt{\pi}\gamma}\exp\left(-\frac{4\ln 2}{\gamma^2}(\nu-\nu_0)^2\right)\right]$$

where $0 \le \mu \le 1$; A_0, ν_0 and γ - the integral intensity, peak position, and full width at half maximum, respectively. The baseline was described as a second-order polynomial. All the parameters (except μ) were allowed to vary without any constrains.

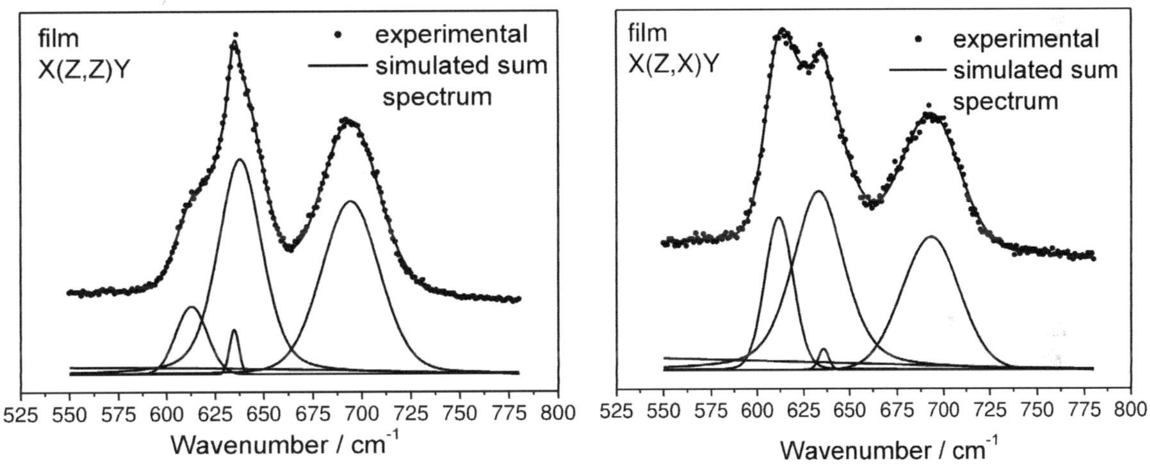

Figure 2. Examples of the deconvolution of the PVC polarized Raman spectra in the region of the C-Cl stretching vibrations: PVC film, prepared from solution of the Aldrich powder in THF, the 532 nm excitation laser line: A – X(Z,Z)Y geometry, B - X(Z,X)Y geometry.

3. Results and Discussion

As a typical example, figure 1 shows Raman spectra of the PVC industrial powder (Aldrich) and the film, prepared from THF solution of this PVC in two most informative spectral regions. We did not observe any Raman-active lines with noticeable intensity in the spectral range from 1650 to 2800 cm^{-1}. All the spectra, presented in figure 1, are recorded using the laser line at 532 nm for excitation. The film was completely isotropic and optically transparent, that allows us to record polarized Raman spectra of very high quality. For the sake of clarity, in each spectral region Raman spectra were normalized to the intensity of the most intense line after subtraction of a linear background. The Raman spectra, recorded in the X(Z,X+Z)Y and X(Z,Z)Y geometry, were normalized to the peak intensity of the line at about 633 and 2914 cm^{-1}, respectively, and the Raman spectra, recorded in the X(Z,X)Y geometry, were normalized to the peak intensity of the line at about 612 and 2940 cm^{-1}, respectively.

The first feature to be noted is obvious difference in the film spectra, recorded with parallel (X(Z,Z)Y geometry) and crossed (X(Z,X)Y geometry) directions of polarization of the laser and scattered radiation (figure 1 (A,B)). The second feature is the presence of additional lines in the film spectra compared with the powder spectrum (two most prominent of them at about 911 and 1030 cm^{-1} are marked in figure 1). We assign these lines to the Raman-active vibrations of THF, remained in the film [27, 28].

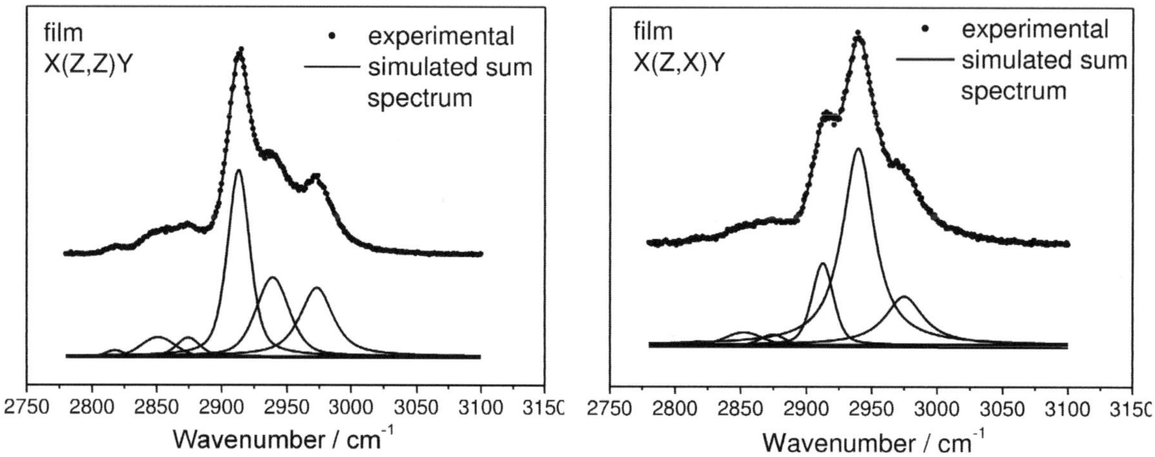

Figure 3. Examples of the deconvolution of the PVC polarized Raman spectra in the region of the C-H and CH_2 stretching vibrations: PVC film, prepared from solution of the Aldrich powder in THF, the 532 nm excitation laser line: A – X(Z,Z)Y geometry, B - X(Z,X)Y geometry.

Table 1. Spectral characteristics of the PVC Raman lines in the C-Cl stretching region.

Sample / geometry	Line N 1			Line N 2			Line N 3			Line N 4		
	ν_0 (cm^{-1})	γ (cm^{-1})	I_0[a]	ν_0 (cm^{-1})	γ (cm^{-1})	I_0	ν_0 (cm^{-1})	γ (cm^{-1})	I_0	ν_0 (cm^{-1})	γ (cm^{-1})	I_0
488 nm												
Aldrich powder / non-polarized	614	18	0.12	639	31	0.44	696	38	0.43	637	7	0.01
C-7059 powder / non-polarized	613	18	0.16	639	30	0.38	696	38	0.45	637	5	0.02
E-62 powder / non-polarized	614	19	0.15	639	29	0.39	696	39	0.45	637	7	0.01
532 nm												
Film (Aldrich/THF) / X(Z,Z)Y	615	19	0.09	640	27	0.46	696	35	0.44	637	6	0.02
Film (Aldrich/THF) / X(Z,X)Y	614	18	0.19	635	33	0.48	695	37	0.33	638	6	0.01
Aldrich powder / non-polarized	613	18	0.14	638	31	0.42	695	38	0.43	636	7	0.01

[a]I_0 - integral intensity, normalized to the total integral intensity of the Raman scattering in the C-Cl stretching region (sum of the integral intensities of four lines).

Two weak features in the Aldrich powder spectrum – the lines at about 1120 and 1510 cm^{-1} (the first is manifested only as a high-frequency shoulder of the PVC line at 1101 cm^{-1}) are related to vibrations of conjugated polyenes [19 - 26], formed in PVC due to dehydrochlorination. Due to the resonance nature of Raman scattering for polyenes, at the 532 nm line excitation, the basic contribution to the Raman scattering at 1510 cm^{-1} comes from polyenes with 15 conjugated double bonds. We did not reveal the line at 1510 cm^{-1} in the Aldrich film spectra, but the additional scattering at about 1120 cm^{-1} was clearly detected. Also, in the spectra of our samples we did not observe the lines at 2143, 2625, 3017, and 3175 cm^{-1}, which were also noticed in Raman spectrum of dehydrochlorinated PVC [20].

Comparison of the spectra of the PVC powders (389293 Aldrich, E-62, and C-7059) did not reveal any significant difference, except presence of the lines at 1120 and 1510 cm^{-1} in the Aldrich powder spectrum. Raman spectra of the films, prepared with THF and acetophenone, differ in presence of the residual THF lines in the spectrum of the first film and in presence of very intense background in the spectrum of the second film.

Table 2. Spectral characteristics of the PVC Raman lines in the region of the C-H and CH_2 stretching vibrations[b].

Sample / geometry	Line N 1			Line N 2			Line N 3		
	ν_0 (cm^{-1})	γ (cm^{-1})	I_0[c]	ν_0 (cm^{-1})	γ (cm^{-1})	I_0	ν_0 (cm^{-1})	γ (cm^{-1})	I_0
488 nm									
Aldrich powder / non-polarized	2940	32	0.41	2914	20	0.35	2974	25	0.19
C-7059 powder / non-polarized	2940	30	0.39	2914	20	0.38	2974	26	0.19
E-62 powder / non-polarized	2940	34	0.45	2914	20	0.36	2975	25	0.14
532 nm									
Film (Aldrich/THF) / X(Z,Z)Y	2941	30	0.24	2915	20	0.41	2975	31	0.25
Film (Aldrich/THF) / X(Z,X)Y	2942	29	0.62	2915	19	0.14	2977	35	0.19
Aldrich powder / non-polarized	2940	32	0.48	2914	20	0.33	2975	23	0.14

Table 2. (continuation).

Sample / geometry	Line N 4			Line N 5		
	ν_0 (cm^{-1})	γ (cm^{-1})	I_0	ν_0 (cm^{-1})	γ (cm^{-1})	I_0
488 nm						
Aldrich powder / non-polarized	2849	25	0.04	2819	10	0.01
C-7059 powder / non-polarized	2847	25	0.04	2819	9	0.01
E-62 powder / non-polarized	2852	27	0.05	2821	12	0.01
532 nm						
Film (Aldrich/THF) / X(Z,Z)Y	2853	30	0.05	2820	15	0.01
Film (Aldrich/THF) / X(Z,X)Y	2854	31	0.03	2819	13	0.003
Aldrich powder / non-polarized	2850	30	0.04	2818	19	0.01

[b]Data for the line N 6 – see text.
[c]I_0 - integral intensity, normalized to the total integral intensity of the Raman scattering in the C-H and CH_2 stretching region (sum of the integral intensities of five or six lines).

Figures 2 and 3 present examples of the deconvolution of the PVC polarized Raman spectra in the region of the C-Cl stretching vibrations (figure 2) and in the region of the C-H and CH_2 stretching vibrations (figure 3) in two scattering geometries. It was revealed that the number of the lines is equal to 4 in the region of the C-Cl stretching vibrations and to 5 or 6 in the region of the C-H and CH_2 stretching vibrations. Tables 1 and 2 summarize spectral characteristics of the PVC Raman lines in these spectral regions, calculated for the PVC powders and films for two excitation wavelengths. Data for the film, prepared using acetophenone, are not included because of low signal-to-noise ratio in this spectrum. It can be seen that the calculated spectral characteristics for the spectra of different samples are in good agreement. It should be noted that all the parameters (except μ) were allowed to vary without any constrains. In the polarized spectra of the film we observed the redistribution of the scattered intensity due to the symmetry properties of the vibrations.

In [5] it was stated that in the C-Cl stretching region of PVC spectra (500 – 800 cm^{-1}) 9 Raman-active lines (+ 4 additional lines with unknown assignment) can be found with pure Lorentz line shapes. Among them are 2 lines, attributed to syndiotactic molecules in the crystalline phase (corresponding to the C-Cl stretching vibration at 608 cm^{-1} with B_{3g} symmetry and the C-Cl stretching vibration at 638 cm^{-1} with A_g symmetry), and 7 lines, attributed to syndiotactic and isotactic molecules in different conformational states in the non-crystalline phase. Authors of the work [5] supposed that integral intensities of the lines in the C-Cl stretching region of PVC spectra can be used to calculate the degree of crystallinity and tacticity of PVC.

Table 3. Depolarization ratios $R=I_{ZZ}/I_{XZ}$ and assignment [5, 9, 13, 15] of the PVC Raman lines.

Wavenumber (cm^{-1})	R, integral intensity	R, peak intensity	Assignment
614	0.42	0.43	non-crystalline phase, syndiotactic molecules in *trans*-conformation
637 (weak)	0.10	0.09	crystalline phase, the C-Cl stretching vibration, A_g
639	0.20	0.16	non-crystalline phase, isotactic molecules in TGTG*TG conformation
696	0.14	0.15	non-crystalline phase, isotactic molecules in TGTGTG conformation
2819	0.09	0.10	the C-H stretching vibration
2851	0.12	0.11	the symmetric CH_2 stretching vibration
2877	0.08	0.09	THF
2914	0.07	0.08	the asymmetric CH_2 stretching vibration
2940	0.52	0.43	the asymmetric CH_2 stretching vibration
2975	0.15	0.12	the C-H stretching vibration

In the Raman spectra of our samples in this region we have found only 4 lines - three strong lines at 614, 639 and 696 cm^{-1} and very weak line at 637 cm^{-1}. The first three lines are clearly visible in the spectra of all the samples under study (figure 1A). The fourth line was introduced to describe additional scattering, which was observed in all the sample spectra and was not described by the model of three lines.

Because of the very low degree of crystallinity of our samples, three strong lines were assigned to the non-crystalline phase. This assumption is confirmed by relatively high values of width of these lines (table 1). The weak and relatively narrow line at 637 cm^{-1} was assigned to the orthorhombic crystalline phase.

Thus, instead of 13 lines, which were found in the C-Cl stretching region of PVC spectra in [5], we observed only four lines. It should be noted that in our deconvolution analysis the shape of each line was described by the weighted sum of Lorentz and Gauss functions, instead of pure Lorentz function, used in [5].

In [5, 6] it was stated that Raman spectra in the region of the C-H and CH_2 stretching vibrations (2800 – 3050 cm^{-1}) weakly depend on the PVC configurational and conformational composition. However, there is little information on the number and assignment of Raman-active lines in this region. Probably, it is due to the fact that analysis of this spectral range is complicated because of strong overlapping of the lines. That is why we used assignment of the IR-active PVC lines for this region (table 3, [9, 13]). This assignment is in good agreement with the results of theoretical calculation of the vibrational spectrum of the syndiotactic PVC chain in *trans*-conformation [7].

In this region we found 5 lines, corresponding to the PVC vibrations, and one THF line [27, 28]. The last line was observed only in the spectra of the film, prepared from solution of the Aldrich powder in THF. The last line has the following spectral characteristics in the X(Z,Z)Y / X(Z,X)Y spectra: peak position – 2877 / 2876 cm^{-1}, width - 21 / 21 cm^{-1}, normalized integral intensity I_0 - 0.04 / 0.02. It is important to note that all the PVC lines are clearly observed in the all sample spectra without spectrum deconvolution (figure 1B).

Using polarized Raman spectra of the isotropic and transparent film, obtained from solution of the Aldrich powder in THF, we calculated the depolarization ratios of the lines in the regions of the C-Cl stretching vibrations and the C-H and CH_2 stretching vibrations. These values as well as the assignment of the PVC lines, taken from literature, are presented in table 3.

Authors of work [6] published values of depolarization ratios of Raman lines in the C-Cl stretching region for a melt-formed isotropic PVC film. We cannot directly compare the values of the depolarization ratios, published in [6] and obtained in this work, because authors of the work [6] used 14 lines in the spectrum deconvolution analysis and, thus, the peak positions and widths of the lines are different for their and our results. However, the line at 614 cm^{-1} with width of about 18 cm^{-1} and depolarization ratio of 0.43, which was found in the spectra of our samples, most probably corresponds to the line at 615 cm^{-1} with width of about 20 cm^{-1} and depolarization ratio of 0.63. Besides, our line at 639 cm^{-1} with width of about 30 cm^{-1} and depolarization ratio of 0.16, is similar to the line at 636 cm^{-1} with width of about 25 cm^{-1} and depolarization ratio of 0.15 from work [6]. In addition, our line at 696 cm^{-1} with width of about 38 cm^{-1} and depolarization ratio of 0.15, can correspond to the line at 698 cm^{-1} with width of about 35 cm^{-1} and depolarization ratio of 0.18.

4. Conclusion

In this work we investigated Raman spectra of a number of industrial grades of PVC powder and films, prepared from solutions in tetrahydrofuran and acetophenone. Raman spectra of three grades of PVC powder, prepared by suspension polymerization and emulsion polymerization and bought from Aldrich Company, turn out to be very similar. Characteristics of the Raman lines, calculated using deconvolution analysis of two most important spectral regions of PVC Raman spectrum – the regions of the C-Cl stretching vibrations and the C-H and CH_2 stretching vibrations, were identical for all powders and films under study. We assigned the lines at 614, 639 and 696 cm^{-1} to vibrations of PVC macromolecules in the non-crystalline phase and the line at 637 cm^{-1} – to macromolecular vibrations in the orthorhombic crystalline phase.

In Raman spectra of the film, prepared from solution of the Aldrich powder in THF, we observed a number of THF lines. It is an indication of presence of residues of the solvent in the film.

Analysis of polarized spectra of the PVC films allows evaluating unambiguously the number and the spectral characteristics of the PVC Raman lines. These data are very useful for more accurate assignment of the PVC Raman lines to particular vibrations and can be applied for evaluation of structure of different PVC-based materials. Analysis of other spectral regions in PVC Raman spectrum is in progress.

Acknowledgments
This work is supported by the Russian Foundation for Basic Research (14-02-00832-a) and the Grant of the President of the Russian Federation for the Support of Leading Scientific Schools (451.2014.2).

Journal of Physics: Conference Series **691** (2016) 012001 doi:10.1088/1742-6596/691/1/012001

References

[1] Robinson M E R, Bower D I and Maddams W F 1976 *Polymer* **17(4)** 355-7
[2] Moore W H and Krimm S 1975 *Die Makromolekul. Chem. Suppl.* **1(S19751)** 491-506
[3] Koenig J L and Druesedow D 1969 *J. Polym. Sci.* A-2 **7(6)** 1075–84
[4] Krimm S 1968 *Pure Appl. Chem.* **16(2-3)** 369-87
[5] Robinson M E R, Bower D I and Maddams W F 1978 *Polymer* **19(7)** 773-84
[6] Jackson R S, Bower D I and Maddams W F 1990 *J. Polym. Sci.* B **28(6)** 837-59
[7] Rubcic A and Zerbi G 1974 *Macromolecules* **7(6)** 754-9
[8] Rubcic A and Zerbi G 1974 *Macromolecules* **7(6)** 759-67
[9] Krimm S 1960 *Adv. Polym. Sci.* **2** 120-7
[10] Scherrenberg R L, Reynaers H, Gondard C and Verluyten J-P 1993 *Macromolecules* **26(16)** 4118-21
[11] Krimm S and Liang C Y 1956 *J. Polym. Sci.* **22(100)** 95-112
[12] Wilkes C E, Folt V L and Krimm S 1973 *Macromolecules* **6(2)** 235-7
[13] Coltro L, Pitta J B and Madaleno E 2013 *Polym. Test.* **32(2)** 272–8
[14] King J, Bower D I and Maddams W F 1988 *J. Appl. Polym. Sci.* **35(3)** 787–96
[15] Bower D I, King J and Maddams W F 1981 *J. Macromol. Sci.* B **20(3)** 305-18
[16] Robinson M E R, Bower D I and Maddams W F 1978 *J. Polym. Sci.* Polym. Phys. Ed. **16(12)** 2115-38
[17] Voyiatzis G A, Andrikopoulos K S, Papatheodorou G N, Kamitsos E I, Chryssikos G D, Kapoutsis J A, Anastasiadis S H and Fytas G 2000 *Macromolecules* **33(15)** 5613-23
[18] Voyiatzis G A and Andrikopoulos K S 2002 *Appl. Spectrosc.* **56(4)** 528-35
[19] Hillemans J P H M, Colemonts C M C J, Meier R J and Kip B J 1993 *Polym. Degrad. Stabil.* **42(3)** 323-33
[20] Liebman S A, Foltz C R, Reuwer J F and Obremski R J 1971 *Macromolecules* **4(1)** 134-8
[21] Dong J, Fredericks P M and George G A 1997 *Polym. Degrad. Stabil.* **58(1-2)** 159-69
[22] Gerrard D L and Maddams W F 1981 *Macromolecules* **14(5)** 1356–62
[23] Baruya A, Gerrard D L and Maddams W F 1983 *Macromolecules* **16(4)** 578–80
[24] Bowden M, Donaldson P, Gardiner D J, Birnie J and Gerrard D L 1991 *Anal. Chem.* **63(24)** 2915–8
[25] Ellahi S, Hester R E and Williams K P J 1995 *Spectrochim. Acta* A **51(4)** 549–53
[26] Gilbert M, Ho K C, Hitt D J and Vrsaljko D 2013 *Polym. Degrad. Stabil.* **98(8)** 1537–47
[27] Ojha A K, Srivastava S K, Peica N, Schlucker S, Kiefer W and Asthana B P 2005 J. Mol. Struct. **735–736** 349–57
[28] Prasad P S R, Prasad K S and Thakur N K 2007 *Spectrochim. Acta* A **68(4)** 1096–100

Unified description of low-order above-threshold ionization on and off axis

W Becker[1] and D B Milošević[1,2,3]

[1]Max-Born-Institut, Max-Born-Str. 2a, 12489 Berlin, Germany

[2]Faculty of Science, University of Sarajevo, Zmaja od Bosne 35, 71000 Sarajevo, Bosnia and Hercegovina

[3]Academy of Sciences and Arts of Bosnia and Herzegovina, Bistrik 7, 71000 Sarajevo, Bosnia and Herzegovina

E-mail: wbecker@mbi-berlin.de,milo@bih.net.ba

Abstract. A recently developed unified description of low-order above-threshold ionization (Becker *et al* 2014 *J. Phys. B: At. Mol. Opt. Phys.* **47** 204022; 2015 *J. Phys. B: At. Mol. Opt. Phys.* **48** 151001) is revisited and extended. By considering the rescattering electron energies and angles at the classical cutoffs and the contributions of particular quantum-orbit solutions, it is shown that summing both the backward- and the forward-scattering contributions, within the low-frequency approximation, it is possible to reproduce the observed features of the ATI spectra both for low and high energies and both on and off the laser-polarization axis in the momentum plane.

1. Introduction

Above-threshold ionization (ATI) of rare-gas atoms had been considered a closed chapter of strong-field physics [1, 2, 3, 4] until, with the availability of mid-infrared lasers, unexpected structures in the low-energy velocity map (at energies well below the ponderomotive energy U_p) were observed [5]. These include a series of energy peaks along the polarization axis below $0.1\,U_\mathrm{p}$ called the low-energy structure (LES) [6, 7, 8], another structure at energies below the former referred to as the very-low-energy structure (VLES) [9, 10], a fork-like structure off the polarization axis [11], and a V structure with the 'V' opening on either side of the origin along the axis perpendicular to the laser polarization [12]. All of these structures came unexpected because they are not afforded by a simple tunneling picture. In calculations based on the time-dependent Schrödinger equation, they require the long-range Coulomb potential for their presence.

In our recent papers [13, 14, 15], for a unified description of these low-energy effects we revisited the classical simple-man model and its quantum-mechanical implementation in terms of the improved strong-field approximation (ISFA), the low-frequency approximation (LFA) [16, 17], and the quantum-orbit expansion [1, 18]. It turned out that rescattering into states with low energy, especially forward scattering, generates all of the afore-mentioned effects. We have shown that the angle-dependent energy cutoffs of the various simple-man and quantum orbits are imprinted in the calculated velocity map [11, 13, 14, 15]. The agreement with the various experimental data is good [11]. In particular, the recently observed V structure [12] is very well developed [14]. The afore-mentioned cut-off pattern is independent of the atomic binding potential. However, via the magnitude of its scattering amplitude it determines whether or not the rescattering contributions can successfully compete with the rather structureless contribution of the direct electrons.

In the present contribution, we revisit and extend some aspects of the above-mentioned results, with special emphasis on the rescattering electron energies and angles at the classical cutoffs and the contributions of particular quantum-orbit solutions to the electron spectra in the momentum plane. We show that summing both the backward- and the forward-scattering contributions within the LFA, it is possible to reproduce the observed features of the ATI spectra both for low and high energies and both on and off the polarization axis. We use the length gauge and dipole approximation and employ the atomic system of units ($\hbar = |e| = m = 4\pi\varepsilon_0 = 1$) if not stated otherwise.

2. Differential ionization rate for above-threshold ionization with rescattering

The differential ionization rate for above-threshold ionization with rescattering from a bound state ψ_i, having the ionization energy $E_i = -I_{\rm p}$, to a final state of an electron with an asymptotic momentum $\mathbf{p}$, is given by [16, 19, 20]

$$w_{\mathbf{p}i} = 2\pi p \left|T^{\rm R}_{\mathbf{p}i}\right|^2, \tag{1}$$

where the rescattering T-matrix element in the saddle-point approximation is proportional to

$$T^{\rm R}_{\mathbf{p}i} \propto \sum_s \Delta(t_{0s}, t_s)^{-1/2} \mathcal{M}_{\mathbf{p}\mathbf{k}_{\rm s}}(t_s) \langle \mathbf{k}_{\rm s} + \mathbf{A}(t_{0s}) | \mathbf{r}\cdot\mathbf{E}(t_{0s}) | \psi_i \rangle \exp[iS(t_{0s}, t_s)]. \tag{2}$$

The sum over s is the sum over the solutions of the system of the two saddle-point equations for the ionization time t_{0s} and the rescattering time t_s, viz.

$$\frac{1}{2}\left[\mathbf{k}_{\rm s} + \mathbf{A}(t_0)\right]^2 = E_i, \quad \frac{1}{2}\left[\mathbf{k}_{\rm s} + \mathbf{A}(t)\right]^2 = \frac{1}{2}\left[\mathbf{p} + \mathbf{A}(t)\right]^2, \tag{3}$$

where $\mathbf{k}_{\rm s} = \int_{t_0}^{t} dt' \mathbf{A}(t')/(t_0 - t)$ is the stationary momentum, $S(t_0, t) = \mathbf{p}^2 t/2 + \mathbf{p}\cdot\boldsymbol{\alpha}(t) + \mathbf{k}_{\rm s}^2(t - t_0)/2 - E_i t_0 + \int^{t_0} dt' \mathbf{A}^2(t')/2$ is the action, and the determinant is given by

$$\Delta(t_0, t) = \left[\frac{i(t - t_0)}{2\pi}\right]^3 \left[\left(\frac{\partial^2 S}{\partial t_0 \partial t}\right)^2 - \frac{\partial^2 S}{\partial t_0^2}\frac{\partial^2 S}{\partial t^2}\right]. \tag{4}$$

The electric field is monochromatic with frequency ω and period $T = 2\pi/\omega$ and linearly polarized in the direction determined by the unit vector $\hat{\mathbf{e}} = \hat{\mathbf{x}}$. In this case, the electric field vector is given by $\mathbf{E}(t) = \hat{\mathbf{e}} E_0 \sin\omega t$ and we have $\mathbf{A}(t) = \hat{\mathbf{e}} A_0 \cos\omega t$, with $A_0 = E_0/\omega$, and $\boldsymbol{\alpha}(t) = \hat{\mathbf{e}}(A_0/\omega)\sin\omega t$. The ponderomotive energy is defined by $U_{\rm p} = \int_0^T \mathbf{A}^2(\tau) d\tau/(2T) = A_0^2/4$.

The rescattering matrix element can be given in the first Born approximation (FBA) or in the low-frequency approximation (LFA) [16]:

$$\mathcal{M}_{\mathbf{p}\mathbf{k}_{\rm s}}(t) = \begin{cases} \langle \mathbf{p} + \mathbf{A}(t) | V_{\rm sh} | \mathbf{k}_{\rm s} + \mathbf{A}(t) \rangle & \text{(FBA)} \\ \langle \psi^{(-)}_{\mathbf{p}+\mathbf{A}(t)} | \; V \; | \mathbf{k}_{\rm s} + \mathbf{A}(t) \rangle & \text{(LFA)} \end{cases}. \tag{5}$$

Here $\psi^{(-)}_{\mathbf{p}}$ is the continuum eigenstate of the Hamiltonian $-\boldsymbol{\nabla}^2/2 + V$ with the atom-specific rescattering potential $V(r)$ modeled by a sum of the Coulomb potential and a short-range potential [16]. The expression in the first Born approximation is simpler since the terms $\mathbf{A}(t)$ cancel. We call this the improved strong-field approximation (ISFA) [20]. In this case, only the short-range part $V_{\rm sh}$ of the potential V is taken into account since the matrix element for the Coulomb potential is singular for forward scattering, i.e. $\propto (\mathbf{p}-\mathbf{k}_{\rm s})^{-2}$.

According to [13, 18, 19] there are two classes of saddle-point solutions of the system of equations (3): the backward-scattering (BS) solutions, which are classified by the multi-index (α, β, m) in [19], and the forward-scattering (FS) solutions, which are classified by the double-index (ν, μ) in [13]. The names BS and FS are related to the case where the liberated electron has its final momentum $\mathbf{p}$ in the direction of the laser polarization axis $\hat{\mathbf{e}}$. In this case, the solutions to the second equation in (3) are given by $\mathbf{p}+\mathbf{A}(t) = \pm[\mathbf{k}_{\rm s}+\mathbf{A}(t)]$. If the electron immediately after the rescattering at the time t moves in the opposite direction, i.e. if $\mathbf{p}+\mathbf{A}(t) = -[\mathbf{k}_{\rm s}+\mathbf{A}(t)]$, we have backward scattering. FS solutions correspond to the '+' sign in the above equation. They are simpler since for these solutions we have $\mathbf{k}_{\rm s} = \mathbf{p}$ and the second equation in (3) is satisfied for all values of t, while the first equation in (3) can be solved analytically over t_0 [13]. Therefore, for $\mathbf{p} \parallel \hat{\mathbf{e}}$, for a fixed value of $\mathbf{p}$, the solution for t_0 is the same for all values of ν and μ. For other electron emission angles θ ($\cos\theta = \mathbf{p}\cdot\hat{\mathbf{e}}/p$), the system of equations (3) should be solved over t_0 and t. New solutions can be easily obtained starting from the known solutions for $\theta = 0$. We keep the name FS for such solutions even for $\theta \neq 0$. On the other hand, the BS solutions $\{t_{0s}, t_s\}$ are different for each $s \equiv (\alpha, \beta, m)$. Similarly to the FS solutions, we will call BS solutions the solutions for $\theta \neq 0$ that are obtained starting from the (α, β, m) solution for $\theta = 0$.

For the BS solutions we can also use a more precise approximation in which the $\alpha = \pm 1$ solutions are combined in the uniform approximation [16, 19]. In this case, the sum over $s \equiv (\alpha, \beta, m)$ in the rescattering T-matrix element is replaced by the sum over (β, m) of the matrix elements which contain the weights and the actions in equation (2) and a linear combination of the Airy function and its first derivative.

3. Rescattering electron energies and angles at the classical cutoff

According to the LFA theory the rescattering ATI rate is determined by the elastic rescattering amplitude (5). This amplitude depends on the electron kinetic energy at the rescattering time $E_s = [\mathbf{k}_{\rm s}+\mathbf{A}(t)]^2/2 = [\mathbf{p}+\mathbf{A}(t)]^2/2$ and on the scattering angle θ_s at the instant of rescattering. For fixed final electron energy $E_{\mathbf{p}} = \mathbf{p}^2/2$ and final emission angle θ in the laboratory frame the rescattering angle is given by [16]

$$\theta_s = \begin{cases} \varphi & \text{for } k_{sx}+A_x(t) \geq 0 \\ |\varphi - \pi| & \text{for } k_{sx}+A_x(t) < 0 \end{cases}, \quad \tan\varphi = \frac{p\sin\theta}{p\cos\theta + A_x(t)}. \tag{6}$$

Let us analyze, as a function of the final electron emission angle θ, the maximum energy $E_{\mathbf{p}}$, the corresponding energy E_s, and the angle θ_s for various backward (βm) and forward (μ) scattering solutions (for fixed βm (μ) the BS (FS) solutions $\alpha = \pm 1$ ($\nu = \pm 1$) approach each other for the corresponding maximum energy $E_{\mathbf{p}}$ so that the curves presented in figure 1 do not depend on the values of α (ν)).

For the BS solutions there are two groups of orbits, which are distinguished by the index β. For all angles $0^\circ \leq \theta \leq 90^\circ$, the orbits having $\beta = -1$ have higher final energies $E_{\mathbf{p}}$, higher energies E_s upon return, and larger scattering angles θ_s than those with $\beta = 1$. Hence, in all three left-hand plots of Fig. 1 the contributions having $\beta = -1$ are located above those with $\beta = 1$. Specifically, for $\theta = 0^\circ$ the energy $E_{\mathbf{p}}$ of the $\beta = -1$ solutions monotonically decreases from $10.01\,U_{\rm p}$ for $m = 0$ to $8\,U_{\rm p}$ for $m \to \infty$, while for the $\beta = 1$ solutions it increases to $8\,U_{\rm p}$ from below. Similarly, for $\beta = -1$ the energies E_s for $\theta = 0^\circ$ decrease from $3.17\,U_{\rm p}$ for $m = 0$ to $2\,U_{\rm p}$ for $m \to \infty$, while for $\beta = 1$ they increase towards the same limit. With increasing angle θ the energies $E_{\mathbf{p}}$ decrease forming a set of almost parallel curves $[E_{\mathbf{p}}(\theta)]_{\beta m}$. For $\theta = 90^\circ$, the curves for $\beta = -1$ approach a nonzero value (equal to $1.8\,U_{\rm p}$ for $m = 0$ and lower for larger values of m), while the curves for $\beta = 1$ all approach the value $E_{\mathbf{p}} = 0$. The energy E_s remains

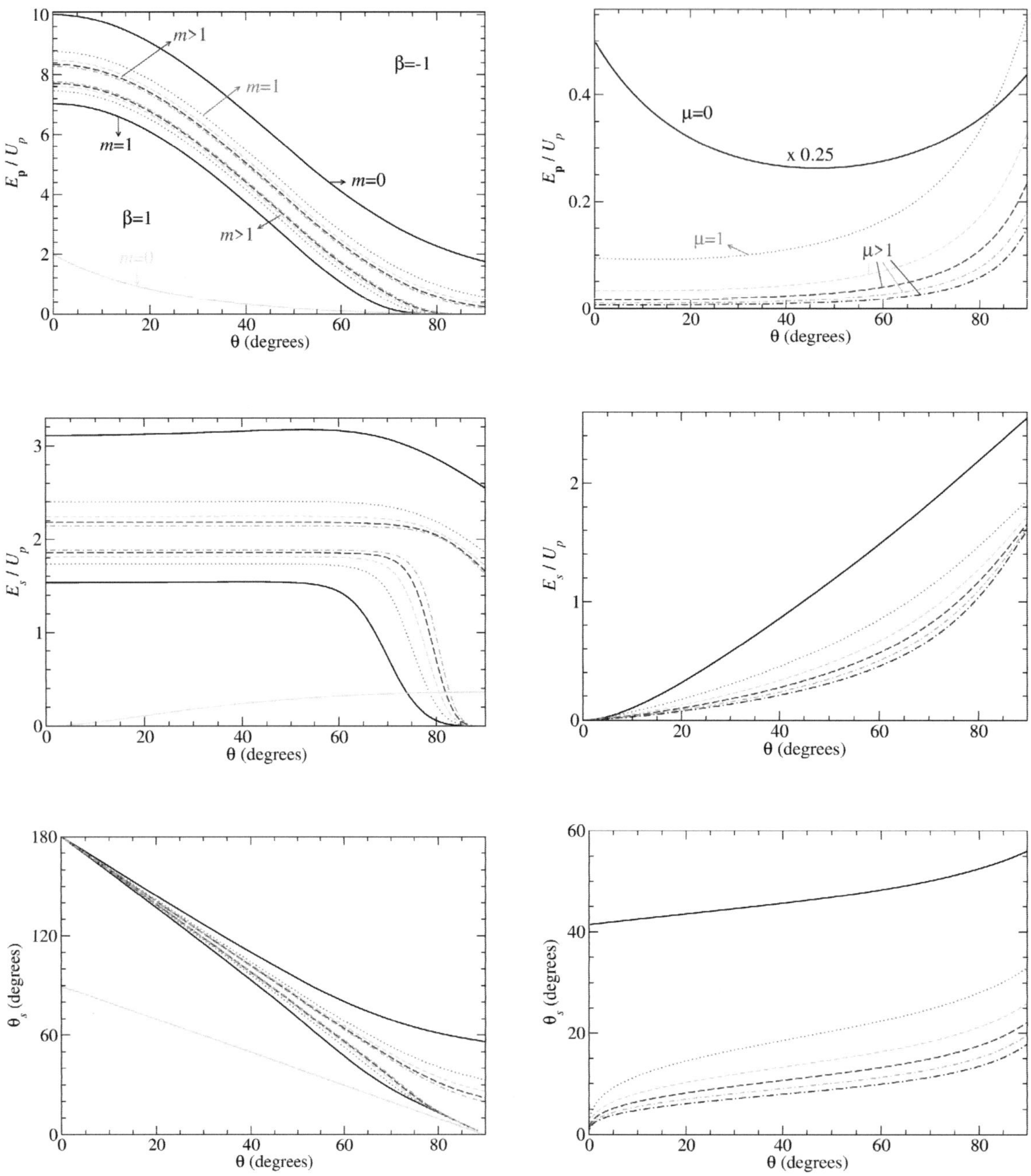

Figure 1. The scattering angle θ_s (lower panels) and the angle-dependent cutoffs of the electron energy E_s at the rescattering time (middle panels) and the final electron energy $E_{\mathbf{p}}$ (upper panels) as functions of the angle θ. The corresponding backward- and forward-scattered quantum-orbit solutions are denoted by the double index (β, m) and the index μ, respectively.

almost constant with increasing θ from zero up to $\approx 60^\circ - 70^\circ$. With a further increase of θ, all energies E_s decrease. For $\beta = -1$ the energies E_s approach nonzero values ($2.55\,U_{\mathrm{p}}$ for $m = 0$ and lower than $1.87\,U_{\mathrm{p}}$ for $m \geq 1$), while for $\beta = 1$ all energies quickly decrease to zero. For all solutions, for $\theta = 0^\circ$ the scattering angle θ_s is equal to 180°, which is in accordance with the fact that the solutions (β, m) are BS solutions (an exception is the recently discovered [18] solution $(\beta, m) = (1, 0)$, presented by a cyan solid line in the left panels of figure 1, which will be commented on at the end of this section). With increasing θ the angle θ_s decreases. For $\beta = -1$ it approaches a nonzero value (for $m = 0$ it is $\theta_s \approx 56°$ for $\theta = 90°$ while for the $m > 0$ solutions we have $\theta_s \approx 20° - 33°$). For $\beta = 1$ the angle θ_s approaches zero for $\theta \to 90°$. Very importantly, according to the Rutherford formula the Coulomb scattering cross section is very large for small values of the angle θ_s and small values of the energy E_s. Now, the orbits characterized by the double index $(\beta, m) = (1, m)$ (which approach the cutoff energy $E_{\mathbf{p}} \approx 8U_{\mathrm{p}}$ from below for increasing values of m and $\theta = 0°$) have $\theta_s \to 0°$ for $\theta \to 90°$. Hence, they are dominant for emission near right angles to the polarization axis, $\theta \to 90°$. In fact, this explains the experimentally observed outer prongs of the fork-like structures in the off-axis LES [11], as well as the V structure [12, 14].

Let us now analyze the FS orbits. In this case we have only one class of solutions, characterized by the index μ. From the right-hand panels of figure 1 we see that the cutoff energy $E_{\mathbf{p}}$ for the solution $\mu = 0$ (black solid curve; the corresponding energy is multiplied by 0.25) decreases from $2\,U_{\mathrm{p}}$ at $\theta = 0°$ to $\theta = 46°$ and then increases from $1.05\,U_{\mathrm{p}}$ to $1.75\,U_{\mathrm{p}}$ at $\theta = 90°$. The corresponding energy E_s increases from zero for $\theta = 0°$ to a large value above $2.54\,U_{\mathrm{p}}$ for $\theta = 90°$, while the corresponding angle θ_s increases from $\theta_s = 42°$ to $\theta_s = 56°$. On the other hand, for the solutions with $\mu \geq 1$ both E_s and θ_s go to zero for $\theta \to 0°$, so that the corresponding differential cross sections are very large and these solutions are responsible for the experimentally observed on-axis LES [6, 7, 8]. From the upper right panel of figure 1 one can see that these LESs appear for energies $E_{\mathbf{p}}$ below $0.1\,U_{\mathrm{p}}$. With increasing angle θ the energies $E_{\mathbf{p}}$ and E_s, as well as the angle θ_s, increase. The angle θ_s remains relatively small even for $\theta = 90°$ (for $\mu = 1, 2, 3, 4, 5, \ldots$ the corresponding values of θ_s are $32.8°, 25.9°, 22.0°, 19.6°, 17.7°, \ldots$). The final electron energy $E_{\mathbf{p}}$ (in units of U_{p}) for $\theta = 90°$ approaches relatively large values (for $\mu = 1, 2, 3, 4, 5, \ldots$ they are $0.54, 0.33, 0.23, 0.18, 0.15, \ldots$; these values are the same as for the BS solution $\beta = -1$ and $m = \mu$ so that the BS and the FS solutions merge). We note that the FS solutions with $\mu \geq 1$ are responsible for the middle prongs of the off-axis LES observed in [11].

It is surprising that for the BS (FS) solution $\beta m = 10$ $(\mu = 0)$ we found that for $\theta = 0°$ the scattering angle θ_s is different from $180°$ $(0°)$. The reason is that for these solutions classical considerations are inadequate. The classical cutoff for these solutions, $E_{\mathbf{p}} = 2\,U_{\mathrm{p}}$ [14, 15], corresponds to the travel time $\tau = t - t_0 = 0$, so that $t = t_0$ and $p + A(t) = k + A(t) = k + A(t_0) = 0$. Therefore, for $\theta = 0°$ both the nominator and the denominator of the second equation in (6) are equal to zero, so that we can obtain a nonzero value for θ_s. Quantum-mechanically we have $p + A(t) \neq 0$ so that we obtain $\theta_s = 0°$ for $\mu = 0$ and the corresponding differential cross section is large. This is important because only the solution $\mu = 0$ from the FS solutions contributes in the energy region $\approx U_{\mathrm{p}}$ and modifies the spectrum of the direct electrons. It should also be mentioned that the solution $\mu = 0$ $(\beta m = 10)$ corresponds to the sign "+" ("−") in equation (11) in [15]. For $\theta > 0$, with increasing τ, the classical solution of the saddle-point equations $E_{\mathbf{p},\mu=0}(\tau)$ [15] increases and has a θ-dependent maximum as can be seen in the lower panel of figure 2 in [15]. On the other hand, for $\theta > 0$, with increasing τ, the classical energy $E_{\mathbf{p},\beta m=10}(\tau)$ continuously decreases from a maximum value towards zero (see the upper panel of figure 2 in [15]). With increasing θ, this maximum value decreases from $E_{\mathbf{p}} = 2\,U_{\mathrm{p}}$ for $\theta = 0°$ to zero for $\theta = 90°$ (see the upper left panel of figure 1).

4. Spectra in the electron momentum plane

In experiments it is common to exhibit the electron spectra in the momentum plane $(p_\perp, p_\parallel) = (p_y, p_x) = (p \sin\theta, p \cos\theta)$. We first present the electron spectra that correspond to the BS quantum orbits. In the left (right) panels of figure 2 we show partial spectra for $\beta = -1$ $(\beta = +1)$ and for $m = 0$ $(m = 1)$, $m = 1$ $(m = 2)$, and $m = 2$ $(m = 3)$ in the upper, middle, and

Journal of Physics: Conference Series **691** (2016) 012002 doi:10.1088/1742-6596/691/1/012002

lower panel, respectively. The results presented are in accordance with the results of section 3 and with the classical analysis in figure 3 in [11] and references [14, 15]. For example, the radii of the rescattering rings of the $\beta = -1$ orbits are all larger than those of the $\beta = 1$ orbits, and the former enclose the origin while the latter all intersect exactly at the origin (the uniform approximation that is used cannot be employed close to the origin). As already mentioned, near the $p_\perp$ axis the $\beta = 1$ rings make up the outer prongs of the fork and also build the V structure. Moreover, quantum interferences of the orbits $\alpha = \pm 1$, which are taken into account in the uniform approximation, are clearly visible in each panel.

Next, in figure 3 we present the ISFA spectra obtained using the saddle-point method (equations (1)–(5) for FBA), for particular FS solutions μ. In the lower right panel we present the coherent sum of the contributions of the solutions $\mu = 0, \ldots, 7$, while in the remaining panels we display the partial contributions of the solutions $\mu = 1, 2$ and 3, as denoted. All spectra exhibit a characteristic diamond-like structure and the cutoff values in p_x and p_y decrease with increasing μ. The cutoffs in p_x are located at the LES momenta whose approximate values are $4\sqrt{U_\mathrm{p}}/[(2\mu + 1)\pi]$ [23, 24]. If we were to sum the partial contributions for $\mu = 1, 2, 3, \ldots$ incoherently, we would obtain nested diamonds (rhombi) similar to those of figure 3 in [11]. Instead, for the coherent sum presented in the lower right panel, we now have a characteristic interference pattern inside the largest diamond (the one corresponding to the solution $\mu = 1$). It should be mentioned that if we were to use the LFA instead of the ISFA the spectra would be enhanced for small values of the momenta and the angle θ, which is caused by the Coulomb effect (large differential cross section). The LES for $\mu = 1, 2$ and 3 would then stand out more distinctly.

In figure 4 we present the coherent sum of both BS and FS quantum-orbit contributions. The three prongs of the fork-like structure of the off-axis LES [11] are clearly visible. The differential cross section depends on the rescattering angle at the instant of the recollision. This dependence not only causes the Coulomb-induced enhancements in the low-energy region, it also modifies the high-energy spectra, which are described by the BS solutions. This is clearly visible in figure 4 where we see characteristic minima for particular values of the angle θ. These minima are atom-specific and similar results have already been obtained for short-wavelength lasers [16]. For a fixed value of θ, from the middle left panel of figure 1 we can read off the electron energy E_s at the rescattering time, while from the lower left panel we obtain the corresponding scattering angle θ_s. For example, we see that for $\theta \approx 20^\circ$ we have $E_s \approx 3U_\mathrm{p}$ and $\theta_s \approx 140^\circ$. For these parameters the differential cross section for elastic scattering of an electron off the xenon positive ion exhibits a minimum. The situation can be reversed in the sense that from a measured ATI electron spectrum in the momentum plane one can extract information about the differential positive-ion cross section [21] (this is particularly useful for the molecular case [22]).

5. Conclusions

In this paper we have shown in detail in figure 1 how the cutoffs of the electron energy (both for the returning electron at the rescattering time and for the final energy at the detector) and the rescattering angle at the time of rescattering depend on the lab-frame electron angle (with respect to the laser polarization axis). This was done both for back-scattering (BS) and forward-scattering (FS) quantum orbits, which are classified by the index (β, m) and μ, respectively. From this figure all aspects of the electron spectra can be extracted. Foe example, from the right panels it is evident that for the $\mu = 1, 2, \ldots$ quantum orbits both the electron energy and the rescattering angle are small for the on-axis spectra. The corresponding differential cross section for electron–parent-ion rescattering is then very large due to the Coulomb potential so that the LES clearly appears in the spectra. This is also in agreement with the soft-recollision model [23, 24] (E_s is small).

In figures 2 and 3 we have shown partial spectra for particular BS and FS quantum orbits.

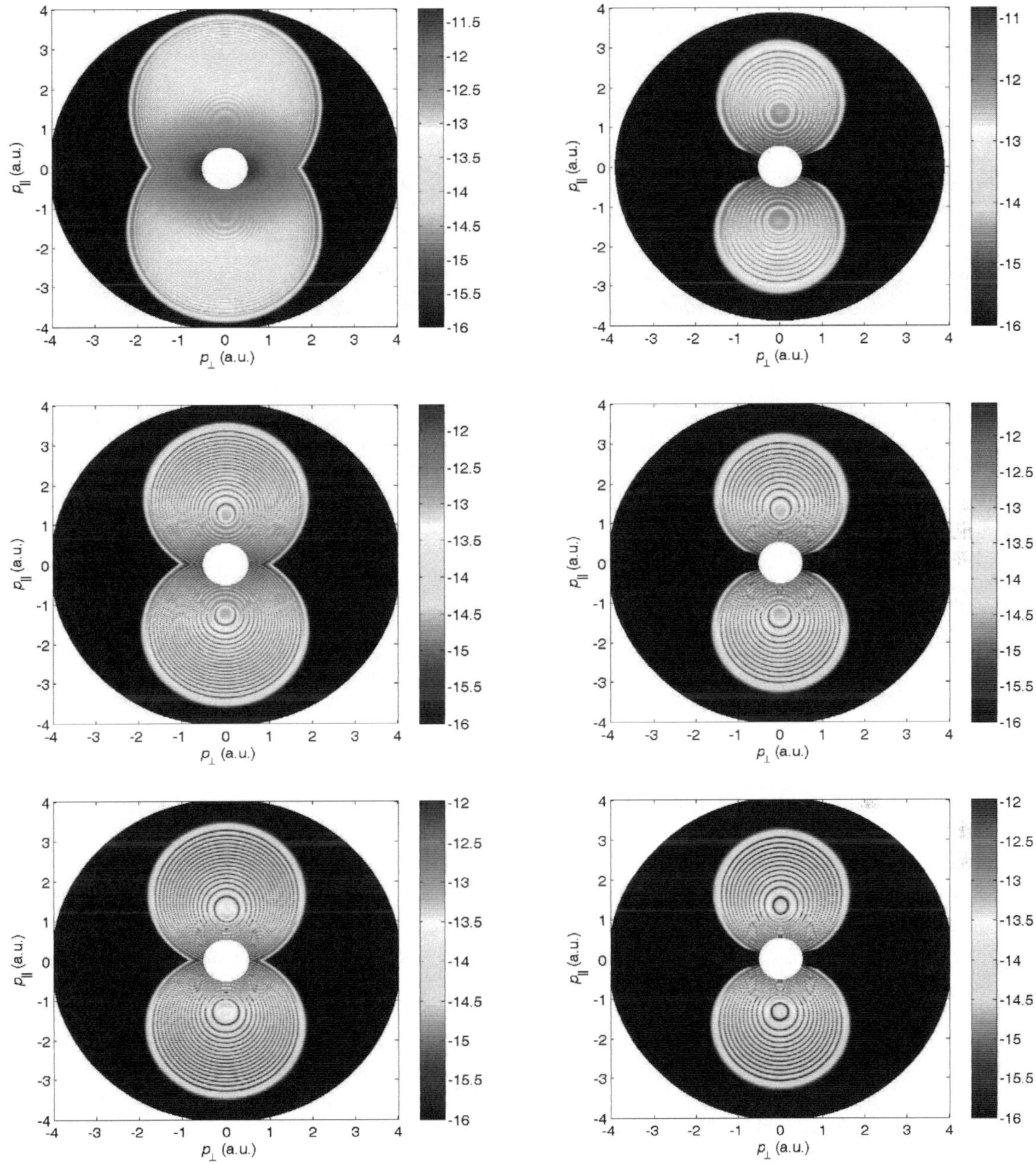

Figure 2. The logarithm of the differential ionization rate of Xe presented in false colors (the color map covers more than four orders of magnitude in arb. units) in the electron momentum plane for ionization by a linearly polarized laser field having the intensity 5×10^{13} W/cm^2 and the wavelength 2000 nm. The results are obtained using the uniform approximation with only one saddle-point solution (β, m). For $\beta = -1$ we present the results for $m = 0$ (upper left panel) $m = 1$ (middle left panel), and $m = 2$ (lower left panel), while for $\beta = +1$ we have $m = 1$ (upper right panel), $m = 2$ (middle right panel), and $m = 3$ (lower right panel).

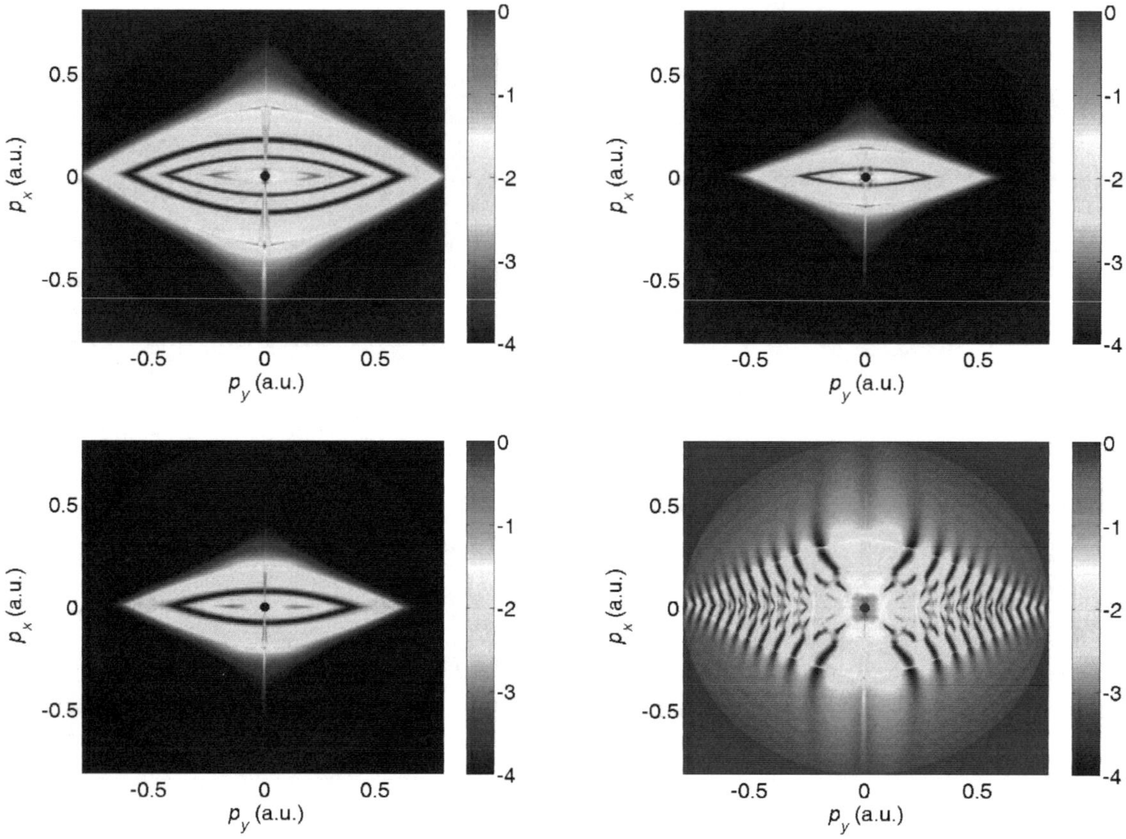

Figure 3. The logarithm of the differential ionization rate of Xe, obtained using the ISFA and the saddle-point method, presented in false colors (the color map covers four orders of magnitude and the rates are normalized to one) in the electron momentum plane for ionization by a linearly polarized laser field having the intensity 4.5×10^{13} W/cm^2 and the wavelength 1800 nm. The results are obtained using the forward-scattering contribution $\mu = 1$ (upper left panel), $\mu = 2$ (lower left panel), $\mu = 3$ (upper right panel) and the coherent sum of the eight solutions $\mu = 0, \ldots, 7$ (lower right panel).

All features of the electron spectra can be explained using these figures. The coherent sum of all quantum-orbit contributions (figure 4) exhibits all the mentioned features. Especially, we retrieved the so-called fork structure, which is located at approximately right angle to and on either side of the polarization axis, both the two outer prongs and the central (split) prong. It should be emphasized that all of these features are already afforded by the ISFA (improved strong-field approximation), which corresponds to the first-order Born approximation for the rescattering electron. The LFA (low-frequency approximation) goes beyond this, essentially by replacing the scattering amplitude in the first-order Born approximation by the exact scattering amplitude (in the absence of the laser field). This way, for higher electron energies the atom-specific minima for particular values of the lab-frame angle also become visible.

References

[1] Becker W, Grasbon F, Kopold R, Milošević D B, Paulus G G and Walther H 2002 Above-threshold ionization: from classical features to quantum effects *Adv. At. Mol. Opt. Phys.* **48** 35

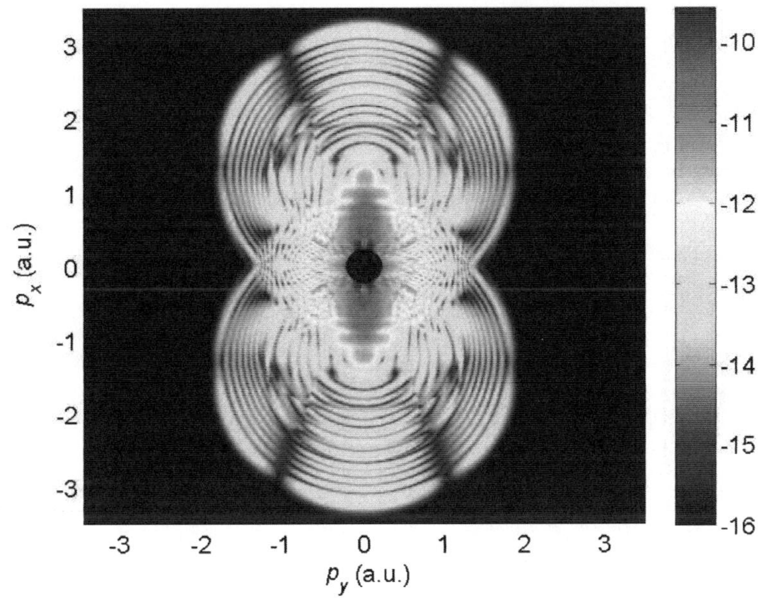

Figure 4. The logarithm of the differential ionization rate of Xe presented in false colors in the electron momentum plane for ionization by a linearly polarized laser field having the intensity 4.5×10^{13} W/cm^2 and the wavelength 1800 nm. The results are obtained as coherent sum of forward-scattering contributions and backward-scattering contributions (taken in the uniform approximation). The rescattering matrix element is calculated in the low-frequency approximation.

[2] Milošević D B, Paulus G G, Bauer D and Becker W 2006 Above-threshold ionization by few-cycle pulses *J. Phys. B: At. Mol. Opt. Phys.* **39** R203
[3] Krausz F and Ivanov M 2009 Attosecond physics *Rev. Mod. Phys.* **81** 163
[4] Agostini P and DiMauro L F 2012 Atomic and molecular ionization dynamics in strong laser fields: from optical to x-rays *Adv. At. Mol. Opt. Phys.* **61** 117
[5] Wolter B, Pullen M G, Baudisch M, Sclafani M, Hemmer M, Senftleben A, Schröter C D, Ullrich J, Moshammer R and Biegert J 2015 Strong-field physics with mid-IR fields *Phys. Rev.* X **5** 021034
[6] Blaga C I, Catoire F, Colosimo P, Paulus G G, Muller H G, Agostini P and DiMauro L F 2009 Strong-field photoionization revisited *Nat. Phys.* **5** 335
[7] Catoire F, Blaga C I, Sistrunk E, Muller H G, Agostini P and DiMauro L F 2009 Mid-infrared strong field ionization angular distributions *Laser Phys.* **19** 1574
[8] Quan W, Lin Z, Wu M, Kang H, Liu H, Liu X, Chen J, Liu L, He X T, Chen S G, Xiong H, Guo L, Xu H, Fu Y, Cheng Y and Zu X X 2009 Classical aspects in above-threshold ionization with a midinfrared strong laser field *Phys. Rev. Lett.* **103** 093001
[9] Wu C Y, Yang Y D, Liu Y Q, Gong Q H, Wu M, Liu X, Hao X L, Li W D, He X T and Chen J 2012 Characteristic spectrum of very-low-energy photoelectrons from above-threshold ionization in the tunneling regime *Phys. Rev. Lett.* **109** 043001
[10] Dura J, Camus N, Thai A, Britz A, Hemmer M, Baudisch M, Senftleben A, Schröter C D, Ullrich J, Moshammer R and Biegert J 2013 Ionization with low-frequency fields in the tunneling regime *Sci. Rep.* **3** 2675
[11] Möller M, Meyer F, Sayler A M, Paulus G G, Kling M F, Schmidt B E, Becker W and Milošević D B 2014 Off-axis low-energy structures in above-threshold ionization *Phys. Rev.* A **90** 023412
[12] Wolter B, Lemell C, Baudisch M, Pullen M G, Tong X-M, Hemmer M, Senftleben A, Schröter C D, Ullrich J, Moshammer R, Biegert J and Burgdörfer J 2014 Formation of very-low-energy states crossing the ionization threshold of argon atoms in strong mid-infrared fields *Phys. Rev.* A **90** 063424

[13] Becker W, Goreslavski S P, Milošević D B and Paulus G G 2014 Low-energy electron rescattering in laser-induced ionization *J. Phys. B: At. Mol. Opt. Phys.* **47** 204022
[14] Becker W and Milošević D B 2015 Above-threshold ionization for very low electron energy *J. Phys. B: At. Mol. Opt. Phys.* **48** 151001
[15] Becker W and Milošević D B 2015 Quantum-orbit theory of low-energy above-threshold ionization on and off axis *Chin. Opt. Lett.* **13** 070006
[16] Čerkić A, Hasović E, Milošević D B and Becker W 2009 High-order above-threshold ionization beyond the first-order Born approximation *Phys. Rev.* A **79** 033413
[17] Milošević D B 2014 Low-frequency approximation for above-threshold ionization by a laser pulse: low-energy forward scattering *Phys. Rev.* A **90** 063423
[18] Milošević D B 2014 Forward- and backward-scattering quantum orbits in above-threshold ionization *Phys. Rev.* A **90** 063414
[19] Milošević D B, Hasović E, Busuladžić M, Gazibegović-Busuladžić A and Becker W 2007 Intensity-dependent enhancements in high-order above-threshold ionization *Phys. Rev.* A **76** 053410
[20] Hasović E, Busuladžić M, Gazibegović-Busuladžić A, Milošević D B and Becker W 2007 Simulation of above-threshold ionization experiments using the strong-field approximation *Laser Phys.* **17** 376
[21] Lin C D, Le A T, Chen Z, Morishita T and Lucchese R 2010 Strong-field rescattering physics–self-imaging of a molecule by its own electrons *J. Phys. B: At. Mol. Opt. Phys.* **43** 122001
[22] Gazibegović-Busuladžić, Hasović E, Busuladžić M, Milošević D B, Kelkensberg F, Siu W K, Vrakking M J J, Lépine F, Sansone G, Nisoli M, Znakovskaya I and Kling M F 2011 Above-threshold ionization of diatomic molecules by few-cycle laser pulses *Phys. Rev.* A **84** 043426
[23] Kästner A, Saalmann U and Rost J M 2012 Electron energy bunching in laser-driven soft recollisions *Phys. Rev. Lett.* **108** 033201
[24] Kästner A, Saalmann U and Rost J M 2012 Energy bunching in soft recollisions revealed with long-wavelength few-cycle pulses *J. Phys. B: At. Mol. Opt. Phys.* **45** 074011

Journal of Physics: Conference Series **691** (2016) 012003 doi:10.1088/1742-6596/691/1/012003

Eikonal perturbation theory in photoionization

F. Cajiao Vélez[1], **K. Krajewska**[1,2], **and J. Z. Kamiński**[1]

[1]Institute of Theoretical Physics, Faculty of Physics, University of Warsaw, Pasteura 5, 02-093 Warszawa, Poland

[2] Department of Physics and Astronomy, University of Nebraska, Lincoln, Nebraska 68588-0299, USA

E-mail: katarzyna.krajewska@fuw.edu.pl

Abstract. The eikonal perturbation theory is formulated and applied to photoionization by strong laser pulses. A special emphasis is put on the first order approximation with respect to the binding potential, which is known as the *generalized eikonal approximation* [2015 *Phys. Rev. A* **91** 053417]. The ordinary eikonal approximation and its domain of applicability is derived from the generalized eikonal approximation. While the former approach is singular for the electron trajectories which return to the potential center, the generalized eikonal avoids this problem. This property makes it a promising tool for further investigations of rescattering and high-order harmonic generation processes.

1. Introduction

The strong-field approximation (SFA), originated from works of Keldysh [1], Faisal [2] and Reiss [3], has had a pivotal role in the development of strong-laser physics. It describes the situation when the laser field is considered to be much more intense than the atomic interaction experienced by ionized electrons with their residual ions. In other words, it neglects the interaction of electrons with their parent ions after ionization. While frequently used, its application to the analysis of photoionization of neutral atoms remains questionable due to the long-range nature of the Coulomb potential. Thus, many efforts have been undertaken in order to include the effects of long-range potentials into this theory. For example, in the improved strong-field approximation, the Coulomb interaction is included as a perturbation, leading to the Born series expansion [4, 5, 6]. In this context, the Coulomb-Volkov ansatz [7, 8, 9, 10], which accounts for the asymptotic phase of the atomic field-free wave function, was also introduced. The latter, combined with the quantum trajectory method [11, 12, 13, 14, 15], have made considerable improvements in the SFA to include the Coulomb interactions [14].

In another effort to introduce the binding potential in the theory, the eikonal approximation was applied to describe photoionization processes. In general, this quasiclassical approximation is extensively used to analyze scattering processes at high energies and small angles [16, 17], regime for which the perturbation theory is not good enough (see, also Ref. [18]). It has been applied, for instance, to study the electron scattering with laser fields in the relativistic and non-relativistic frameworks [19]. The application of the so-called Molière-Glauber eikonal [20, 21] to ionization provided a new tool to treat photoelectron dynamics in the presence of strong laser fields [22, 23]. In this context, the eikonal-Volkov approximation was also introduced [24, 25]. This consisted in including the laser field in full extent by means of the Volkov wave function,

Journal of Physics: Conference Series **691** (2016) 012003 doi:10.1088/1742-6596/691/1/012003

together with the ionic potential treated within the eikonal approximation. This approach was further developed to describe molecular ionization [26].

Following our recent work [27], we present the eikonal perturbation theory and its application to strong-field ionization. This approach in the first order with respect to the binding potential, which we call the generalized eikonal approximation, avoids a singularity at the potential center. Thus, in contrast to the original eikonal approximation [20, 21], it is not singular for returning electron trajectories. This is illustrated in this paper numerically by direct comparison between the generalized and ordinary eikonals calculated along the complex-time quantum trajectories. As a result, we avoid an unphysical behavior of probability distributions of ionization, which is observed when the ordinary eikonal approximation is used [27].

Throughout this paper, the atomic units (a.u.) are used, in which $\hbar = 1$ but we explicitly keep the electron charge ($e < 0$) and mass (m) in the equations. In numerical calculations, both quantities are set to $|e| = 1$ and $m = 1$.

2. Generalized eikonal approximation

Consider an electron of charge $e < 0$ and mass m placed in an external potential $V(\boldsymbol{r}, t)$ and in a laser field; the latter being characterized by the vector potential $\boldsymbol{A}(t)$. The propagator that describes the electron dynamics in the velocity gauge, $K_V(\boldsymbol{r}, t; \boldsymbol{r}', t')$, satisfies the equation,

$$\Big(-\mathrm{i}\frac{\partial}{\partial t'} - \frac{1}{2m}[\mathrm{i}\boldsymbol{\nabla}' - e\boldsymbol{A}(t')]^2 - V(\boldsymbol{r}', t')\Big) K_V(\boldsymbol{r}, t; \boldsymbol{r}', t') = \mathrm{i}\delta(t - t')\delta(\boldsymbol{r} - \boldsymbol{r}'). \tag{1}$$

We seek for the solution of this equation using the proper time representation [19],

$$K_V(\boldsymbol{r}, t; \boldsymbol{r}', t') = \int_0^\infty \mathrm{d}s \int \frac{\mathrm{d}\Omega\,\mathrm{d}^3k}{(2\pi)^4} \exp\Big[-\mathrm{i}\Omega(t - t') + \mathrm{i}\boldsymbol{k}\cdot(\boldsymbol{r} - \boldsymbol{r}') + \mathrm{i}s(\Omega - \frac{\boldsymbol{k}^2}{2m} + \mathrm{i}\varepsilon) \\ + \mathrm{i}\Phi_{\boldsymbol{k}}(t', s) + \mathrm{i}\chi_{\boldsymbol{k}}(\boldsymbol{r}', t', s)\Big], \tag{2}$$

where $\varepsilon > 0$ is introduced for convergence. Plugging this ansatz into Eq. (1), we obtain two independent equations satisfied by the unknown functions, $\Phi_{\boldsymbol{k}}(t', s)$ and $\chi_{\boldsymbol{k}}(\boldsymbol{r}', t', s)$,

$$\Big(\frac{\partial}{\partial t'} - \frac{\partial}{\partial s}\Big)\Phi_{\boldsymbol{k}}(t', s) = -\frac{e}{m}\boldsymbol{A}(t')\cdot\big[\boldsymbol{k} - \frac{e}{2}\boldsymbol{A}(t')\big], \tag{3}$$

$$\Big(\frac{\partial}{\partial t'} - \frac{\partial}{\partial s}\Big)\chi_{\boldsymbol{k}}(\boldsymbol{r}', t', s) = -\frac{1}{m}\big[\boldsymbol{k} - e\boldsymbol{A}(t')\big]\cdot\boldsymbol{\nabla}'\chi_{\boldsymbol{k}}(\boldsymbol{r}', t', s) + V(\boldsymbol{r}', t') \\ + \frac{1}{2m}\big(\boldsymbol{\nabla}'\chi_{\boldsymbol{k}}(\boldsymbol{r}', t', s)\big)^2 - \frac{\mathrm{i}}{2m}\Delta'\chi_{\boldsymbol{k}}(\boldsymbol{r}', t', s). \tag{4}$$

Eq. (3) can be easily solved, leading to

$$\Phi_{\boldsymbol{k}}(t', s) = \int_{t'}^{t'+s} \mathrm{d}\tau\, \frac{e}{m}\boldsymbol{A}(\tau)\cdot\big[\boldsymbol{k} - \frac{e}{2}\boldsymbol{A}(\tau)\big]. \tag{5}$$

Eq. (4), on the other hand, is a nonlinear second order differential equation which can be solved explicitly only for particular potentials. Other than that, one can formulate a general approach of treating Eq. (4) perturbatively.

As shown in Ref. [27], the solution of Eq. (4), the so-called *eikonal*, can be represented in the form

$$\chi_{\boldsymbol{k}}(\boldsymbol{r}', t', s) = -\int_{t'}^{t'+s} \mathrm{d}\sigma \int \mathrm{d}^3\rho \Big(\frac{m}{2\pi\mathrm{i}(\sigma - t')}\Big)^{3/2} \exp\Big(\frac{\mathrm{i}m}{2(\sigma - t')}[\boldsymbol{R}_{\boldsymbol{k}}(\boldsymbol{r}', t', \sigma) - \boldsymbol{\rho}]^2\Big) \\ \times W_{\boldsymbol{k}}(\boldsymbol{\rho}, \sigma, t' + s - \sigma). \tag{6}$$

Here,

$$\boldsymbol{R}_{\boldsymbol{k}}(\boldsymbol{r}',t',\sigma)=\boldsymbol{r}'+\frac{1}{m}\int_{t'}^{\sigma}\mathrm{d}\tau[\boldsymbol{k}-e\boldsymbol{A}(\tau)] \tag{7}$$

is interpreted as a free electron trajectory in the laser field and $W_{\boldsymbol{k}}(\boldsymbol{r}',t',s)$ is defined as

$$W_{\boldsymbol{k}}(\boldsymbol{r}',t',s)=V(\boldsymbol{r}',t')+\frac{1}{2m}\big(\boldsymbol{\nabla}'\chi_{\boldsymbol{k}}(\boldsymbol{r}',t',s)\big)^2. \tag{8}$$

This is the starting point for developing the eikonal perturbation theory. Specifically, in the first approximation, when

$$W_{\boldsymbol{k}}^{(1)}(\boldsymbol{r}',t',s)\approx V(\boldsymbol{r}',t'), \tag{9}$$

one obtains

$$\chi_{\boldsymbol{k}}^{(1)}(\boldsymbol{r}',t',s)=-\int_{t'}^{t'+s}\mathrm{d}\sigma\int\mathrm{d}^3\rho\left(\frac{m}{2\pi\mathrm{i}(\sigma-t')}\right)^{3/2}\exp\Big(\frac{\mathrm{i}m}{2(\sigma-t')}[\boldsymbol{R}_{\boldsymbol{k}}(\boldsymbol{r}',t',\sigma)-\boldsymbol{\rho}]^2\Big)V(\boldsymbol{\rho},\sigma). \tag{10}$$

In the following, we shall call $\chi_{\boldsymbol{k}}^{(1)}(\boldsymbol{r}',t',s)$ the *generalized eikonal*. Note that it also defines the propagator (2) in the first order eikonal approximation,

$$K_V^{(1)}(\boldsymbol{r},t;\boldsymbol{r}',t')=\int\frac{\mathrm{d}^3k}{(2\pi)^3}\exp\Big[\mathrm{i}\boldsymbol{k}\cdot(\boldsymbol{r}-\boldsymbol{r}')-\mathrm{i}(t-t')\frac{\boldsymbol{k}^2}{2m}+\mathrm{i}\Phi_{\boldsymbol{k}}(t',t-t')+\mathrm{i}\chi_{\boldsymbol{k}}^{(1)}(\boldsymbol{r}',t',t-t')\Big], \tag{11}$$

where, in Eq. (2), we have explicitly performed the integrals over s and Ω.

For our further purpose, let us represent the generalized eikonal (10) such that

$$\chi_{\boldsymbol{k}}^{(1)}(\boldsymbol{r}',t',s)=-\int_{t'}^{t'+s}\mathrm{d}\sigma\, V_{\mathrm{eff}}^{(1)}\big(\boldsymbol{R}_{\boldsymbol{k}}(\boldsymbol{r}',t',\sigma),t',\sigma\big), \tag{12}$$

where we have introduced the effective potential,

$$V_{\mathrm{eff}}^{(1)}\big(\boldsymbol{R}_{\boldsymbol{k}}(\boldsymbol{r}',t',\sigma),t',\sigma\big)=\int\mathrm{d}^3\rho\left(\frac{m}{2\pi\mathrm{i}(\sigma-t')}\right)^{3/2}\exp\Big(\frac{\mathrm{i}m}{2(\sigma-t')}[\boldsymbol{R}_{\boldsymbol{k}}(\boldsymbol{r}',t',\sigma)-\boldsymbol{\rho}]^2\Big)V(\boldsymbol{\rho},\sigma). \tag{13}$$

As argued in Ref. [27], the effective potential differs from the classical one, $V(\boldsymbol{r},t)$, by quantum corrections which vanish in the limit $\hbar\to 0$. This means that the effective potential accounts for the spreading of the electron wave packet during the time evolution. Thus, it eliminates the problem with rescattering trajectories, as it will be illustrated numerically in Sec. 4.

In closing this Section we note that, for the generalized eikonal approximation [Eq. (9)] to be applicable, it must hold that $\big(\boldsymbol{\nabla}'\chi_{\boldsymbol{k}}(\boldsymbol{r},t,s)\big)^2\ll 2m|V(\boldsymbol{r},t)|$. This condition is very well satisfied for numerical illustrations presented in Sec. 4. Following the first order approximation, we note that Eq. (8) allows us to define the recurrence relation,

$$W_{\boldsymbol{k}}^{(n+1)}(\boldsymbol{r}',t',s)=V(\boldsymbol{r}',t')+\frac{1}{2m}\big(\boldsymbol{\nabla}'\chi_{\boldsymbol{k}}^{(n)}(\boldsymbol{r}',t',s)\big)^2 \tag{14}$$

and, as a consequence, to define the generalized eikonal (6) in subsequent orders. Investigations of higher order terms in the eikonal perturbation theory are, however, beyond the scope of this paper.

3. Generalized eikonal for the Coulomb potential

For the Coulomb potential describing the interaction of an electron and a nucleus of charge $-Ze$, where $Z = 1, 2, ...$ is the atomic number, we have $V(\boldsymbol{r}, t) = -Z\alpha c/r$. Here, $\alpha = e^2/(4\pi\varepsilon_0 c)$ is the fine structure constant (in a.u., $\alpha c = 1$). Thus, the generalized eikonal defined in (10) becomes

$$\chi_{\boldsymbol{k}}^{(1)}(\boldsymbol{r}', t', s) = \int_{t'}^{t'+s} \mathrm{d}\sigma \int \mathrm{d}^3\rho \left(\frac{m}{2\pi\mathrm{i}(\sigma - t')}\right)^{3/2} \frac{Z\alpha c}{\rho} \exp\left(\frac{\mathrm{i}m}{2(\sigma - t')}[\boldsymbol{R}_{\boldsymbol{k}}(\boldsymbol{r}', t', \sigma) - \boldsymbol{\rho}]^2\right). \tag{15}$$

Performing the integral over $\boldsymbol{\rho}$, we arrive at

$$\chi_{\boldsymbol{k}}^{(1)}(\boldsymbol{r}', t', s) = Z\alpha c \int_{t'}^{t'+s} \mathrm{d}\sigma \frac{1}{|\boldsymbol{R}_{\boldsymbol{k}}(\boldsymbol{r}', t', \sigma)|} \mathrm{erf}\left(\sqrt{\frac{m}{2\mathrm{i}(\sigma - t')}}\,|\boldsymbol{R}_{\boldsymbol{k}}(\boldsymbol{r}', t', \sigma)|\right), \tag{16}$$

where $\mathrm{erf}(z)$ is the error function.

The originally used eikonal [20, 21, 19, 26],

$$\chi_{\boldsymbol{k},\mathrm{original}}(\boldsymbol{r}', t', s) = Z\alpha c \int_{t'}^{t'+s} \mathrm{d}\sigma \frac{1}{|\boldsymbol{R}_{\boldsymbol{k}}(\boldsymbol{r}', t', \sigma)|}, \tag{17}$$

is recovered from Eq. (16), if $\sqrt{\frac{m}{2(\sigma - t')}}\,|\boldsymbol{R}_{\boldsymbol{k}}(\boldsymbol{r}', t', \sigma)| \gg 1$. Note that this condition is satisfied for short-time intervals $\sigma \approx t'$, given that $|\boldsymbol{R}_{\boldsymbol{k}}(\boldsymbol{r}', t', \sigma)| \neq 0$. For $\sigma \not\approx t'$, it requires that $|\boldsymbol{R}_{\boldsymbol{k}}(\boldsymbol{r}', t', \sigma)| \not\to 0$. As it was shown in Ref. [27], in the limit when $|\boldsymbol{R}_{\boldsymbol{k}}(\boldsymbol{r}', t', \sigma)| \to 0$, the generalized eikonal (16) presents an integrable singularity. This suggests that the generalized eikonal approximation is applicable even to cases when the electron trajectory can return back to the origin of the Coulomb potential. This is in contrast to the original eikonal (17), which diverges logarithmically for such trajectories.

4. Application of the generalized eikonal approximation to ionization

In this paper, we apply the generalized eikonal approximation to ionization of a hydrogen-like atom by a modulated laser pulse. We assume that the pulse lasts for time T, i.e., it is described by the electric field $\boldsymbol{\mathcal{E}}(t)$ which vanishes for $t < 0$ and $t > T$. The pulse duration T defines the fundamental frequency of field oscillations, $\omega = 2\pi/T$. One can also introduce the field phase, $\phi = \omega t$, which allows us to rewrite the above condition such that $\boldsymbol{\mathcal{E}}(\phi)$ vanishes for $\phi < 0$ and $\phi > 2\pi$. We also assume that the driving pulse is linearly polarized along the z-axis.

4.1. Model of the laser pulse

In the dipole approximation, we describe the laser field by the electric field vector

$$\boldsymbol{\mathcal{E}}(\phi) = \mathcal{E}_0 f_{\mathcal{E}}(\phi)\boldsymbol{e}_z, \tag{18}$$

where $\mathcal{E}_0$ is related to the amplitude of field oscillations. We consider the shape function $f_{\mathcal{E}}(\phi)$ such that

$$f_{\mathcal{E}}(\phi) = \begin{cases} \sin^2(N_{\mathrm{rep}}\frac{\phi}{2})\sin(N_{\mathrm{rep}}\phi), & \phi \in [0, 2\pi], \\ 0, & \text{otherwise}. \end{cases} \tag{19}$$

This means that the pulse consists of N_{rep} modulations ($N_{\mathrm{rep}} = 1, 2, 3, ...$), each one being a single-cycle pulse. We also define the shape function for the vector potential $\boldsymbol{A}(\phi)$ such that

$$f_A(\phi) = -\int_0^\phi f_{\mathcal{E}}(\varphi)\mathrm{d}\varphi, \tag{20}$$

which leads to

$$\boldsymbol{A}(\phi) = \frac{\mathcal{E}_0}{\omega} f_A(\phi)\boldsymbol{e}_z. \tag{21}$$

Note that, for our choice of the shape function (19), the condition $\boldsymbol{A}(2\pi) = \boldsymbol{A}(0) = \boldsymbol{0}$ holds.

Journal of Physics: Conference Series **691** (2016) 012003 doi:10.1088/1742-6596/691/1/012003

4.2. Probability amplitude of ionization using the generalized eikonal approximation

As derived in Ref. [27], under the generalized eikonal approximation, the probability amplitude to ionize a hydrogen-like atom equals

$$\mathcal{A}^{(1)}(\boldsymbol{p}) = -\,\mathrm{i}\int_0^T \mathrm{d}t' \mathrm{e}^{\mathrm{i}\frac{\boldsymbol{p}^2}{2m}T-\mathrm{i}E_0t'}\int \mathrm{d}^3r\int \mathrm{d}^3r'\psi_{\boldsymbol{p}}^{(-)*}(\boldsymbol{r})K_L^{(1)}(\boldsymbol{r},T;\boldsymbol{r}',t')(-e\boldsymbol{\mathcal{E}}(t')\cdot\boldsymbol{r}')\psi_0(\boldsymbol{r}'). \quad (22)$$

Here, we assume that $\psi_0(\boldsymbol{r}) = \lambda\sqrt{\frac{\lambda}{\pi}}\mathrm{e}^{-\lambda r}$ is the ground state of a hydrogen-like atom with energy $E_0 = -\frac{\lambda^2}{2m}$ [$\lambda = (Za_0)^{-1}$, a_0 is the Bohr radius] whereas $\psi_{\boldsymbol{p}}^{(-)}(\boldsymbol{r})$ is the scattering state describing the ionized electron of momentum $\boldsymbol{p}$. Note that the probability amplitude (22) is expressed in the length gauge. Thus, $K_L^{(1)}(\boldsymbol{r},T;\boldsymbol{r}',t')$ represents the electron propagator in the length gauge as well. More specifically,

$$K_L^{(1)}(\boldsymbol{r},t;\boldsymbol{r}',t') = \int\frac{\mathrm{d}^3k}{(2\pi)^3}\exp\Big[\mathrm{i}(\boldsymbol{k}-e\boldsymbol{A}(t))\cdot\boldsymbol{r}-\mathrm{i}(\boldsymbol{k}-e\boldsymbol{A}(t'))\cdot\boldsymbol{r}'-\mathrm{i}(t-t')\frac{\boldsymbol{k}^2}{2m} + \mathrm{i}\Phi_{\boldsymbol{k}}(t',t-t')+\mathrm{i}\chi_{\boldsymbol{k}}^{(1)}(\boldsymbol{r}',t',t-t')\Big], \quad (23)$$

which follows from Eq. (11) after applying the gauge transformation [27].

In order to proceed, it was assumed in Ref. [27] that the final electron state $\psi_{\boldsymbol{p}}^{(-)}(\boldsymbol{r})$ is essentially the plane wave. This made it possible to explicitly perform the integral over $\boldsymbol{r}$ in Eq. (22). The remaining integrals were reformulated such that

$$\mathcal{A}^{(1)}(\boldsymbol{p}) = -\mathrm{i}\frac{\mathrm{e}^{\mathrm{i}(\boldsymbol{p}^2/2m-E_0)T}}{\omega}\int_0^{2\pi}\mathrm{d}\phi\int \mathrm{d}^3r'\mathrm{e}^{-\mathrm{i}(\boldsymbol{p}-e\boldsymbol{A}(\phi))\cdot\boldsymbol{r}'}(-e\boldsymbol{\mathcal{E}}(\phi)\cdot\boldsymbol{r}')\psi_0(\boldsymbol{r}')\mathrm{e}^{\mathrm{i}W[\phi,\boldsymbol{p}|\boldsymbol{r}_{\mathrm{cl}}]}. \quad (24)$$

Here, the functional $W[\phi,\boldsymbol{p}|\boldsymbol{r}_{\mathrm{cl}}]$, which depends on the classical electron trajectory,

$$\boldsymbol{r}_{\mathrm{cl}}(\sigma;\boldsymbol{r}',\phi,\boldsymbol{p}) \equiv \boldsymbol{r}_{\mathrm{cl}}(\sigma) = \boldsymbol{R}_{\boldsymbol{p}}\Big(\boldsymbol{r}',\frac{\phi}{\omega},\frac{\sigma}{\omega}\Big), \quad (25)$$

has been introduced. More specifically,

$$W[\phi,\boldsymbol{p}|\boldsymbol{r}_{\mathrm{cl}}] = S[\phi|\boldsymbol{r}_{\mathrm{cl}}] + m\omega\boldsymbol{r}_{\mathrm{cl}}(\phi)\cdot\boldsymbol{r}'_{\mathrm{cl}}(\phi) - \boldsymbol{p}\cdot\boldsymbol{r}_{\mathrm{cl}}(2\pi), \quad (26)$$

with the action and the effective Lagrangian equal to

$$S[\phi|\boldsymbol{r}_{\mathrm{cl}}] = \frac{1}{\omega}\int_\phi^{2\pi}\mathrm{d}\sigma\mathcal{L}_{\mathrm{eff}}(\boldsymbol{r}_{\mathrm{cl}}(\sigma),\boldsymbol{r}'_{\mathrm{cl}}(\sigma),\sigma), \quad (27)$$

$$\mathcal{L}_{\mathrm{eff}}(\boldsymbol{r}_{\mathrm{cl}}(\sigma),\boldsymbol{r}'_{\mathrm{cl}}(\sigma),\sigma) = \frac{m\omega^2}{2}[\boldsymbol{r}'_{\mathrm{cl}}(\sigma)]^2 + e\boldsymbol{\mathcal{E}}(\sigma)\cdot\boldsymbol{r}_{\mathrm{cl}}(\sigma) - V_{\mathrm{eff}}^{(1)}(\boldsymbol{r}_{\mathrm{cl}}(\sigma),\sigma-\phi)+E_0, \quad (28)$$

respectively. Since the effective potential describing the interaction of the photoelectron and its parent ion oscillates rapidly, in Eq. (24) we apply the saddle-point method. While the details of the derivations are presented in Ref. [27], we remind that, instead of classical trajectories, the complex-time quantum trajectories are used in these derivations.

4.3. Complex-time quantum trajectories

The complex-time quantum trajectory is the solution of the classical Newton equation in the laser field,

$$\boldsymbol{r}''_q(\phi;\boldsymbol{p},\phi_s) = \frac{e}{m\omega^2}\boldsymbol{\mathcal{E}}(\phi), \quad (29)$$

where the '*prime*' means the derivative with respect to ϕ. Here, ϕ_s defines the initial phase for which the following conditions are satisfied,

$$\mathrm{Re}[\boldsymbol{r}_q(\phi_s;\boldsymbol{p},\phi_s)]=\boldsymbol{0},\qquad [m\omega\boldsymbol{r}'_q(\phi_s;\boldsymbol{p},\phi_s)]^2=-\lambda^2. \tag{30}$$

It turns out that the initial phase is the solution of the equation,

$$\frac{\partial}{\partial\phi}G(\boldsymbol{p},\phi)\Big|_{\phi=\phi_s}=0, \tag{31}$$

with

$$G(\boldsymbol{p},\phi)=\frac{1}{\omega}\int_0^{\phi}\mathrm{d}\phi'\Big\{\frac{[\boldsymbol{p}-e\boldsymbol{A}(\phi')]^2}{2m}-E_0\Big\}. \tag{32}$$

The quantum trajectories, satisfying the conditions (30), can be written in the form [27]

$$\boldsymbol{r}_q(\phi;\boldsymbol{p},\phi_s)=\frac{\boldsymbol{p}}{m\omega}[\phi-\mathrm{Re}(\phi_s)]+\frac{e}{m}\boldsymbol{\alpha}(\phi)-\frac{e}{m}\mathrm{Re}[\boldsymbol{\alpha}(\phi_s)], \tag{33}$$

where $\boldsymbol{\alpha}(\phi)=-\frac{1}{\omega}\int_0^{\phi}\boldsymbol{A}(\varphi)\mathrm{d}\varphi$. In addition, the trajectories (33) satisfy the conditions,

$$\mathrm{Im}[\boldsymbol{r}_q(\mathrm{Re}\,\phi;\boldsymbol{p},\phi_s)]=\boldsymbol{0},\qquad \mathrm{Im}[\boldsymbol{r}'_q(\mathrm{Re}\,\phi;\boldsymbol{p},\phi_s)]=\boldsymbol{0}. \tag{34}$$

Since we deal with the Coulomb-free trajectories, if the laser pulse is switched off the electron will carry the momentum $\boldsymbol{p}$. In other words,

$$m\omega\boldsymbol{r}'_q(2\pi;\boldsymbol{p},\phi_s)=\boldsymbol{p}. \tag{35}$$

This agrees with the assumption that the electron final state is approximated by the plane wave solution.

4.4. Probability amplitude of ionization using the saddle-point approach

After applying the saddle-point method along with the complex-time quantum trajectories with respect to Eq. (24), the probability amplitude of ionization of a hydrogen-like atom by a finite laser pulse becomes

$$\mathcal{A}^{(1)}_{\mathrm{saddle}}(\boldsymbol{p})=-2\sqrt{\frac{\lambda}{\pi}}\Big(\frac{\pi\lambda}{m\omega}\Big)^2\mathrm{e}^{\mathrm{i}\Phi_0(\boldsymbol{p})}\sum_s\frac{\mathrm{e}^{\mathrm{i}G(\boldsymbol{p},\phi_s)+\mathrm{i}\chi_q^{(1)}(\boldsymbol{p},\phi_s)}}{G''(\boldsymbol{p},\phi_s)}. \tag{36}$$

The summation is over all solutions of Eq. (31) with $\mathrm{Im}[\phi_s]>0$ (since now on, called the *saddle points*). Let us note, however, that the most important contributions to the sum in (36) come from these saddle points which satisfy the conditions,

$$\mathrm{Im}[G(\boldsymbol{p},\phi_s)]>0,\quad \mathrm{Im}[G''(\boldsymbol{p},\phi_s)]>0, \tag{37}$$

and result in the smallest value of $\mathrm{Im}[G(\boldsymbol{p},\phi_s)]$. Therefore, only those points will be considered in our numerical illustrations presented in the next Section. In Eq. (36), we have also introduced the abbreviation $\Phi_0(\boldsymbol{p})=(\frac{\boldsymbol{p}^2}{2m}-E_0)T-G(\boldsymbol{p},2\pi)$. Moreover, $\chi_q^{(1)}(\boldsymbol{p},\phi_s)$ present there is the generalized eikonal (16) calculated along the quantum trajectory (33) starting at ϕ_s,

$$\chi_q^{(1)}(\boldsymbol{p},\phi_s)=\frac{Z\alpha c}{\omega}\int_{\phi_s}^{2\pi}\mathrm{d}\sigma\frac{1}{|\boldsymbol{r}_q(\sigma;\boldsymbol{p},\phi_s)|}\mathrm{erf}\Big(\sqrt{\frac{m\omega}{2\mathrm{i}(\sigma-\phi_s)}}|\boldsymbol{r}_q(\sigma;\boldsymbol{p},\phi_s)|\Big). \tag{38}$$

Following Eq. (17), one can also define the original eikonal,

$$\chi_{q,\mathrm{original}}(\boldsymbol{p},\phi_s)=\frac{Z\alpha c}{\omega}\int_{\phi_s}^{2\pi}\mathrm{d}\sigma\frac{1}{|\boldsymbol{r}_q(\sigma;\boldsymbol{p},\phi_s)|}. \tag{39}$$

It defines the respective probability amplitude of ionization, which is essentially the same as Eq. (36) except that $\chi_q^{(1)}(\boldsymbol{p},\phi_s)$ has to be replaced by $\chi_{q,\mathrm{original}}(\boldsymbol{p},\phi_s)$.

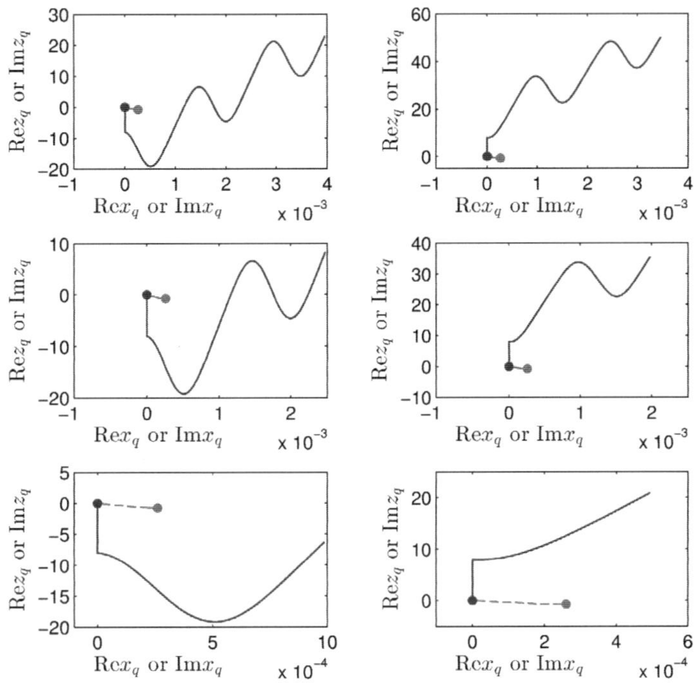

Figure 1. Real (solid blue lines) and imaginary (dashed red lines) parts of the quantum trajectories (in a.u.), calculated from Eq. (33), for an electron with the kinetic energy $E_{\boldsymbol{p}} = 3.13\text{eV}$. The electron is ionized by a modulated pulse defined by Eqs. (18) and (19) where $N_{\text{rep}} = 3$, $\omega_{\text{L}} = 1.55\text{eV}$ and $I = 3.125 \times 10^{13}\text{W/cm}^2$. Each panel represents a trajectory starting at a different saddle point ϕ_s. Only the essential saddle points are considered.

4.5. Numerical illustrations

To illustrate numerically the introduced theory, we calculate first the complex-time quantum trajectories defined by Eq. (33). For this purpose, we consider a Ti:sapphire laser working at frequency $\omega_{\text{L}} = 1.55\text{eV}$. We assume that it produces a modulated pulse, as defined by Eqs. (18) and (19), which consists of three single-cycle modulations ($N_{\text{rep}} = 3$). The averaged intensity carried out by the pulse, which in general is defined as $I = \langle c\varepsilon_0 \boldsymbol{\mathcal{E}}^2 \rangle = \frac{1}{2\pi}\int_0^{2\pi} c\varepsilon_0 \boldsymbol{\mathcal{E}}^2(\phi)\mathrm{d}\phi$, equals $I = 3.125 \times 10^{13}\text{W/cm}^2$. Let us also remind that the pulse is linearly polarized in the z-direction. As introduced in Ref. [27], one can define the ponderomotive energy of electron free oscillations in such a laser field, U_p. For the current parameters, $U_p = 1.024\omega_{\text{L}}$.

In Fig. 1, we present the quantum trajectories starting at saddle points ϕ_s for a photoelectron that is detected with energy $E_{\boldsymbol{p}} = 3.13\text{eV} \approx 3U_p$ at the polar angle $\theta_{\boldsymbol{p}} = 0.2\pi$. The same but for the electron with energy $E_{\boldsymbol{p}} = 12.22\text{eV} \approx 12U_p$ is shown in Fig. 2. Since the laser field is linearly polarized the results are independent of the azimuthal angle of the electron, $\varphi_{\boldsymbol{p}}$. Actually, for the laser field consider in this example there are 24 saddle points. However, only those saddle points which satisfy the conditions (37) and posses the smallest $\text{Im}[G(\boldsymbol{p}, \phi_s)]$ have been chosen for numerical illustrations. As we have checked this numerically, the remaining saddle points have a marginally small contribution to the probability amplitude of ionization (36). Therefore, we do not consider them for the purpose of these and subsequent figures. Here, we plot the real (solid blue lines) and imaginary (dashed red lines) parts of the quantum trajectories which originate from the essential saddle points. As we have checked this, these saddle points relate

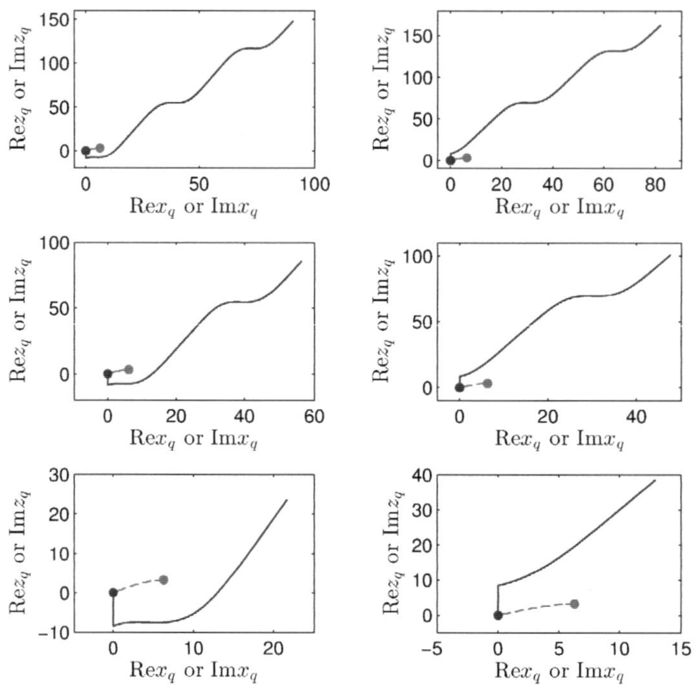

Figure 2. The same as in Fig. 1 but for $E_{\boldsymbol{p}} = 12.22\text{eV}$.

to either the maximum or minimum of the electric field. The trajectories starting at the former points are plotted in the left column of Figs. 1 and 2, whereas the trajectories starting at the latter points are plotted in the right column of Figs. 1 and 2. Note that these trajectories are parametrized by Cartesian coordinates x_q and z_q. When comparing Figs. 1 and 2, we note a significant difference of the scale of x_q.

It can be seen in Fig. 1 that, for $E_{\boldsymbol{p}} = 3.13\text{eV} \approx 3U_p$, the real part of some of the quantum trajectories approach the ion after the electron is ionized. This is shown in the top and middle panels of the left column. For $E_{\boldsymbol{p}} = 12.22\text{eV} \approx 12U_p$ (Fig. 2), already shortly after ionization, x_q acquires large values and the electron never comes back near the potential center. As it will become clear shortly, the trajectories which come back to its center lead to singularities in the formulation based on the original eikonal approximation. Our generalized eikonal approximation avoids this problem.

In Fig. 3, we plot the real and imaginary parts of the original eikonal, $\chi_{q,\,\text{original}}(\boldsymbol{p}, \phi_s)$, which is defined by Eq. (39). The eikonal is calculated along the quantum trajectories which start at the saddle points at either the maximum (solid blue lines) or minimum (dashed red lines) of the laser field. As before, these saddle points fulfill the conditions (37) and have the smallest $\text{Im}[G(\boldsymbol{p}, \phi_s)]$. As we can observe, the real parts of $\chi_{q,\text{original}}(\boldsymbol{p}, \phi_s)$ exhibit the sharp peaks at two instances. This happens for those trajectories which return very close to the origin of the Coulomb potential, and which are plotted in Fig. 1. Note that the fact that the original eikonal is singular for the returning trajectories is a consequence of its definition, i.e., Eq. (39). Moreover, we see in Fig. 3 that the rapid change of $\text{Re}\,\chi_{q,\,\text{original}}(\boldsymbol{p}, \phi_s)$ happens when the photoelectron energy passes through the value $3U_p$. This explains why in Fig. 10 of Ref. [27] the corresponding energy distribution of ionization shows unphysical wiggles around this energy. In contrast, such a behavior is not observed in [27] when the generalized eikonal approximation is used. The reason

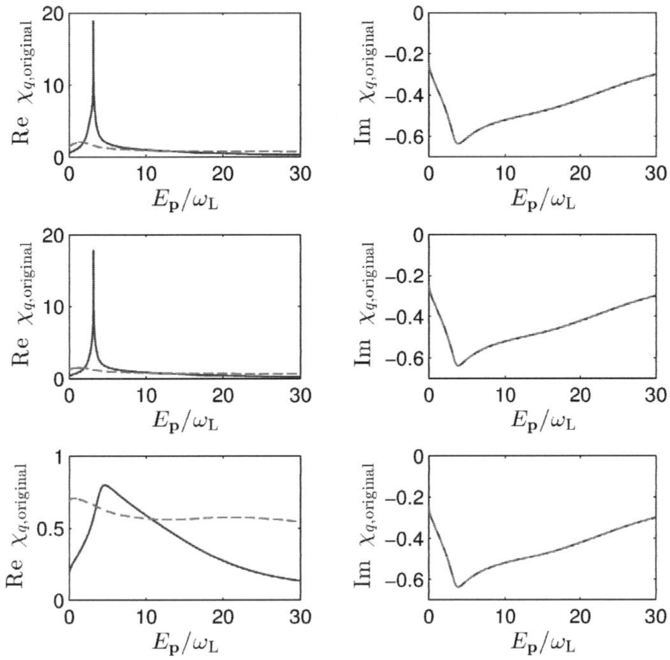

Figure 3. Imaginary and real parts of the original eikonal, $\chi_{q,\,\text{original}}(\boldsymbol{p}, \phi_s)$, calculated according to Eq. (39), as a function of the electron energy. Solid blue lines represent the trajectories starting at the essential saddle points which appear at the maximum of the electric field whereas the dashed red lines are for the essential saddle points which appear at the minimum of the electric field. The laser field parameters are $\omega_{\text{L}} = 1.55\text{eV}$, $I = 3.125 \times 10^{13}\text{W/cm}^2$ and $N_{\text{rep}} = 3$.

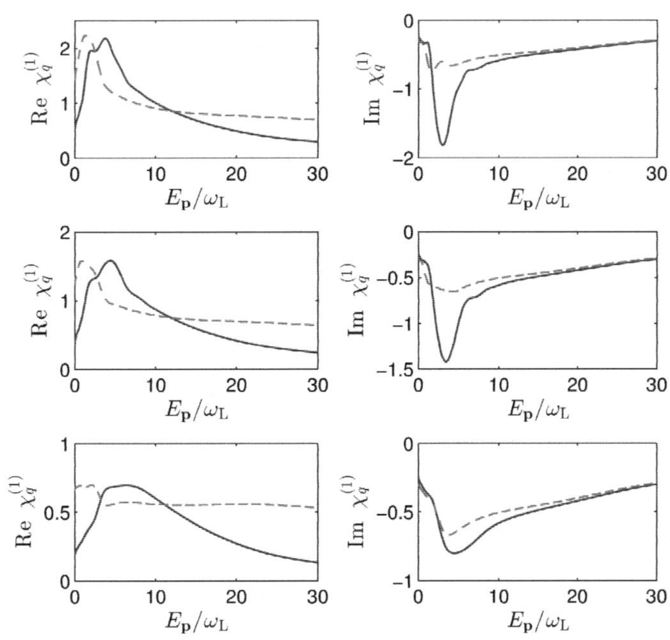

Figure 4. The same as in Fig. 3 but for the generalized eikonal, $\chi_q^{(1)}(\boldsymbol{p}, \phi_s)$, given by Eq. (38).

being that the generalized eikonal, $\chi_q^{(1)}(\boldsymbol{p}, \phi_s)$, defined by Eq. (38), is not singular even for those trajectories which come back to the potential center (see, Fig. 4). Thus, the lack of spurious behavior of trajectories returning back to the vicinity of the parent ion makes the approach presented here an attractive tool for investigations of ionization, rescattering and high-order harmonic generation by strong laser pulses. This includes also more complex systems such as diatomic molecules or fullerenes.

5. Conclusions

In this paper, we have presented the basics of the eikonal perturbation theory. We have focused on its first term, the generalized eikonal approximation [27], when applied to strong-field ionization. A comparison with the original eikonal approximation, introduced in Refs. [20, 21], was performed. It turned out that, while the original eikonal is singular for electron trajectories which return to the Coulomb potential center, the generalized eikonal does not share this property. Thus, in contrast to the original eikonal, the newly developed approach does not lead to unphysical behavior of the probability distributions of ionization [27]. Its further applications to rescattering and high-order harmonic generation are being considered.

Acknowledgments

This work was supported by the Polish National Science Center (NCN) under Grant No. 2012/05/B/ST2/02547. Moreover, F.C.V. acknowledges the support from the Foundation for Polish Science International PhD Projects Programme co-financed by the EU European Regional Development Fund.

References

[1] Keldysh L V 1965 *Sov. Phys. JETP* **20** 1307
[2] Faisal F H M 1973 *J. Phys. B* **6** L89
[3] Reiss H R 1980 *Phys. Rev. A* **22** 1786
[4] Becker W, Lohr A and Kleber M 1994 *J. Phys. B* **27** L325
[5] Lewenstein M, Kulander K C, Schafer K J and Bucksbaum P H 1995 *Phys. Rev. A* **51** 1495
[6] Milošević D B 2013 *Phys. Rev. A* **88** 023417
[7] Jain M and Tzoar N 1978 *Phys. Rev. A* **18** 538
[8] Cavaliere P, Ferrante G and Leone C 1980 *J. Phys. B* **13** 4495
[9] Kamiński J Z 1986 *Phys. Scr.* **34** 770
[10] Kamiński J Z 1988 *Phys. Rev. A* **37** 622
[11] Popov V S 2004 *Phys. Usp.* **47** 855
[12] Popov V S 2005 *Phys. At. Nucl.* **68** 686
[13] Gribakin G F and Kuchiev M Yu 1997 *Phys. Rev. A* **55** 3760
[14] Popruzhenko S V, Paulus G G and Bauer D 2008 *Phys. Rev. A* **77** 053409
[15] Popruzhenko S V 2014 *J. Phys. B* **47** 204001
[16] Saha B C, Sarkar K and Ghosh A S 1973 *J. Phys. B* **6** 2303
[17] Byron Jr. F W and Joachain Ch J 1977 *Phys. Rev. A* **15** 128
[18] Joachain Ch J 1975 *Quantum Collision Theory* (Amsterdam: North Holland)
[19] Kamiński J Z 1984 *Acta Phys. Pol. A* **66** 517
[20] Molière G 1947 *Z. Naturforsch.* **2** 133
[21] Glauber R J 1959 *Lectures in Theoretical Physics* ed. W E Brittin and L G Dunham (New York: Interscience) vol 1 pp 315-414
[22] Banerji J and Mittelman M H 1981 *J. Phys. B* **14** 3717
[23] Krstić P and Mittleman M H 1982 *Phys. Rev. A* **25** 1568
[24] Reiss H R and Krainov V P 1996 *Proc. SPIE* **39** 2796
[25] Krainov V P 1997 *J. Opt. Soc. Am. B* **14** 425
[26] Smirnova O, Spanner M and Ivanov M 2008 *Phys. Rev. A* **77** 033407
[27] Cajiao Vélez F, Krajewska K and Kamiński J Z 2015 *Phys. Rev. A* **91** 053417

Journal of Physics: Conference Series **691** (2016) 012004 doi:10.1088/1742-6596/691/1/012004

Atomic processes in bicircular fields

S Odžak[1], E Hasović[1], W. Becker[2] and D B Milošević[1,2,3]

[1] Faculty of Science, University of Sarajevo, Zmaja od Bosne 35, 71000 Sarajevo, Bosnia and Herzegovina

[2] Max-Born-Institut, Max-Born-Str. 2a, 12489 Berlin, Germany

[3] Academy of Sciences and Arts of Bosnia and Herzegovina, Bistrik 7, 71000 Sarajevo, Bosnia and Herzegovina

E-mail:
senad.odzak@gmail.com,ahasovic1@yahoo.com,wbecker@mbi-berlin.de,milo@bih.net.ba

Abstract. We investigate laser-assisted electron-ion recombination (LAR), high-order harmonic generation (HHG) and above-threshold ionization (ATI) of argon atoms by a bicircular laser field, which consists of two coplanar counter-rotating circularly polarized fields of frequencies $r\omega$ and $s\omega$. The energy of soft x rays generated in the LAR process is analyzed as a function of the incident electron angle and numerical results of direct recombination of electrons with Ar^+ ions are presented. We also present the results of HHG by a bicircular field and confirm the selection rules derived earlier for inert-gas atoms in a p ground state. We show that the photoelectron spectra in the ATI process, presented in the momentum plane, as well as the LAR spectra exhibit the same discrete rotational symmetry as the applied field.

1. Introduction

Various nonlinear processes can occur when atoms or molecules are exposed to intense laser fields [1, 2]. Generally, these processes can be categorized into two groups: laser-assisted processes and laser-induced processes. The first group includes phenomena that can occur without the presence of a laser field, such as laser-assisted electron-atom scattering and laser-assisted electron-ion recombination (LAR) [3]. In contrast, laser-induced processes cannot happen except in the presence of a laser field. Examples of such processes are high-order harmonic generation (HHG) and above-threshold ionization (ATI) [4]. In the last two decades, all of these processes have been investigated in detail by considering atoms or molecules interacting with linearly polarized laser fields.

It was almost generally accepted that HHG and high-order above-threshold ionization (HATI) will be inefficient unless the laser polarization is linear. Otherwise, an electron that starts its orbit with zero velocity will almost never revisit its parent ion. For linear polarization, the electron trajectories are one-dimensional and along the direction of the linearly polarized driving laser field. However, already 20 years ago a scheme for efficient HHG in two dimensions was realized [5, 6], and 15 years ago its properties were investigated in detail and explained [7, 8, 9, 10, 11, 12]. This scheme employed a bicircular field, which consists of two coplanar counterrotating circularly polarized fields with frequencies that are integer multiples of a fundamental laser frequency. Very recently, this scheme has attracted great attention [13, 14, 15, 16, 17, 18, 19, 20, 21]. This is because a bicircular field generates harmonics that are circularly polarized (in addition to being bright). Related schemes employ a circularly polarized field plus a static magnetic field [22, 23] or crossed-beam two-color fields [24]. Bright circularly polarized harmonics open up possibilities for various applications, for example, for photoelectron circular dichroism in chiral molecules or x-ray magnetic circular dichroism spectroscopy, two important areas of chemistry and solid state physics.

In the present contribution we consider various atomic processes driven by a bicircular field. From the first-order processes we consider (direct) ATI and LAR and present a symmetry analysis of the angle-resolved ATI electron spectra and the emitted x-ray spectra. We will also consider higher-order processes such as HHG and HATI. Our underlying theory, which is based on the S-matrix formalism, is presented in Sec. 2 while the bicircular field and the corresponding symmetry analysis are presented in Sec. 3. Our numerical results are presented and discussed in Sec. 4 and finally our conclusions are summarized in Sec. 5. The atomic system of units ($\hbar = e = m_e = 4\pi\varepsilon_0 = 1$) is used throughout the paper.

2. Theory

In the present paper we use an S-matrix approach to describe the above-mentioned processes occurring in the presence of intense laser fields. In order to model the interaction of the strong laser field with atoms we use the single-active-electron approximation and the dipole approximation. The corresponding Hamiltonian can be written as

$$H(t) = -\frac{\boldsymbol{\nabla}^2}{2} + \mathbf{r}\cdot\mathbf{E}(t) + V(\mathbf{r}), \tag{1}$$

where $\mathbf{r}\cdot\mathbf{E}(t)$ is the laser field-electron interaction in the length gauge and $V(\mathbf{r})$ is the interaction of the electron with the rest of the atom in the absence of the laser field. The S-matrix element for a transition from an initial state $|\psi_i\rangle$ to a final state $|\psi_f\rangle$ is given by

$$S_{fi} = i \lim_{\substack{t\to\infty \\ t'\to-\infty}} \langle\psi_f(t)|U(t,t')|\psi_i(t')\rangle, \tag{2}$$

where $U(t,t')$ is the time evolution operator, which is a solution of the Schrödinger equation with the Hamiltonian (1). For the LAR (HHG, HATI) process the initial state is the unbound laser-free state $|\psi_i\rangle = |\mathbf{p}\rangle$ (atomic laser-free ground state $|\psi_B\rangle$) with $\langle\mathbf{r}|\mathbf{p}\rangle = (2\pi)^{-3/2}\exp(i\mathbf{p}\cdot\mathbf{r})$. The final state of the HATI (LAR, HHG) process is the continuum state $|\mathbf{p}\rangle$ (atomic bound state $|\psi_B\rangle$). In order to calculate the transition amplitude from the initial to the final state we use the strong-field approximation (SFA) in all of the above-mentioned processes [4, 1].

In the case of the direct LAR process the incident electron with momentum $\mathbf{p}$ recombines with the positive ion and an atomic bound state with energy $E_B < 0$ is formed. This process happens in the presence of a strong periodic laser field, which is defined by the electric field vector $\mathbf{E}(t)$, with the period T and frequency $\omega = 2\pi/T$, followed by emission of an x-ray photon with wave vector $\mathbf{K}$, frequency $\omega_{\mathbf{K}}$ and unit complex polarization vector $\hat{\mathbf{e}}_{\mathbf{K}}$. The LAR process is characterized by the differential power spectrum [3]

$$S(\mathbf{K},\mathbf{p}) = \frac{p\omega_{\mathbf{K}}^4}{2\pi c^3}\left|T_n\right|^2, \tag{3}$$

where $T_n = \mathbf{T}(\mathbf{K},\mathbf{p})\cdot\hat{\mathbf{e}}_{\mathbf{K}}^*$ is the T-matrix element. The number of photons n exchanged with the laser field is obtained from the energy-conserving condition

$$n = (\omega_{\mathbf{K}} + E_B - E_{\mathbf{p}} - U_{\mathrm{P}})/\omega, \tag{4}$$

where $E_{\mathbf{p}} = \mathbf{p}^2/2$ is the energy of the incident electron, $U_{\mathrm{P}} = \int_0^T \mathbf{A}^2(t)dt/(2T)$ is the ponderomotive energy and $\mathbf{A}(t) = -\int^t \mathbf{E}(t')dt'$ the vector potential. In length gauge and dipole approximation we have

$$\mathbf{T}(\mathbf{K},\mathbf{p}) = \int_0^T \frac{dt}{T} \langle \psi_B | \mathbf{r} e^{-i\mathbf{K}\cdot\mathbf{r}} | \mathbf{p} + \mathbf{A}(t) \rangle \exp\left\{ i(E_B + \omega_{\mathbf{K}})t - \frac{i}{2}\int^t dt' \left[\mathbf{p} + \mathbf{A}(t')\right]^2 \right\}. \tag{5}$$

In the case of the HHG process, the harmonic power (intensity) of the nth harmonic is defined by [16, 19]

$$P_n = \frac{(n\omega)^4}{2\pi c^3} |T_n|^2 , \tag{6}$$

where $T_n = \mathbf{T}_n \cdot \hat{\mathbf{e}}_n^*$ is the T-matrix element for the transition from the initial to the final state and $\hat{\mathbf{e}}_n = \hat{\mathbf{e}}_{\mathbf{K}}$ the complex unit polarization vector of the nth harmonic photon. Here

$$\mathbf{T}_n = \int_0^T \frac{dt}{T} \mathbf{d}_{fi}(t) e^{in\omega t} \tag{7}$$

is the Fourier component of the time-dependent dipole matrix element $\mathbf{d}_{fi}(t)$. Within the SFA, the time-dependent dipole can be approximated by [19]

$$\mathbf{d}_{fi}(t) \approx \sum_a \mathbf{d}_{fi}^a(t), \tag{8}$$

with the time-dependent dipole moment for the active electron, denoted with the index a,

$$\mathbf{d}_{fi}^a(t) \approx -i \left(\frac{2\pi}{i}\right)^{3/2} \int_0^\infty \frac{d\tau}{\tau^{3/2}} e^{iS_s} \langle \psi_{af} | \mathbf{r}_a | \mathbf{k}_s + \mathbf{A}(t) \rangle \langle \mathbf{k}_s + \mathbf{A}(t-\tau) | \mathbf{r}_a \cdot \mathbf{E}(t-\tau) | \psi_{ai} \rangle, \tag{9}$$

where $\mathbf{k}_s \equiv -\int_{t-\tau}^t dt' \mathbf{A}(t')/\tau$ is the stationary momentum, $S_s \equiv E_f t - \int_{t-\tau}^t dt' [\mathbf{k}_s + \mathbf{A}(t')]^2/2 - E_i(t-\tau)$ the action and $E_i = E_f = -I_{\mathrm{P}}$, where I_{P} is the ionization potential.

The ground-state atomic wave function is modeled by a linear combination of Slater-type orbitals obtained using the Hartree-Fock-Roothaan method [25]. For atoms with a p ground state the magnetic quantum number is $m = 0, \pm 1$. Taking into account the fact that for atoms with closed electron shells (Ne, Ar, Kr, Xe) m-changing transitions are forbidden by the Pauli exclusion principle, only three electrons are active. In this case we have [19]

$$\mathbf{d}_{fi}(t) \approx \sum_{a=1,2,3} \mathbf{d}_{fi}^a(t)\Big|_{m_i = m_f = m} = \sum_m \mathbf{d}_m(t), \tag{10}$$

where $\mathbf{d}_m(t)$ $(m = 0, \pm 1)$ is given by equation (9).

The SFA theory of the (H)ATI process is described in detail in [4]. The observable quantity is the averaged differential ionization rate given by

$$\overline{w}_{\mathbf{p}i}(n) = \frac{N_e}{2l+1} \sum_{m=-l}^{l} w_{\mathbf{p}ilm}(n), \tag{11}$$

where N_e is the number of equivalent electrons in the ionizing shell and $w_{\mathbf{p}ilm}(n) = 2\pi p |T_{\mathbf{p}i}(n)|^2$. Here n is the number of photons absorbed from the laser field such that $n\omega = E_{\mathbf{p}} + U_{\mathrm{P}} + I_{\mathrm{P}}$. The T-matrix element $T_{\mathbf{p}i}(n)$ can be written in the framework of the SFA as the sum

$$T_{\mathbf{p}i}(n) = \int_0^T \frac{dt}{T} e^{i[\mathbf{p}\cdot\boldsymbol{\alpha}(t) + U_1(t)]} \left[\mathcal{T}_{\mathbf{p}i}^{(0)}(t) + \mathcal{T}_{\mathbf{p}i}^{(1)}(t) \right] e^{in\omega t}, \tag{12}$$

where

$$\mathcal{T}^{(0)}_{\mathbf{p}i}(t) = \langle \mathbf{p} + \mathbf{A}(t) | \mathbf{r} \cdot \mathbf{E}(t) | \psi_i \rangle \tag{13}$$

describes the direct ATI process, while

$$\mathcal{T}^{(1)}_{\mathbf{p}i}(t) = -i \int_0^\infty d\tau \left(\frac{2\pi}{i\tau} \right)^{3/2} \langle \mathbf{p} | V | \mathbf{k}_s \rangle \langle \mathbf{k}_s + \mathbf{A}(t-\tau) | \mathbf{r} \cdot \mathbf{E}(t-\tau) | \psi_i \rangle e^{i[S_{\mathbf{k}_s i}(t-\tau) - S_{\mathbf{k}_s i}(t)]} \tag{14}$$

allows for one additional interaction of the liberated electron with its parent ion (HATI process). The phase factor in the previous equation is given by $S_{\mathbf{k}_s i}(t) = \int^t dt' \{ [\mathbf{k}_s + \mathbf{A}(t')]^2/2 + I_{\mathrm{P}} \} = (\mathbf{k}_s^2/2 + U_{\mathrm{P}} + I_{\mathrm{P}})t + \mathbf{k}_s \cdot \boldsymbol{\alpha}(t) + U_1(t)$, with $U_1(t) = \int^t \mathbf{A}^2(t')dt'/2 - U_{\mathrm{P}}t$ and $\boldsymbol{\alpha}(t) = \int^t \mathbf{A}(t')dt'$.

3. Bicircular field and dynamical symmetries

We use a bichromatic circularly polarized field with coplanar counterrotating components having the angular frequencies $r\omega$ and $s\omega$, which are integer multiples of the same fundamental frequency ω. The corresponding electric-field vector in the xy plane is defined by [7, 8, 9, 26]

$$\mathbf{E}(t) = \frac{i}{2} \left[E_r \hat{\mathbf{e}}_+ e^{-ir\omega t} + E_s \hat{\mathbf{e}}_- e^{-is\omega t} \right] + \text{c.c.}, \tag{15}$$

where $\hat{\mathbf{e}}_\pm = (\hat{\mathbf{e}}_x \pm i\hat{\mathbf{e}}_y)/\sqrt{2}$. In equation (15) E_j and $I_j = E_j^2$ are the amplitude and intensity of the jth field component of helicity h_j ($h_r = 1$, $h_s = -1$). The components of the bicircular field (15) are

$$E_x(t) = \left[E_r \sin(r\omega t) + E_s \sin(s\omega t) \right] / \sqrt{2}, \tag{16}$$

$$E_y(t) = \left[-E_r \cos(r\omega t) + E_s \cos(s\omega t) \right] / \sqrt{2}. \tag{17}$$

Denoting $A_r = E_r/(r\omega)$ and $A_s = E_s/(s\omega)$, for the ponderomotive energy we obtain $U_{\mathrm{P}} = (A_r^2 + A_s^2)/4$. The normalized electric-field vector $\mathbf{E}(t)$ and the pertinent normalized vector potential $\mathbf{A}(t)$ for various combinations of $(r,s) \in (1,2), (1,3), (2,3)$ and equal intensities of the field components are presented in figure 1.

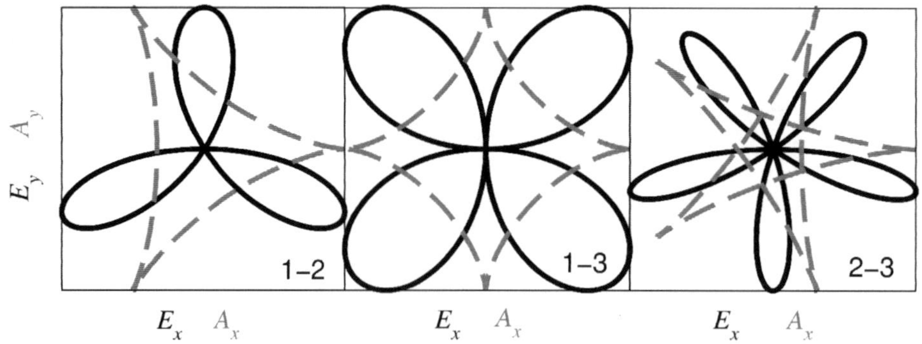

Figure 1. (Color online) Normalized electric-field vector $\mathbf{E}(t)$ (solid black lines) and vector potential $\mathbf{A}(t)$ (dashed red lines) of the $r\omega - s\omega$ bicircular laser field given by equation (15) for equal laser field component intensities and $(r,s) \in (1,2), (1,3), (2,3)$, plotted for $0 \le t \le T$, $T = 2\pi/\omega$.

Since the electric field vector lies in the xy plane, similarly as in references [7, 17], we define the degree of circular polarization $\xi_n = \mathrm{Im}\,(2T_{nx}^* T_{ny})/|T_n|^2$. Then the ellipticity of the emitted harmonic is given by

$$\varepsilon_n = \mathrm{sgn}(\xi_n)\left(\frac{1-\sqrt{1-\xi_n^2}}{1+\sqrt{1-\xi_n^2}}\right)^{1/2}. \tag{18}$$

The bicircular field $\mathbf{E}(t)$, as well as the vectors $\mathbf{A}(t)$ and $\boldsymbol{\alpha}(t)$, obey the following dynamical symmetry: the rotation by the angle $\alpha_j = -2\pi jr/(r+s)$ about the z axis is equivalent to the translation in time by $\tau_j = jT/(r+s)$ [16, 19, 27, 28], i.e.

$$R_z(\alpha_j)\mathbf{E}(t) = \mathbf{E}(t+\tau_j), \tag{19}$$

where j is an integer. The diagonal matrix elements of the rotation matrix $R_z(\alpha_j)$ are $\cos\alpha_j$, while the off diagonal elements are $\pm\sin\alpha_j$. We expect that the spectra of HHG and LAR photons as well as the spectra of (H)ATI electrons exhibit the same type of the symmetry [19, 27, 28, 29]. In addition to the discrete rotational symmetry (19), the bicircular fields observe a reflection symmetry about any of the various obvious symmetry axes.

4. Numerical results

In this section, we will present numerical results for the processes HHG, HATI and LAR. As an example, we choose the Ar atom (Ar^+ ion) with a $2p$ ground state for the HHG and HATI (LAR) process whose bound-state energy is -15.75 eV. We use the bicircular field (15) with $r=1$ and $s=2$, equal component intensities $I_r = I_s = 1.5\cdot 10^{14}$ W/cm^2 and the fundamental wavelength 800 nm. For the LAR process the momentum of the incident electron is in the xy plane, such that $\mathbf{p}\cdot\hat{\mathbf{e}}_x = p\cos\theta$.

In figure 2 we present the differential power spectrum coded in false color, as a function of the incident-electron angle (horizontal axis) and the x-ray energy (vertical axis) for the incident-electron energy $E_{\mathbf{p}} = 50$ eV and the magnetic quantum $m=1$ number of the ground state (qualitatively similar results are obtained for the $m=-1$ case). The figure is dominated by the superposition of three sine-like wave patterns per cycle, which create conical-like shapes. In [27] these structures are explained by interference of different numbers of classical solutions. Additionally, for the chosen combination $r=1$ and $s=2$, the effects of the dynamical symmetry (19) are clearly visible, as explained in section 2. The afore-mentioned reflection symmetry is also obvious, as the pattern is symmetric about any angle that is a multiple of 60°.

In figure 3 we present the harmonic power as a function of the harmonic order for Ar atoms modeled by a realistic $2p$ state with four orbitals [25]. Our numerical results presented in figure 3 confirm the selection rules [5, 7] where the harmonics $n=3q$ (q integer) are absent, while the harmonics $n=3q\pm1$ have the ellipticity $\varepsilon_n = \pm1$. Additionally, from the left panel of figure 3 it is clearly visible that the intensities of the harmonics $n=3q-1$, having the ellipticity $\varepsilon_n=-1$, are different from those of the harmonics $n=3q+1$, for which $\varepsilon_n = 1$. Which ones of the harmonics with positive or negative ellipticities are stronger depends on their position in the plateau. The same feature has also been observed in spectra obtained from solutions of the time-dependent Schrödinger equation [18]. The partial spectra for particular values $m=-1$ and $m=+1$ are shown in the right panel of figure 3. The asymmetry between the $\varepsilon_n=-1$ and $\varepsilon_n=1$ harmonics is also visible for all four possible combinations $(\varepsilon_n, m) = (\pm1,\pm1)$ and more pronounced than in the m-averaged case.

The logarithm of the differential ionization rate of Ar atoms is presented as a function of the electron momentum in figure 4. We show the spectra of the so-called direct electrons (electrons that reach the detector without interacting with the parent ion after ionization) as well as those of the rescattered electrons (electrons that undergo exactly one scattering off

Journal of Physics: Conference Series **691** (2016) 012004 doi:10.1088/1742-6596/691/1/012004

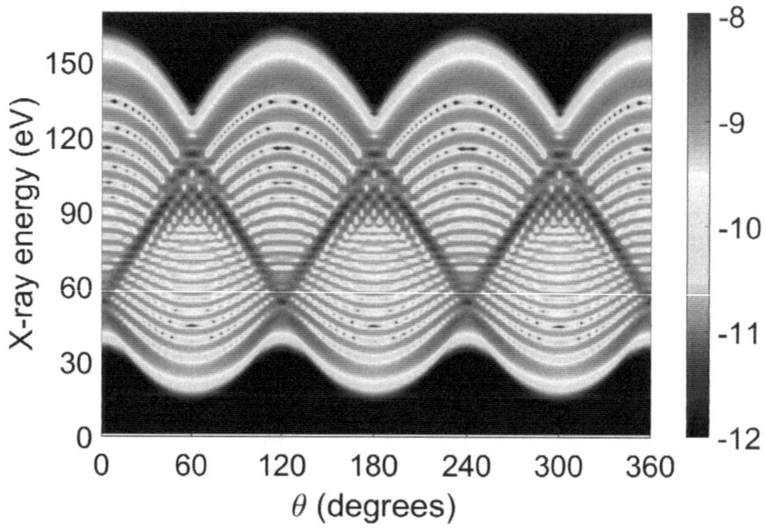

Figure 2. (Color online) Logarithm of the differential power spectrum presented in false color as a function of the incident electron angle and the x-ray energy for laser-assisted radiative recombination of an electron with an Ar^+ ion in the presence of the bicircular laser field (15) with the fundamental wavelength of 800 nm, $I_r = I_s = 1.5 \cdot 10^{14}$ W/cm^2 and $r = 1$, $s = 2$. The incident electron energy is $E_{\mathbf{p}} = 50$ eV and the Ar bound state is $2p$ with the magnetic quantum number $m = 1$.

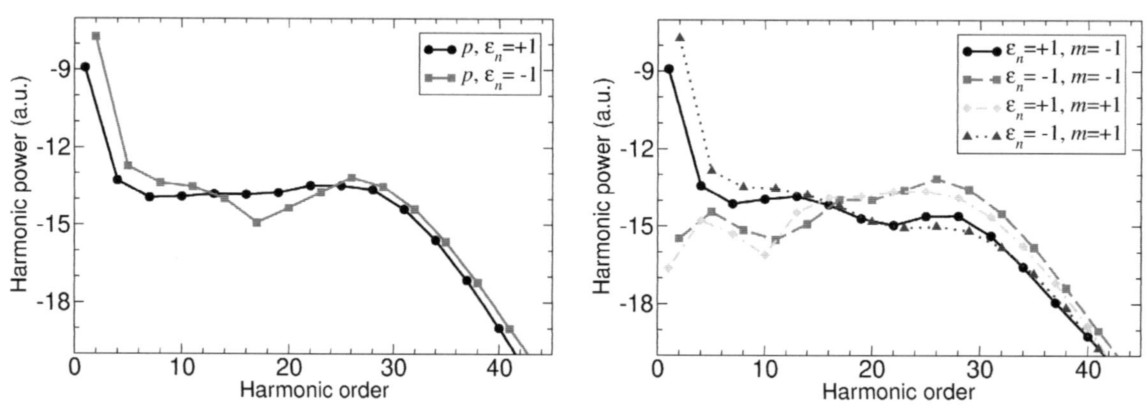

Figure 3. (Color online) Harmonic power as a function of the harmonic order for HHG by the bicircular field (15) for Ar atoms and the field-component wavelengths $\lambda_1 = 800$ nm and $\lambda_2 = 400$ nm ($r = 1$, $s = 2$) and intensities $I_r = I_s = 1.5 \cdot 10^{14}$ W/cm^2. The nth harmonic ellipticity is $\varepsilon_n = \pm 1$ for $n = 3q \pm 1$, where q is an integer. The left panel exhibits the results obtained using the realistic $2p$ ground state, while the right panel displays separate results for specific values of the initial- and final-state magnetic quantum numbers $m_i = m_f = m$.

the parent ion after ionization). As one can see the contribution of the direct electrons to the ionization rate is more then one order of magnitude larger than that of the rescattered electrons. Furthermore, as expected from the streaking condition $\mathbf{p} + \mathbf{A}(t) = \mathbf{0}$ the direct electrons are emitted predominantly in the direction opposite to the vector potential at the time of ionization (i.e. in the directions determined by the emission angles $\theta = 60°, 180°$ and $300°$). The structure of the rescattered electron spectrum is more complex. It consists of three broad peaks at the emission angles $\theta \approx 60^\circ, 180^\circ$ and 300° from which extensions emanate in the clockwise direction, which while becoming weaker and weaker proceed to form concentric rings similar to the ATI rings that are familiar from ionization by linear polarized fields; see, e.g., [30].

It is obvious that both spectra are symmetric with respect to rotations about the z axis by multiples of an angle of 120°. However, in contrast to the spectrum of the rescattered electrons, the direct electron spectrum obeys an additional reflection symmetry about one of the three axes at the angles of $60^\circ, 180^\circ$ and 300°. The presence or absence of the reflection symmetry allows one to assess whether or not rescattering plays a role in a given spectrum. It should be mentioned that ATI spectra of Ar atoms generated by a bicircular field having an intensity three times lower than the one used in figure 4 have recently been examined in Ref. [31]. The three-fold symmetry as well as traces of rescattered electrons are clearly visible in this experiment.

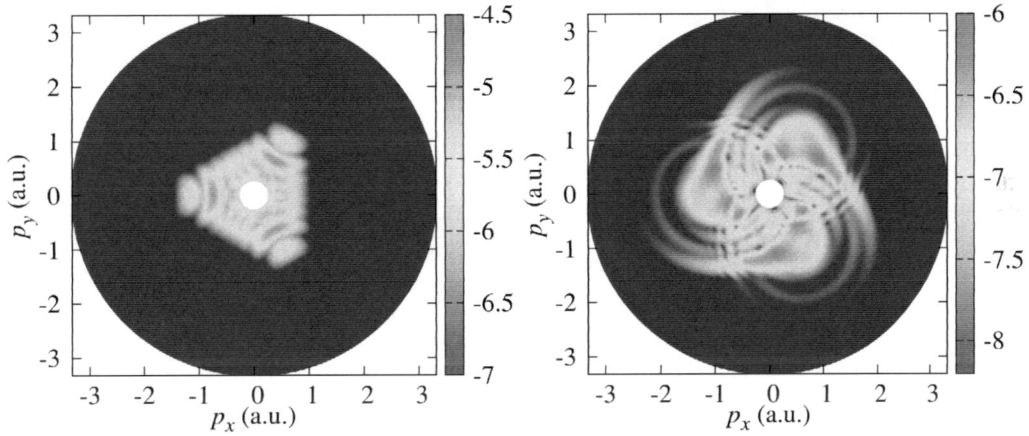

Figure 4. (Color online) Logarithm of the differential ionization rate of Ar atoms presented in the momentum plane for the laser-field parameters as in figure 2. The contribution of the direct (rescattered) electrons is shown in the left (right) panel.

5. Conclusions

We applied the SFA theory to investigate the LAR, HHG and HATI processes in the presence of a bicircular laser field. For the LAR process we numerically calculated the differential power spectrum of the emitted x rays as a function of the incident-electron angle θ for a fixed value of the incoming electron energy $E_{\mathbf{p}}$. The spectrum exhibits symmetry under rotations by an angle of $2\pi/(r+s) = 120^\circ$, which reflects the symmetry of the applied laser field.

Considering the HHG process of argon atoms with a p ground state, we found strong asymmetry in the emission of left- and right-circularly polarized harmonics. According to the selection rules the harmonics $n = 3q \pm 1$, with q an integer, are emitted. The HHG spectra display a strong dependence on the magnetic quantum number m. The asymmetry between emission of circularly polarized harmonics with opposite helicities ($\varepsilon_n = \pm 1$) is important for the investigation of photoelectron circular dichroism in chiral molecules and for x-ray magnetic circular dichroism spectroscopy. These circularly polarized harmonics provide a much-desired table-top source of coherent soft x rays.

We also investigated above-threshold ionization generated by a bicircular field. The photoelectron spectrum presented in the momentum plane is symmetric with respect to rotations by multiples of 120° about the z axis, both for the direct and the rescattered electrons. The direct electron spectrum exhibits an additional symmetry: reflection about one of three different axes in the xy plane. Since this kind of symmetry is absent for the rescattered electrons, it allows one to assess whether or not rescattering plays a role in a given spectrum.

The combination of attoclock features with the discrete rotational symmetry makes the bicircular field a powerful tool to investigate the electron dynamics on the attosecond time scale. Since a bicircular field develops in a plane, the electron trajectories are two-dimensional and the direct and rescattering ATI structures are well separated. We expect that bicircular fields can be especially useful for the investigation of the dynamics of planar molecules.

We gratefully acknowledge support by the Federal Ministry of Education and Science, Bosnia and Herzegovina.

References

[1] Milošević D B and Ehlotzky F 2003 *Adv. At. Mol. Opt. Phys.* **49** 373
[2] Lein M 2007 *J. Phys. B: At. Mol. Opt. Phys.* **40** R135
[3] Milošević D B and Ehlotzky F 2002 *Phys. Rev.* A **65** 042504
[4] Becker W, Grasbon F, Kopold R, Milošević D B, Paulus G G and Walther H 2002 *Adv. At. Mol. Opt. Phys.* **48** 35
[5] Eichmann H, Egbert A, Nolte S, Momma C, Wellegehausen B, Becker W, Long S and McIver J K 1995 *Phys. Rev.* A **51** R3414(R)
[6] Long S, Becker W and McIver J K 1995 *Phys. Rev.* A **52** 2262
[7] Milošević D B, Becker W and Kopold R 2000 *Phys. Rev.* A **61** 063403
[8] Milošević D B and Sandner W 2000 *Opt. Lett.* **25** 1532
[9] Milošević D B, Becker W, Kopold R and Sandner W 2001 *Laser Phys.* **11** 165
[10] Milošević D B and Becker W 2000 *Phys. Rev.* A **62** 011403(R)
[11] Ceccherini F, Bauer D and Cornolti F 2003 *Phys. Rev.* A **68** 053402
[12] Milošević D B and Becker W 2005 *J. Mod. Opt.* **52** 233
[13] Fleischer A, Kfir O, Diskin T, Sidorenko P and Cohen O 2014 *Nature Photon.* **8** 543
[14] Pisanty E, Sukiasyan S and Ivanov M 2014 *Phys. Rev.* A **90** 043829
[15] Kfir O *et al* 2015 *Nature Photon.* **9** 99
[16] Milošević D B 2015 *Opt. Lett.* **40** 2381
[17] Milošević D B 2015 *J. Phys. B: At. Mol. Opt. Phys.* **48** 171001
[18] Medišauskas L, Wragg J, van der Hart H and Ivanov M Yu 2015 *Phys. Rev. Lett.* **115** 153001
[19] Milošević D B 2015 *Phys. Rev.* A **92** 043827
[20] Fan T *et al* 2015 *Proc. Natl. Acad. Sci. USA* **112** 14206
[21] Yuan K-J and Bandrauk A D 2015 *Phys. Rev.* A **92** 063410
[22] Zuo T and Bandrauk A D 1995 *J. Nonlinear Opt. Phys. Mat.* **4** 533
[23] Bandrauk A D and Lu H Z 2003 *Phys. Rev.* A **68** 043408
[24] Tong X M and Chu S I 1998 *Phys. Rev.* A **58** R2656
[25] Radzig A A and Smirnov B M 1985 *Reference Data on Atoms, Molecules and Ions* (Berlin: Springer)
[26] Milošević D B, Becker W and Kopold R 2001 *Atoms, Molecules and Quantum Dots in Laser Fields: Fundamental Processes* (Conference Proceedings vol 71) ed N Bloembergen, N Rahman and A Rizzo (Bologna: Societá Italiana di Fisica) pp 239–252
[27] Odžak S and Milošević D B 2015 *Phys. Rev.* A **92** 053416
[28] Kramo A, Hasović E, Milošević D B and Becker W 2007 *Laser Phys. Lett.* **4** 279
[29] Hasović E, Kramo A and Milošević D B 2008 *Eur. Phys. J. Spec. Top.* **160** 205
[30] Möller M, Meyer F, Sayler A M, Paulus G G, Kling M F, Schmidt B E, Becker W and Milošević D B 2014 *Phys. Rev.* A **90** 023412
[31] Mancuso C A *et al* 2015 *Phys. Rev.* A **91** 031402(R)

Coherence in laser-induced Compton scattering

F. Cajiao Vélez[1], **K. Krajewska**[1,2], **and J. Z. Kamiński**[1]
[1]Institute of Theoretical Physics, Faculty of Physics, University of Warsaw, Pasteura 5, 02-093 Warszawa, Poland
[2] Department of Physics and Astronomy, University of Nebraska, Lincoln, Nebraska 68588-0299, USA
E-mail: katarzyna.krajewska@fuw.edu.pl

Abstract. The concept of the electron mass dressing by a powerful laser pulse is discussed. It is shown, by considering the coherent frequency combs generated out of the Compton radiation, how the electron dressed mass can be determined experimentally. This also opens a possibility to measure properties of extremely intense pulses for which the previously developed methods, working at moderate intensities, are not applicable. Namely, one can determine these properties from the properties of coherent Compton radiation.

1. Motivation

The concept of the electron mass dressing by a laser field is almost as old as the laser itself. It was V. I. Ritus, A. I. Nikishov and their collaborators (see, e.g., [1, 2, 3] and references therein) who thoroughly investigated the importance of this concept by considering fundamental processes of quantum electrodynamics (QED), modified or induced by an action of intense electromagnetic plane wave (for recent reviews see, e.g., Refs. [4, 5, 6, 7, 8]). The prominent example of such a process is the Compton scattering in which an electron, interacting with a laser field, emits a nonlaser photon. Its classical analogue is the so-called Thomson scattering in which an electron, accelerated or decelerated while interacting with a laser field, emits an electromagnetic radiation.

If an electron interacts with a monochromatic electromagnetic plane wave of the frequency $\omega_{\rm L}$ and of the time-averaged intensity I, the energy spectrum of Compton photons is defined by the so-called Klein-Nishina formula [9, 10],

$$\omega_{\boldsymbol{K},N} = \frac{N\omega_{\rm L}}{\frac{p_{\rm i}\cdot n_{\boldsymbol{K}}}{p_{\rm i}\cdot n} + \frac{U}{c}\frac{n\cdot n_{\boldsymbol{K}}}{p_{\rm i}\cdot n} + \frac{N\omega_{\rm L}}{\omega_{\rm cut}}}, \tag{1}$$

with positive integers N. In this formula, the symbol $a\cdot b$ means the relativistic scalar product of four-vectors, $a\cdot b = a^0 b^0 - \boldsymbol{a}\cdot\boldsymbol{b}$, with the four-vector $a=(a^0,\boldsymbol{a})$ and similarly for b, where $\boldsymbol{a}\cdot\boldsymbol{b}$ is the three-dimensional scalar product. Here, it is assumed that the plane wave propagates in the space direction $\boldsymbol{n}$, meaning that $n=(1,\boldsymbol{n})$ and $n\cdot n=0$. Similarly, the Compton photon is generated in the direction $\boldsymbol{n_K}$ [$n_{\boldsymbol{K}}=(1,\boldsymbol{n_K})$ and $n_{\boldsymbol{K}}\cdot n_{\boldsymbol{K}}=0$] with the four-momentum $K_N=(\omega_{\boldsymbol{K},N}/c)n_{\boldsymbol{K}}$, and $p_{\rm i}=(p_{\rm i}^0,\boldsymbol{p}_{\rm i})$ is the electron initial momentum on the free-electron mass shell, $p_{\rm i}\cdot p_{\rm i}=(m_{\rm e}c)^2$. The so-called ponderomotive energy of electrons in the laser field equals

$$U = \frac{1}{4}\mu^2\frac{(m_{\rm e}c^2)^2}{cp_{\rm i}\cdot n}. \tag{2}$$

Here,

$$\mu = \frac{|e|\mathcal{E}_0}{m_{\rm e} c \omega_{\rm L}}, \tag{3}$$

where the electric field amplitude of the plane wave, $\mathcal{E}_0$, is related to the time-averaged intensity of the laser field such that (ε_0 means the vacuum permittivity)

$$I = \frac{1}{2}\varepsilon_0 c \mathcal{E}_0^2. \tag{4}$$

Moreover,

$$\omega_{\rm cut} = \frac{c}{\hbar}\frac{p_{\rm i}\cdot n}{n\cdot n_{\boldsymbol{K}}} \tag{5}$$

defines the maximum value of $\omega_{\boldsymbol{K},N}$ (i.e., $\omega_{\boldsymbol{K},N} \to \omega_{\rm cut}$ when $N \to \infty$), and it represents the only parameter that contains the Planck constant $\hbar$. Note that, in the limit $\hbar \to 0$, we recover the classical Thomson formula for frequencies of generated radiation,

$$\omega^{\rm Th}_{\boldsymbol{K},N} = \frac{N\omega_{\rm L}}{\frac{p_{\rm i}\cdot n_{\boldsymbol{K}}}{p_{\rm i}\cdot n} + \frac{U}{c}\frac{n\cdot n_{\boldsymbol{K}}}{p_{\rm i}\cdot n}}, \tag{6}$$

which are not limited from above as, in the classical limit, $\omega_{\rm cut} \to \infty$. Hence, one can find the scaling law,

$$\omega^{\rm Th}_{\boldsymbol{K},N} = \frac{\omega_{\boldsymbol{K},N}}{1-\frac{\omega_{\boldsymbol{K},N}}{\omega_{\rm cut}}}, \tag{7}$$

which relates the Compton and Thomson frequencies. The consequences of this scaling law for the plane waves and long laser pulses have been studied in [11, 12, 13], whereas its extension to arbitrary short laser pulses with the discussion of the electron spin and laser pulse polarization effects has been investigated in [14]. In our further analysis we use units in which $\hbar = 1$.

It follows from the Klein-Nishina formula [Eq. (1)] that, for a given kinematics of the Compton scattering, the measurement of positions of two consecutive peaks in the Compton spectrum, $\omega_{\boldsymbol{K},N}$ and $\omega_{\boldsymbol{K},N+1}$, allows one to determine the ponderomotive energy U. Indeed, after some algebra, we arrive to the expression,

$$\frac{p_{\rm i}\cdot n_{\boldsymbol{K}}}{p_{\rm i}\cdot n} + \frac{U}{c}\frac{n\cdot n_{\boldsymbol{K}}}{p_{\rm i}\cdot n} = \frac{\omega_{\rm L}}{\omega_{\boldsymbol{K},N+1}-\omega_{\boldsymbol{K},N}}\Big(1-\frac{\omega_{\boldsymbol{K},N+1}}{\omega_{\rm cut}}\Big)\Big(1-\frac{\omega_{\boldsymbol{K},N}}{\omega_{\rm cut}}\Big). \tag{8}$$

Note that the right-hand-side of this expression has to be independent of N. This means that the proposed procedure of measuring U does not require the knowledge of the order of the Compton frequency, N, but only positions of arbitrary chosen two adjacent peaks. In general, one could select peaks in the spectrum corresponding to frequencies $\omega_{\boldsymbol{K},N}$ and $\omega_{\boldsymbol{K},N+M}$, $M = 1, 2, \ldots$, and arrive at the similar conclusion.

In order to relate the ponderomotive energy U to the electron mass dressing let us recall that the Volkov solution of the Dirac equation [15] for an electron with the four-momentum $p = (p^0, \boldsymbol{p})$, $p\cdot p = m_{\rm e}^2 c^2$, is

$$\psi^{(+)}_{\boldsymbol{p}\lambda}(x) = \sqrt{\frac{m_{\rm e}c^2}{VE_{\boldsymbol{p}}}}\Big[1 - \frac{e}{2k\cdot p}\not{A}(k\cdot x)\not{k}\Big] u^{(+)}_{\boldsymbol{p}\lambda} {\rm e}^{-{\rm i}S^{(+)}_p(x)}, \tag{9}$$

where

$$S^{(+)}_p(x) = p\cdot x + \int_0^{k\cdot x}\Big[e\frac{A(\phi)\cdot p}{k\cdot p} - e^2\frac{A^2(\phi)}{2k\cdot p}\Big]{\rm d}\phi, \tag{10}$$

$E_{\boldsymbol{p}} = cp^0$, and $u^{(+)}_{\boldsymbol{p}\lambda}$ is the free-electron bispinor normalized such that $\bar{u}^{(+)}_{\boldsymbol{p}\lambda} u^{(+)}_{\boldsymbol{p}\lambda'} = \delta_{\lambda\lambda'}$, with $\lambda = \pm$ labeling its spin degrees of freedom. Moreover, for the linearly polarized plane wave, the vector potential has the form,

$$A(\phi) = \varepsilon \frac{\mathcal{E}_0}{\omega_{\rm L}} \cos\phi, \tag{11}$$

where ε is the polarization four-vector such that $\varepsilon = (0, \boldsymbol{\varepsilon})$ and $\varepsilon \cdot \varepsilon = -\boldsymbol{\varepsilon} \cdot \boldsymbol{\varepsilon} = -1$. Note also that $k = (\omega_{\rm L}/c)(1, \boldsymbol{n})$, with the unit vector $\boldsymbol{n}$ defined above. Hence, the function $S^{(+)}_p(x)$ can be splitted into two parts

$$S^{(+)}_p(x) = \bar{p} \cdot x + G_p(k \cdot x), \tag{12}$$

where $G_p(\phi)$ is a periodic function of ϕ with the period of 2π, whereas

$$\bar{p} = p + \frac{1}{4}(\mu m_{\rm e} c)^2 \frac{k}{p \cdot k} \tag{13}$$

is the so-called electron dressed momentum by the laser field. One can easily verify that

$$\bar{p} \cdot \bar{p} = (\overline{m}_{\rm e} c)^2, \quad \overline{m}_{\rm e} = m_{\rm e} \sqrt{1 + \frac{1}{2}\mu^2}. \tag{14}$$

This means that the electron momentum in the laser field is on the mass shell with the dressed mass $\overline{m}_{\rm e}$. By determining the ponderomotive energy U and, hence, the parameter μ [see, Eq. (2)], the electron dressed mass [Eq. (14)] can be directly determined from the Compton spectrum.

The procedure described above allows one to determine the electron mass dressing in the presence of a monochromatic plane wave or, after some modifications, in the presence of polychromatic plane waves with commensurate frequencies. In real experiments, however, one deals with finite-in-time and focused-in-space laser pulses. The requirement of space focusing can be soften by considering the head-on collision of the laser pulse with relativistic electrons. Under such conditions, in the electron reference frame, the laser focus is largely oblate. Thus, the laser pulse can be modeled as a plane-wave-fronted pulse, which is finite in its propagation direction but it is infinite in the perpendicular directions [16]. For such laser pulses the following questions arise:

(i) Is it still meaningful to consider the electron mass dressing by an arbitrary short laser pulse?

(ii) If yes, can we measure the electron dressed mass for an arbitrary chosen laser pulse?

The purpose of our paper is to answer these two questions.

2. Momentum dressing by short laser pulses

Assume that the laser pulse lasts from $t = 0$ up to $t = T_{\rm p}$ and vanishes beyond this time interval. Note that its time duration defines the fundamental frequency, $\omega = 2\pi/T_{\rm p}$, which is the smallest frequency in the spectral decomposition of the pulse. If $k = (\omega/c)(1, \boldsymbol{n})$, then the electromagnetic vector potential has the form,

$$A(\phi) = A_0(\varepsilon_1 f_1(\phi) + \varepsilon_2 f_2(\phi)), \quad A_0 = \frac{\mathcal{E}_0}{\omega}, \tag{15}$$

where $\phi = k \cdot x$. Moreover, $\varepsilon_j = (0, \boldsymbol{\varepsilon}_j)$ for $j = 1, 2$ denote two linear polarizations of the laser field such that $\varepsilon_{j'} \cdot \varepsilon_j = -\delta_{j'j}$ and $k \cdot \varepsilon_j = 0$. The two functions $f_j(\phi)$, $j = 1, 2$, which we call the shape functions, are arbitrary and sufficiently smooth functions that vanish for $\phi < 0$ and $\phi > 2\pi$. We also define their mean values as

$$\langle f_j^m \rangle = \frac{1}{2\pi} \int_0^{2\pi} {\rm d}\phi \, [f_j(\phi)]^m, \quad j, m = 1, 2. \tag{16}$$

Journal of Physics: Conference Series **691** (2016) 012005 doi:10.1088/1742-6596/691/1/012005

With these definitions, the function $S_p^{(+)}(x)$ [Eq. (10)], can be decomposed as in (12). This time, however, the electron dressed momentum becomes

$$\bar{p} = p - \mu m_{\rm e}c\Big(\frac{p\cdot\varepsilon_1}{p\cdot k}\langle f_1\rangle + \frac{p\cdot\varepsilon_2}{p\cdot k}\langle f_2\rangle\Big)k + \frac{1}{2}(\mu m_{\rm e}c)^2\frac{\langle f_1^2\rangle + \langle f_2^2\rangle}{p\cdot k}k. \tag{17}$$

Is it the most general form of the momentum dressing? The answer to this question depends on the QED process under considerations. As it follows from the analysis presented in [17], for the Compton scattering, the scattering amplitude depends only on the difference between the final and initial dressed momenta, $\bar{p}_{\rm f} - \bar{p}_{\rm i}$. This means that the physical observables do not change if we redefine the laser field dressing such that

$$\bar{p}_{\rm f,i} \to \bar{p}_{\rm f,i} + P, \tag{18}$$

where P is an arbitrary four-vector. (Another transformation has to be introduced, for instance, for the Breit-Wheeler process, as it has been discussed in [18].)

In general, the four-vector P can be represented as

$$P = g_1\varepsilon_1 + g_2\varepsilon_2 + g_0 k, \tag{19}$$

where, in the absence of the laser field, $g_j = 0$ ($j = 0, 1, 2$). It appears that the shifted dressed momentum (18) is on the mass-shell (i.e., $\bar{p}_{\rm f,i}\cdot\bar{p}_{\rm f,i}$ is independent of $p_{\rm f,i}$) for a particular choice of g_j such that

$$g_1 = \mu m_{\rm e}c\langle f_1\rangle,\quad g_2 = \mu m_{\rm e}c\langle f_2\rangle,\quad g_0 = 0. \tag{20}$$

In this case,

$$\bar{p}_{\rm f,i}\cdot\bar{p}_{\rm f,i} = (\overline{m}_{\rm e}c)^2 = (m_{\rm e}c)^2\Big[1 + \frac{2U_0}{m_{\rm e}c^2}\Big], \tag{21}$$

where $\overline{m}_{\rm e}$ can be called the electron dressed mass in the laser pulse, whereas

$$U_0 = \frac{1}{2}\mu^2 m_{\rm e}c^2[\langle f_1^2\rangle - \langle f_1\rangle^2 + \langle f_2^2\rangle - \langle f_2\rangle^2]. \tag{22}$$

Note that, for the monochromatic plane wave, when $\langle f_j\rangle = 0$, $j = 1, 2$, and $\langle f_1^2\rangle + \langle f_2^2\rangle = 1/2$, the last two equations lead to the previous result (14). This answers our first question raised in Sec. 1: *if one accepts the above definition of the momentum dressing* (17) *then the measurement of the electron dressed mass induced by a pulse is equivalent to determining its three characteristics* [19],

$$\mu\langle f_j\rangle,\ j = 1, 2, \quad \text{and} \quad \mu^2[\langle f_1^2\rangle + \langle f_2^2\rangle]. \tag{23}$$

Actually, this task belongs to a broader topic which is the diagnosis of powerful laser pulses. The point being that, for such pulses, the conventional diagnostic methods developed for moderately intense fields are not applicable. Therefore, one may determine the peak intensity of the laser pulse by studying the properties of radiation generated in either the laser-modified recombination process [20, 21, 22] or in the Thomson scattering [23]. It seems that this goal can be realized even more effectively by analyzing the ionization spectrum of photoelectrons [24] (see, also Ref. [25]). Another possibility is to exploit the chirp of the laser pulse in the Thomson or Compton scattering [26, 27, 28]. Moreover, the problem of the relative change of the electron mass dressing by two laser pulses of different shapes has been studied in [29]. The latter, however, does not provide a method for a direct measurement of the electron dressed mass. The aim of the next Sections is to discuss how the electron dressed mass can be determined directly (and, from the theoretical point of view, very precisely even for extremely short one-cycle laser pulses) from the spectrum of radiation generated during the Compton scattering.

3. Coherent frequency combs

It is known that the coherent properties of radiation, which manifest themselves in interference and diffraction experiments, can lead to very precise measurements of physical quantities. This approach was initiated by T. Young and his famous double-slit experiment [30], which then triggered the development of other experimental techniques (like diffraction gratings). Among them, the frequency comb generation is of a particular interest. Generation of optical frequency combs [31, 32] has become an invaluable experimental tool for precise measurements of optical frequencies, that also enabled the creation of optical atomic clocks [33]. A particular realization of this technique, called the high-order harmonic generation [34], has led to the development of attoscience [35, 36]. This newly growing field is related to the interaction of matter with attosecond pulses of coherent radiation. Also, it has been proposed that the interaction of modulated laser pulses with relativistic and nearly monoenergetic electron beams can create x- and γ-ray frequency combs in the keV or MeV region [37, 38]. This may extend the applicability of frequency comb generation to nuclear and elementary particle physics. The same phenomenon was studied for matter and anti-matter waves [37, 39].

It has been demonstrated for the Compton scattering stimulated by a finite train of identical pulses [37, 38] that the following interference/diffraction formula, which describes the probability amplitude of the process, $\mathcal{A}_\sigma(\omega_{\boldsymbol{K}},\lambda_{\rm i},\lambda_{\rm f})$, holds

$$\mathcal{A}_\sigma(\omega_{\boldsymbol{K}},\lambda_{\rm i},\lambda_{\rm f}) = \exp\big[{\rm i}\Phi_\sigma(\omega_{\boldsymbol{K}},\lambda_{\rm i},\lambda_{\rm f})\big]\frac{\sin(\pi\bar{Q}^+/k^0)}{\sin(\pi\bar{Q}^+/k^0N_{\rm rep})}|\mathcal{A}_\sigma^{(1)}(\omega_{\boldsymbol{K}},\lambda_{\rm i},\lambda_{\rm f})|. \tag{24}$$

Here, $N_{\rm rep}$ is the number of laser pulses in the train, $\mathcal{A}_\sigma^{(1)}(\omega_{\boldsymbol{K}},\lambda_{\rm i},\lambda_{\rm f})$ is the probability amplitude of Compton scattering by a single pulse from the train, and $\Phi_\sigma(\omega_{\boldsymbol{K}},\lambda_{\rm i},\lambda_{\rm f})$ is the global phase. Moreover, σ labels the polarization of the Compton photon, and $\lambda_{\rm i}$ and $\lambda_{\rm f}$ define the spin degrees of freedom of the initial and final electron states, respectively. The coherent enhancement of the Compton amplitude $\mathcal{A}_\sigma(\omega_{\boldsymbol{K}},\lambda_{\rm i},\lambda_{\rm f})$ is observed for such frequencies of emitted photons $\omega_{\boldsymbol{K},N}$ (with integer N), that satisfy the condition

$$\bar{Q}^+/k^0N_{\rm rep} = -N, \quad \bar{Q}^+ = \bar{p}_{\rm i}^+ - \bar{p}_{\rm f}^+ - K^+, \tag{25}$$

provided that $N_{\rm rep} > 1$. Here, the light-cone variable $\bar{Q}^+ = (\bar{Q}^0 + \boldsymbol{n}\cdot\bar{\boldsymbol{Q}})/2$ (and similarly for the remaining four-momenta of electron and photon) has been used. It follows from Eq. (25) that the probability distribution of Compton scattering ($\sim|\mathcal{A}_\sigma(\omega_{\boldsymbol{K}},\lambda_{\rm i},\lambda_{\rm f})|^2$) is enhanced by a factor of $N_{\rm rep}^2$. Meaning that the energy spectrum of radiated photons consists of a sequence of well separated peaks, which is in contrast to the Compton process induced by a single pulse (for which $N_{\rm rep} = 1$). Note also that, in contrast to the Thomson scattering, these peaks are not *exactly* equally separated from each other over the whole interval of allowed frequencies $(0,\omega_{\rm cut})$. When $\omega_{\boldsymbol{K}}$ approaches the cut-off value, the distribution of $\omega_{\boldsymbol{K},N}$ becomes increasingly denser. Thus, one can get the regular frequency combs out of the Compton radiation only within limited frequency intervals.

4. Generalized Klein-Nishina formula and laser pulse diagnosis

Eq. (25) allows us to determine the peak frequencies in the energy spectrum of emitted Compton photons. After algebraic manipulations, we arrive at the *generalized Klein-Nishina formula*,

$$\omega_{\boldsymbol{K},N}^{(\rm GKN)} = \frac{(N/N_{\rm osc})\omega_{\rm L}}{\frac{p_{\rm i}\cdot n_{\boldsymbol{K}}}{p_{\rm i}\cdot n} + \frac{\nu n\cdot n_{\boldsymbol{K}} + g_1 p_{{\rm i},1} + g_2 p_{{\rm i},2}}{(p_{\rm i}\cdot n)^2} + \frac{(N/N_{\rm osc})\omega_{\rm L}}{\omega_{\rm cut}}}, \tag{26}$$

which for the plane wave (i.e., when $N_{\rm osc} = 1$ and $g_j = 0$) reduces to the original one (1). Here, g_1 and g_2 are defined in Eq. (20),

$$\nu = \frac{1}{2}(\mu m_{\rm e}c)^2(\langle f_1^2\rangle + \langle f_2^2\rangle), \tag{27}$$

and (for $j = 1, 2$)

$$p_{\mathrm{i},j} = (p_\mathrm{i} \cdot n)(n_{\boldsymbol{K}} \cdot \varepsilon_j) - (p_\mathrm{i} \cdot \varepsilon_j)(n \cdot n_{\boldsymbol{K}}). \tag{28}$$

Similar to the original Klein-Nishina formula, the quantum signature of Eq. (26) is hidden in the definition of ω_cut. The frequencies determined by Eq. (26) mark the positions of main peaks in the Compton spectrum. Similarly, one can find the location of secondary peaks (if $N_\mathrm{rep} > 2$) and zeros (if $N_\mathrm{rep} > 1$) in the angular-resolved frequency distributions. It is also worth noting that the polarization-dependent terms appear now in the definition of the directly measurable quantity; in other words, they affect the positions of peak frequencies in the Compton spectrum. Next, we define the quantity

$$\mathcal{N}_\mathrm{GKN} = \frac{N_\mathrm{osc}\omega_{\boldsymbol{K}}\omega_\mathrm{cut}}{\omega_\mathrm{L}(\omega_\mathrm{cut} - \omega_{\boldsymbol{K}})}\Big(\frac{p_\mathrm{i} \cdot n_{\boldsymbol{K}}}{p_\mathrm{i} \cdot n} + \frac{\nu n \cdot n_{\boldsymbol{K}} + g_1 p_{\mathrm{i},1} + g_2 p_{\mathrm{i},2}}{(p_\mathrm{i} \cdot n)^2}\Big), \tag{29}$$

which, according to (26), acquires integer values for peak frequencies $\omega_{\boldsymbol{K},N}^{(\mathrm{GKN})}$.

The positions of the interference peaks allow us to determine parameters of the driving laser field. In order to demonstrate this, we introduce the quantities,

$$\eta = \nu n \cdot n_{\boldsymbol{K}} + g_1 p_{\mathrm{i},1} + g_2 p_{\mathrm{i},2}, \tag{30}$$

$$A(\omega_{\boldsymbol{K}}) = \frac{N_\mathrm{osc}\omega_{\boldsymbol{K}}\omega_\mathrm{cut}}{\omega_\mathrm{L}(\omega_\mathrm{cut} - \omega_{\boldsymbol{K}})(p_\mathrm{i} \cdot n)^2}, \tag{31}$$

$$B(\omega_{\boldsymbol{K}}) = \frac{N_\mathrm{osc}\omega_{\boldsymbol{K}}\omega_\mathrm{cut}}{\omega_\mathrm{L}(\omega_\mathrm{cut} - \omega_{\boldsymbol{K}})}\frac{p_\mathrm{i} \cdot n_{\boldsymbol{K}}}{p_\mathrm{i} \cdot n}. \tag{32}$$

They depend on the initial electron energy and the geometry of the process. Except of η, they also depend on the frequency of emitted photons. The latter means that, for a given frequency distribution, η remains constant. Moreover, only η depends on the laser pulse parameters such as $\mu^2(\langle f_1^2\rangle + \langle f_2^2\rangle)$ and $\mu\langle f_j\rangle$, $j = 1, 2$ (note that only these parameters are necessary for determining the electron dressed mass). Using the above definitions, Eq. (29) can be rewritten as

$$A(\omega_{\boldsymbol{K}})\eta = \mathcal{N}_\mathrm{GKN} - B(\omega_{\boldsymbol{K}}). \tag{33}$$

This equation determines η provided that we can unambiguously prescribe an integer number $\mathcal{N}_\mathrm{GKN}$ to an arbitrary chosen peak. In general, this is impossible. Therefore, we choose instead two arbitrary, consecutive peaks from the spectrum, i.e., $\omega_{\boldsymbol{K},N}^{(\mathrm{GKN})}$ and $\omega_{\boldsymbol{K},N+1}^{(\mathrm{GKN})}$. By solving the system of two equations (33) evaluated at $\omega_{\boldsymbol{K},N}^{(\mathrm{GKN})}$ and $\omega_{\boldsymbol{K},N+1}^{(\mathrm{GKN})}$, we obtain that

$$\eta = \frac{1 - B(\omega_{\boldsymbol{K},N+1}^{(\mathrm{GKN})}) + B(\omega_{\boldsymbol{K},N}^{(\mathrm{GKN})})}{A(\omega_{\boldsymbol{K},N+1}^{(\mathrm{GKN})}) - A(\omega_{\boldsymbol{K},N}^{(\mathrm{GKN})})}. \tag{34}$$

This quantity is already independent of N. Thus, in order to determine three independent parameters ν, g_1, and g_2 present in the definition of η, we should repeat this procedure for three different geometries. This finally leads to the system of three linear equations for the unknown ν, g_1, and g_2.

5. Illustration of the proposed diagnostic method

In the following, we shall consider the linearly polarized laser pulse such that, for $0 \leqslant \phi = k \cdot x \leqslant 2\pi$, the four-vector potential has the form $A(\phi) = A_0\varepsilon f(\phi)$ [i.e., in Eq. (15), we take $\varepsilon = \varepsilon_1$,

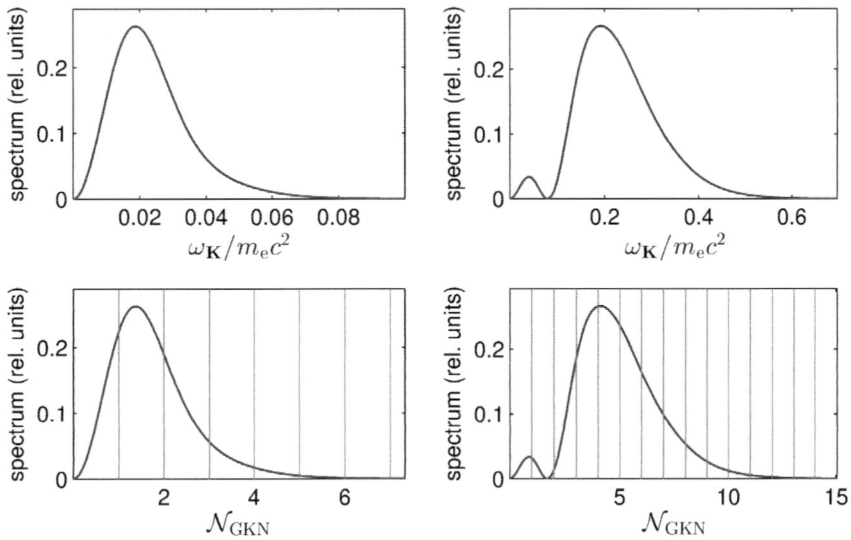

Figure 1. Spectra of Compton radiation resulting from the head-on collision of a linearly polarized laser pulse and an electron of momentum $\boldsymbol{p}_{\rm i} = -10^2 m_{\rm e}c\boldsymbol{e}_z$. The laser pulse ($\mu = 1$ and $\omega_{\rm L} = 3\times10^{-6}m_{\rm e}c^2$) propagates along the z direction and it is polarized along the x direction. These two axes define the scattering plane. The pulse has a $\sin^2$ envelope (35) with $N_{\rm osc} = 1$ and $N_{\rm rep} = 1$. The Compton photon is emitted in the direction determined by the polar angle $\theta_{\boldsymbol{K}} = 0.99\pi$ (left column) or $\theta_{\boldsymbol{K}} = 0.995\pi$ (right column) and by the azimuthal angle $\varphi_{\boldsymbol{K}} = \pi$, with the polarization vector parallel to the scattering plane. In the upper row, the energy spectra of Compton radiation are plotted as functions of the frequency $\omega_{\boldsymbol{K}}$ whereas, in the lower row, as functions of $\mathcal{N}_{\rm GKN}$. The vertical lines mark the integer values of the respective argument.

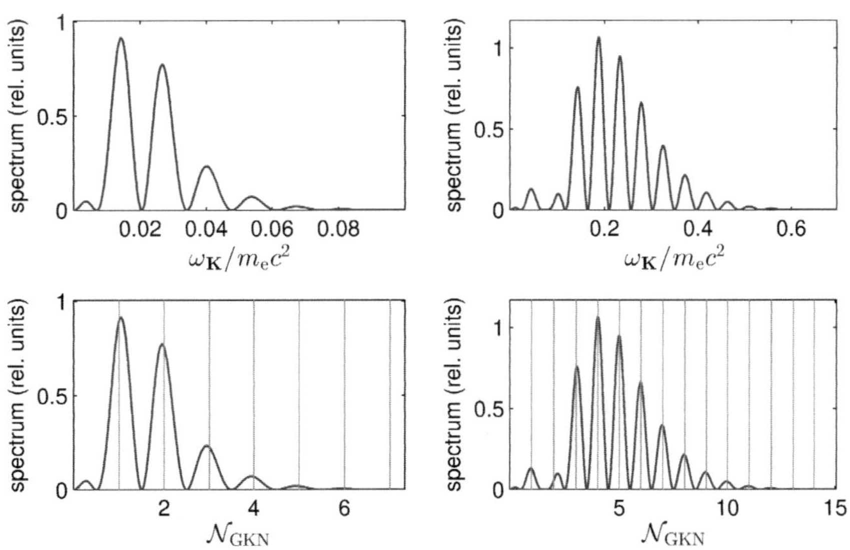

Figure 2. The same as in Fig. 1 but for $N_{\rm rep} = 2$. We clearly see that the peaks appear for integer values of $\mathcal{N}_{\rm GKN}$.

Journal of Physics: Conference Series **691** (2016) 012005 doi:10.1088/1742-6596/691/1/012005

$f(\phi) = f_1(\phi)$, $f_2(\phi) = 0$] and the electric field vector is $\boldsymbol{\mathcal{E}}(\phi) = -\omega A_0 \boldsymbol{\varepsilon} f'(\phi)$. The shape function $f(\phi)$ is defined via its derivative,

$$f'(\phi) = \begin{cases} 0, & \phi < 0, \\ N_f' \sin^2\left(N_{\rm rep}\frac{\phi}{2}\right)\sin(N_{\rm rep}N_{\rm osc}\phi), & 0 \leqslant \phi \leqslant 2\pi, \\ 0, & \phi > 2\pi, \end{cases} \tag{35}$$

where we assume that $f(0) = 0$. The pulse described by this shape function consists of $N_{\rm rep}$ identical subpulses, with $N_{\rm osc}$ cycles each. In other words, it can be treated as a finite train of $N_{\rm rep}$ pulses. Regardless of this interpretation, it is justified to apply the theory derived in Ref. [17] for the process driven by a single pulse. Since the modulated pulse (35) lasts for time $T_{\rm p}$ we can define its fundamental, $\omega = 2\pi/T_{\rm p}$, and its central frequency, $\omega_{\rm L} = N_{\rm rep}N_{\rm osc}\omega$. The latter is supposed to be fixed and equal to $\omega_{\rm L} = 3\times10^{-6} m_{\rm e}c^2$ in all calculations performed below. In the following, we assume also that, in Eq. (35), we have $N_f' = N_{\rm rep}N_{\rm osc}$. This guarantees that the time-averaged intensity carried out by the laser field is independent of $N_{\rm rep}$ and $N_{\rm osc}$, as for this particular choice of N_f' the amplitude of the electric field scales as $\mu\omega_{\rm L}$.

To illustrate the diagnostic method described in the previous Section, we choose the extremely short laser pulse consisting of only one cycle, i.e., in Eq. (35), we take $N_{\rm osc} = 1$ and $N_{\rm rep} = 1$. In Fig. 1, we present the energy spectrum of radiation generated by such a pulse as functions of either the Compton photon frequency $\omega_{\boldsymbol{K}}$ or the parameter $\mathcal{N}_{\rm GKN}$, for two different kinematics. For the polar angle $\theta_{\boldsymbol{K}} = 0.99\pi$ (left column), one can hardly attribute an integer value of $\mathcal{N}_{\rm GKN}$ to the broad peak which appears in the spectrum. On the other hand, for $\theta_{\boldsymbol{K}} = 0.995\pi$ (right column), one can prescribe integer values $\mathcal{N}_{\rm GKN} = 1$ and $\mathcal{N}_{\rm GKN} = 4$ to both peaks, even though there are no additional peaks in between them. Let us also note that, if one tried to associate to both peaks the consecutive integers $\mathcal{N}_{\rm GKN} = 1$ and $\mathcal{N}_{\rm GKN} = 2$, one would obtain the incorrect estimation of the laser pulse characteristics $\langle f\rangle$ and $\langle f^2\rangle$.

The situation changes if we apply such a pulse twice, meaning that $N_{\rm osc} = 1$ and $N_{\rm rep} = 2$ in Eq. (35). Now, as it is demonstrated in Fig. 2 for both polar angles, we observe peaks in the spectrum to which one can prescribe unambiguously integer values of $\mathcal{N}_{\rm GKN}$. This is particularly evident for the most pronounced peaks. Next, choosing for both cases two adjacent peaks from these spectra, we acquire from the numerical (or experimental) data the corresponding Compton frequencies (without knowing what the integer values of $\mathcal{N}_{\rm GKN}$ are prescribed to them). Applying the method described in Sec. 4 one can estimate now the laser pulse characteristics. Namely, for the considered shape function (35), they are very close to the exact values $\langle f\rangle = -0.375$ and $\langle f^2\rangle = 0.1328125$. Note that, for larger $N_{\rm rep}$, the peaks in the Compton spectrum practically do not change their positions and they become sharper. Thus, for larger $N_{\rm rep}$, the determination of the laser pulse properties can be made with an increased precision.

6. Conclusions

The aim of this paper was to discuss the concept of the electron mass dressing by a short and intense laser pulse. We have shown that this concept seems to be not as fundamental as for the monochromatic plane wave (in general, for the polychromatic plane waves with commensurate frequencies), as it does not follow directly from the Floquet-Bloch analysis of the Volkov solution. However, with some modifications, it is possible to redefine the electron momentum dressing such that it is on the mass shell with the dressed mass, independent of the asymptotic electron momentum. If such a modification is accepted, then the analysis of the frequency combs in the Compton scattering allows one to measure the dressed electron mass for arbitrarily short (even one-cycle) and intense laser pulses. This is achieved by determining characteristics of the laser pulse. From the practical point of view, the procedure described above provides a very precise method of diagnosis of extremely intense pulses in cases when the conventional methods fail.

Note also that this method is not only limited to the Compton or Thomson scattering, but can be extended to other quantum processes as well (for instance, to the Breit-Wheeler process [39] or to ionization of atoms and molecules [25, 40]).

References

[1] Ritus V I and Nikishov A I 1979 *Trudy FIAN* **111** 5
[2] Ritus V I 1985 *J. Sov. Laser Res.* **6** 497
[3] Ritus V I and Nikishov A I 1986 *Trudy FIAN* **168** 5
[4] Ehlotzky F, Krajewska K and Kamiński J Z 2009 *Rep. Prog. Phys.* **72** 046401
[5] Ruffini R, Vereshchagin G and Xue S S 2010 *Phys. Rep.* **487** 1
[6] Di Piazza A, Müller C, Hatsagortsyan K Z and Keitel C H 2012 *Rev. Mod. Phys.* **84** 1177
[7] Nedoreshta V N, Roshchupkin S P and Voroshilo A I 2013 *Laser Phys.* **23** 055301
[8] Narozhny N B and Fedotov A M 2015 *Contemp. Phys.* **56** 249
[9] Klein O and Nishina Y 1929 *Z. Physik* **52** 853
[10] Brown L S and Kibble T W B 1964 *Phys. Rev.* **133** A705
[11] Heinzl T, Seipt D and Kämpfer B 2010 *Phys. Rev. A* **81** 022125
[12] Seipt D and Kämpfer B 2011 *Phys. Rev. A* **83** 022101
[13] Seipt D and Kämpfer B 2011 *Phys. Rev. ST Accel. Beams* **14** 040704
[14] Krajewska K and Kamiński J Z 2014 *Phys. Rev. A* **90** 052117
[15] Volkov D M 1935 *Z. Phys.* **94** 250
[16] Lee K, Chung S Y, Park S H, Jeong Y U and Kim D 2010 *Europhys. Lett.* **89** 64006
[17] Krajewska K and Kamiński J Z 2012 *Phys. Rev. A* **85** 062102
[18] Krajewska K and Kamiński J Z 2012 *Phys. Rev. A* **86** 052104
[19] Krajewska K, Cajiao Vélez F and Kamiński J Z 2015 *Phys. Rev. A* **91** 062106
[20] Jaroń A, Kamiński J Z and Ehlotzky F 2000 *Phys. Rev. A* **61** 023404
[21] Jaroń A, Kamiński J Z and Ehlotzky F 2001 *Phys. Rev. A* **63** 055401
[22] Müller C, Voitkiv A B and Najjari B 2009 *J. Phys. B* **42** 221001
[23] Har-Shemesh O and Di Piazza A 2012 *Opt. Lett.* **37** 1352
[24] Kalashnikov M, Andreev A, Ivanov K, Galkin A, Korobkin V, Romanovsky M, Shiryaev O, Schnuerer M, Braenzen J and Trofimov V 2015 *Laser Part. Beams* **33** 361
[25] Krajewska K and Kamiński J Z 2016 *Phys. Lett. A*, http://dx.doi.org/10.1016/j.physleta.2015.11.038
[26] Ghebregziabher I, Shadwick B A and Umstadter D 2013 *Phys. Rev. ST Accel. Beams* **16** 030705
[27] Terzić B, Deitrick K, Hofler A S and Krafft G A 2014 *Phys. Rev. Lett.* **112** 074801
[28] Seipt D, Rykovanov S G, Surzhykov A and Fritzsche S 2015 *Phys. Rev. A* **91** 033402
[29] Harvey C, Heinzl T, Ilderton A and Marklund M 2012 *Phys. Rev. Lett.* **109** 100402
[30] Young T 1804 *Phil. Trans. R. Soc. Lond.* **94** 1
[31] Hall J L 2006 *Rev. Mod. Phys.* **78** 1279
[32] Hänsch T W 2006 *Rev. Mod. Phys.* **78** 1297
[33] Margolis H S 2009 *J. Phys. B* **42** 154017
[34] Agostini P and DiMauro L F 2004 *Rep. Prog. Phys.* **67** 813
[35] Farkas Gy and Tóth Cs 1992 *Phys. Lett. A* **168** 447
[36] Krausz F and Ivanov M 2009 *Rev. Mod. Phys.* **81** 163
[37] Krajewska K and Kamiński J Z 2014 *Laser Phys. Lett.* **11** 035301
[38] Krajewska K, Twardy M and Kamiński J Z 2014 *Phys. Rev. A* **89** 052123
[39] Krajewska K and Kamiński J Z 2014 *Phys. Rev. A* **90** 052108
[40] Cajiao Vélez F, Krajewska K and Kamiński J Z 2015 *Phys. Rev. A* **91** 053417

Numerical method for calculating atomic spectra in strong laser fields

E V Koryukina
Plasma Physics Department, National Research Tomsk State University, 36 Lenin Avenue, Tomsk 634050, Russia

E-mail: evk@phys.tsu.ru

Abstract. In this paper, a new theoretical method is suggested for calculating atomic emission spectra in strong laser fields. This numerical method based on diagonalization of the energy matrix of an atom in the electric field is free from limitations of perturbation theory and valid for a wide range of changes in the electric field strength and frequency. Within the framework of the suggested method, the ground-state shifts of He, Ne, Ar, and Kr atoms in a circularly polarized electric field are computed. Calculations are carried out for electric fields with strengths of 0.06 and 0.15 a.u. and frequencies from 10^{-4} to 0.057 a.u. In addition, the effect of ordering of the transition probabilities with respect to the magnetic quantum number M in the electric field is also investigated. The results obtained can be useful for correct interpretation of the experimental data and prediction of the behaviour of atomic emission spectra in strong laser fields.

1. Introduction

The influence of laser fields on atoms leads to an appearance of emission spectra specified by bound-bound transitions and electronic spectra arising due to ionization of atoms. For correct interpretation of these spectra, one needs accurate calculation methods, and non-stationary perturbation theory is one of them. In spite of limitations of perturbation theory (not very strong electric fields, isolated atomic energy levels, and Rydberg atoms [1]), most of the experimental data obtained for low-power optical lasers were explained within this theory ([2] and the references therein). However, by the present time, a lot of other excitation sources (in particular, electrodeless high-frequency lamps, light-emitting diodes, and superpower lasers) have appeared. These sources generate electric fields with parameters essentially different from those of optical lasers. In such cases, when non-stationary perturbation theory is unsuitable by virtue of its limitations, one needs a new reliable and efficient theoretical method for calculating atomic emission spectra in electric fields.

As for theoretical study of influence of strong laser fields on atoms and ions, the pioneer work in this brunch of physics was Keldysh's work [3]. In this work, the theory and methods for calculating tunnel and multiphoton ionizations in strong laser fields were suggested. After this work, many scientific papers were devoted to these problems (see, for example, reviews [4–6] and the references therein). It should be noted that the main attention in these works was paid to ionization processes in atoms (e.g., to the bound-free transitions) under the influence of alternating electric fields, but processes caused by the bound-bound transitions, as a rule, were neglected, because these probabilities were considered to be too small. However, at present time, calculating the discrete atomic spectra and the bound-bound transition probabilities in strong laser fields is also a vital problem. Firstly, shifts and splitting of atomic energy levels induced by the non-resonant dynamic Stark effect always lead to an appearance of the so-called dynamic resonances [7]. Secondly, the ground-state energy shifts in the electric fields are necessary for the refinement of formulas for calculating the ionization rates of the ground states of rare gas atoms derived in [8, 9]. Moreover, because the ground states of rare gas atoms are not Rydberg ones, calculations of their shifts in the electric field necessitate the consideration of interactions between the Stark states [10]. Finally, from the theoretical viewpoint, it is very interesting to reveal some regularities in the behaviour of the ionization rates and the bound-bound transition probabilities in an electric field and to understand reasons for these regularities.

In the present work, a theoretical method free from limitations of perturbation theory is used for calculating rare gas emission spectra in strong circularly polarized laser fields that allows one to take into account the Stark state interactions. This numerical method suggested and developed by us in [11, 12] is based on diagonalization of the energy matrix of an atom in electric fields. It gives one the possibility of calculating emission spectra of any atoms in alternating circularly polarized electric fields with strength and frequency changing in wide ranges.

Previously, the suggested numerical method was applied to calculate emission spectra of rare gas atoms in alternating circularly polarized electric fields with the electric field strength changing from 0 to 150 kV/cm and electric field frequency changing in the range 10^2–10^8 MHz [13–15]. The calculation results are in excellent agreement with the available theoretical and experimental data, in particular, with [16]. The results obtained stimulated our interest to calculation of the ground-state shifts and the bound-bound transition probabilities of rare gas atoms in strong low-frequency laser fields by the method of energy matrix diagonalization of an atom in the electric field, because the conditions of applicability of perturbation theory are violated for such electric fields.

In Section 2 of the present work we describe the method of energy matrix diagonalization of an atom in the electric field. In Section 3 we check the validity of the suggested numerical method and investigate its convergence in calculations of the ground-state shifts of rare gas atoms in laser fields with different values of the electric field parameters. Then, in Section 3, we study a very interesting effect of ordering of the bound-bound transition probabilities with respect to the magnetic quantum number M and discuss the presence of the same M-ordering for the bound-free transition probabilities.

2. Theoretical method

In a circularly polarized electric field, the non-stationary Schrödinger equation is written as

$$\mathrm{i}\frac{\partial \psi_n(\mathbf{r},t)}{\partial t} = (H_0(\mathbf{r}) - F(x\cos\omega t \pm y\sin\omega t))\psi_n(\mathbf{r},t), \tag{1}$$

where ψ_n is the wave function of the n-th state of the system, H_0 is the unperturbed Hamiltonian, and the operator $-F(x\cos\omega t \pm y\sin\omega t)$ describes perturbation induced by the interaction of an atom with a circularly polarized electric field of frequency ω and strength F. The "+" and "–" signs correspond to the right and left polarization of the field, respectively. For the case of a circularly polarized field, the non-stationary Schrödinger equation can be reduced to the stationary one due to separation of spatial and time variables within the framework of the rotating-wave approximation [17].

Within this approximation, the wave function in the coordinate system rotating about the Z-axis with the frequency ω, has the form

$$\varphi(\mathbf{r},t) = \exp(i\omega t J_z)\psi(\mathbf{r},t), \tag{2}$$

where J_z is the z-component of the total angular momentum operator. On substituting equation (2) in equation (1), we get

$$\mathrm{i}\frac{\partial \varphi(\mathbf{r},t)}{\partial t} = Q\varphi(\mathbf{r},t), \qquad Q = (H_0 - \omega J_z \pm F D_x). \tag{3}$$

As seen from equation (3), the operator Q is time-independent. Hence, in the rotating-wave approximation, it is possible to go from the non-stationary Schrödinger equation (1) to the stationary one, and we have

$$Q\varphi(\mathbf{r}) = \varepsilon\varphi(\mathbf{r}), \qquad \varphi(\mathbf{r},t) = \exp(-i\varepsilon t)\varphi(\mathbf{r}), \tag{4}$$

where Q is the energy operator of an atom in the electric field, and ε and $\varphi(\mathbf{r},t)$ are the energy and wave function of an atom in the electric field in the rotating coordinate system. It is obvious that the Schrödinger equation (4) can be solved within stationary perturbation theory, but this way of solution has been found too complicated [18, 19]. We found that instead of solving the Schrödinger equation (4) within perturbation theory, it is much more convenient to solve it using diagonalization procedure. It was shown that the wave functions and energies of an atom, being solutions to the Schrödinger equation (4), are found from diagonalization of the energy matrix of an atom in the electric field with the following elements

$$Q_{mn} = E_n^{(0)}\delta_{mn} - \omega < \varphi_m^{(0)}(\mathbf{r})|J_z|\varphi_n^{(0)}(\mathbf{r}) > \pm F < \varphi_m^{(0)}(\mathbf{r})|D_x|\varphi_n^{(0)}(\mathbf{r}) >. \tag{5}$$

This matrix is obtained in the coordinate system rotating about the *Z*-axis with frequency ω. In equation (5), $\varphi_n^{(0)}$ and $E_n^{(0)}$ are the wave function and energy of the *n*-th state of an atom in the absence of the electric field; J_z is the *z*-component of the total angular momentum operator; and D_x is the *x*-component of the dipole moment operator. Upon diagonalization of the energy matrix with elements (5), we get the energies ε_n and wave functions φ_n as

$$\varphi_n(\mathbf{r},t) = \exp(-i\varepsilon_n t)\sum_k C_{nk}(F,\omega)\varphi_k^{(0)}(\mathbf{r}) \tag{6}$$

for the *n* atomic states in the electric field in the rotating coordinate system. The coefficients C_{nk} in the wave function (6) are the expansion coefficients of the wave functions of an atom under the action of the electric field in terms of unperturbed wave functions. These coefficients depend on the electric field strength and frequency.

To go to the initial coordinate system and find the average atomic energies in the electric field, it is necessary to perform averaging over the oscillation period. After averaging over the oscillation period, the wave functions and average energies of an atom in the electric field in the initial coordinate system are written as

$$\psi_n(\mathbf{r},t) = \exp(-i(\omega J_z + \varepsilon_n)t)\sum_k C_{nk}(F,\omega)\varphi_k^{(0)}(\mathbf{r}), \tag{7}$$

$$\bar{E}_n = < \psi_n(\mathbf{r},t)|H(\mathbf{r},t)\psi_n|\psi_n(\mathbf{r},t) > = \varepsilon_n + \omega < \varphi_n(\mathbf{r})|J_z|\varphi_n(\mathbf{r}) >. \tag{8}$$

It follows from equation (8) that $\bar{E}_n$ is time-independent.

The wave functions and energies of an atom in the electric field specified by equations (7) and (8) are further used for calculating spontaneous transition probabilities between the Stark states *JM* and *J'M'*. These transition probabilities are computed using the formula

$$A(JM \to J'M') = \frac{4\omega^3_{JM,J'M'}}{3\hbar c^3}|D_{JM,J'M'}|^2, \tag{9}$$

$$|D_{JM,J'M'}|^2 = \left|\sum_q \sum_{ij} C_i^{(JM)*} C_j^{(J'M')} (-1)^{J_i - M_i} \begin{pmatrix} J_i & 1 & J_j \\ -M_i & q & M_j \end{pmatrix} \cdot < \gamma_i J_i \| D \| \gamma_j J_j > \right|^2, \tag{10}$$

where $C_i^{(JM)}$ and $C_j^{(J'M')}$ are the expansion coefficients from equation (7), and $\omega_{JM,J'M'}$ is the frequency of the $JM \to J'M'$ transition.

The matrix elements of the D_x operator in equations (5) and (10) are determined as

$$< \varphi_m^{(0)}|D_x|\varphi_n^{(0)} > = < \gamma JM|D_x|\gamma' J'M' > = \frac{(-1)^{J-M}}{\sqrt{2}}\left[\begin{pmatrix} J & 1 & J' \\ -M & -1 & M' \end{pmatrix} - \begin{pmatrix} J & 1 & J' \\ -M & 1 & M' \end{pmatrix}\right] < \gamma J \| D \| \gamma' J' >, \tag{11}$$

where the reduced matrix elements $< \gamma J \| D \| \gamma' J' >$ are calculated depending on a coupling scheme. The reduced matrix elements $< \gamma J \| D \| \gamma' J' >$ are computed using the following formulas

$$<\gamma J\|D\|\gamma' J'>=A(-1)^{\varphi}Q(T_1T_2J,U_1'U_2'J')<l\|r\|l'>, \tag{12}$$

$$<l\|r\|l'>=(-1)^{\frac{3l+l'+3}{2}}\sqrt{\max(l,l')}<R_{nl}|r|R_{n'l'}>, \tag{13}$$

$$<R_{nl}|r|R_{n'l'}>=\int_0^{\infty}R_{nl}(r)rR_{n'l'}(r)r^2dr, \tag{14}$$

where T_1T_2J and $U_1'U_2'J'$ are the quantum numbers of the atomic energy level within the corresponding coupling scheme; *nl* and *n'l'* are the quantum numbers of the optical electron, which goes from the state *nl* to the state *n'l'*. The calculation formulas for the multipliers $Q(T_1T_2J,U_1'U_2'J')$ from equation (12) were reported in [20]. The radial integral $<R_{nl}|r|R_{n'l'}>$ in equation (14) is calculated using a semiempirical formula, which is an improved modification of the Bates-Damgaard formula [21]. The specific form of our semiempirical formula for computation of the radial integrals and details of its derivation were reported in [12].

It follows from the above reasoning that the proposed theoretical approach is free from limitations of perturbation theory and can be used for calculating the dynamic Stark effect in a circularly polarized electric field with the strength and frequency changing in wide ranges. Since the energies and wave functions in the electric field are determined by the diagonalization procedure, it allows us to take into account the interaction of the Stark states. The algorithm of the developed theoretical approach is implemented in a special software package StarkD written in FORTRAN.

3. Results and discussion

In this work, the suggested theoretical approach was applied to calculate the energy spectra of rare gas atoms in strong alternating circularly polarized electric fields of different strengths and frequencies. First of all, it is necessary to check the validity of our numerical method for strong laser fields.

It is well known that the use of any theoretical methods with expansions in some basis functions requires an investigation of the convergence of the calculation results with increasing basis. The convergence problem is very complicated, and it must be solved in each individual case. In particular, in perturbation theory, the energy level shifts induced by the electric field are proportional to the atomic dynamic polarizability [1]. The convergence of the basis in calculating atomic polarizabilities was discussed in classical monographs [1, 19]. The detailed investigation of the basis convergence was performed in [10], where atomic polarizabilities were computed by the method of the reduced-added Green function.

The method of energy matrix diagonalization of an atom in the electric field used in the present work is also based on calculating atomic characteristics with the wave functions being the expansions in the basis set of unperturbed wave functions. It was shown in [11], for the electric fields with parameters F = 0–10 kV/cm and $\omega = 10^2$–10^8 cm^{-1}, the inclusion of unperturbed wave functions $\varphi_{nl}^{(0)}$ with l = 0–3 and $n \le 10$ in the basis set (the basis N_0) allows us to achieve convergence of calculations of atomic energies in the electric field. In case of strong laser fields, our calculation results showed that in order to achieve the calculation convergence, it is necessary to increase the basis set. Firstly, we have found that it is necessary to include the one-electron functions $\varphi_{nl}^{(0)}$ with $l = 4$ in the basis set N_0 (the basis N_1). Secondly, the one-electron functions $\varphi_{nl}^{(0)}$ with the principal quantum number $n \le 15$ also must be added to the basis set N_1 (the basis N_2).

As an illustration of the convergence of our calculation method with increasing basis set, the computed ground-state energy shifts for rare gas atoms in comparison with other theoretical data [1] are listed in table 1. These calculations were carried out for the shifts in the electric field of the neodymium laser (laser intensity of 30 MW/cm^2 and ω = 18800 cm^{-1}).

As follows from table 1, the sign of our calculated shifts is negative that agrees with the computation results [1] obtained within the framework of perturbation theory. In addition, the calculation results from table 1 demonstrate the convergence of our numerical method with increasing basis set. It should be noted that a further increase in the basis set has a weak influence on the shift value. As a whole, our computation data are in good agreement with the results presented in [1] except the ground-state shift of the Ne atom. The above-indicated difference is caused by the fact that in [1] the calculated value of the polarizability of the neon atom was $\alpha_n(\omega) = 5.25$ a.u., whereas at $\omega = 0.0$ a.u., the experimental value of the polarizability was 2.669 a.u. [22], and at $\omega = 0.2$ a.u., $\alpha_n(\omega)$ equals 2.814 a.u. [10] or 2.794 a.u. [23].

Table 1. Ground-state energy shifts ΔE (in 10^{-3} cm^{-1}) for rare gas atoms in the Nd-laser field.

ΔE	He	Ne	Ar	Kr
N_0	–0.032	–0.061	–0.262	–0.391
N_1	–0.054	–0.098	–0.491	–0.674
N_2	–0.059	–0.108	–0.544	–0.793
[1]	–0.067	–0.240	–0.602	–0.850

Then, our suggested theoretical approach was applied to calculate the energy spectra of rare gas atoms in alternating circularly polarized electric fields of strengths F=0.06 a.u. and F=0.15 a.u. at different values of the electric field frequency. First of all, the calculations of these shifts were computed at the electric field frequency $\omega = 10^{-4}$ a.u. to verify the convergence of our method. The results of these calculations are listed in table 2. As follows from this table, an increase in the set functions leads to the convergence of our calculation results. The signs of all ground-state energy shifts are negative and these shifts increase with atomic nuclear charge. The results obtained also agree with estimations of perturbation theory, and this fact confirms once more the validity of our numerical method.

Table 2. Ground-state energy shifts ΔE (in a.u.) for rare gas atoms in the circularly polarized electric fields ($\omega = 10^{-4}$ a.u.).

Electric field strength	Basis	Ne	Ar	Kr
	N_0	–0.0025	–0.0075	–0.0117
$F = 0.06$ a.u.	N_1	–0.0038	–0.0112	–0.0249
	N_2	–0.0041	–0.0127	–0.0303
	N_0	–0.0191	–0.0702	–0.0867
$F = 0.15$ a.u.	N_1	–0.0261	–0.0903	–0.1613
	N_2	–0.0287	–0.1091	–0.1842

In addition, we also calculated the ground state shifts of rare gas atoms in strong laser field with the electric field strength $F = 0.15$ a.u. and frequency $\omega = 0.05$ a.u. The results obtained with the basis set N_2 are listed in table 3.

Table 3. Ground-state energy shifts ΔE (in a.u.) for rare gas atoms in the circularly polarized electric field with parameters $F = 0.15$ a.u. and $\omega = 0.05$ a.u.

ΔE	Ne	Ar	Kr
This work	–0.0381	–0.1229	–0.2277
[10]	–0.0308	–0.1212	–0.2029

As follows from this table, the values of shifts of atomic ground states computed by our method are slightly greater than the calculation results presented in [10]; however, as a whole, our results are in good agreement with the calculation data [10].

It is well known that in the case of one-photon ionization, the ionization rate of the bound state depends on its quantum number m. For transitions with increasing angular momentum of the electron, the states with $m > 0$ are depleted faster than those with $m < 0$. Since both the bound-free and bound-

bound transitions are induced by the electric field, this suggests that the bound-bound transition probabilities between the Stark states JM will demonstrate a similar dependence on the magnetic quantum number M. Actually, we have found that switching on the electric field leads to the M-ordering of the transition probabilities, and this effect has a global character.

In relatively weak circularly polarized electric fields, the bound-bound transition probabilities are ordered with respect to the magnetic quantum number M of the optical electron, and this regularity is valid in a wide range of changes in the electric field strength and frequency. As an illustration, figure 1 shows an evolution of probabilities of the $nd\,^3D_2$–$2p\,^1P_1$ transitions with increasing electric field strength for the He atom. In this figure and following figures, the magnetic quantum numbers M, M' of the $JM \rightarrow J'M'$ transitions are pointed in the frame. As can be seen from this figure, switching on the electric field leads to ordering of transition probabilities with respect to the magnetic quantum number M, e.g., probabilities of the max$|M| \rightarrow \pm M'$ transitions are maximal, and these probabilities diminish with M. This effect is observed for the electric field strength from 0 to 120 kV/cm (see figure 1).

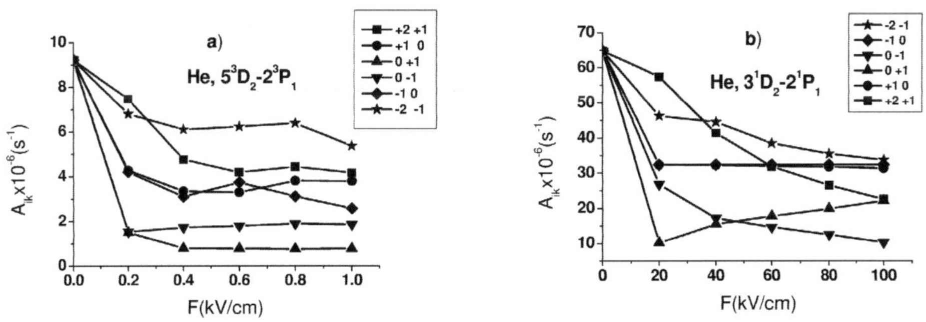

Figure 1. Dependence of transition probabilities between the Stark states on the electric field strength for the He atom: a) $\omega = 10^2$ MHz, b) $\omega = 10^3$ MHz.

The M-ordering of transition probabilities is also detected for other rare gas atoms in the electric field. For example, the behaviour of the transition probabilities depending on the electric field strength for neon and argon atoms is shown in figure 2.

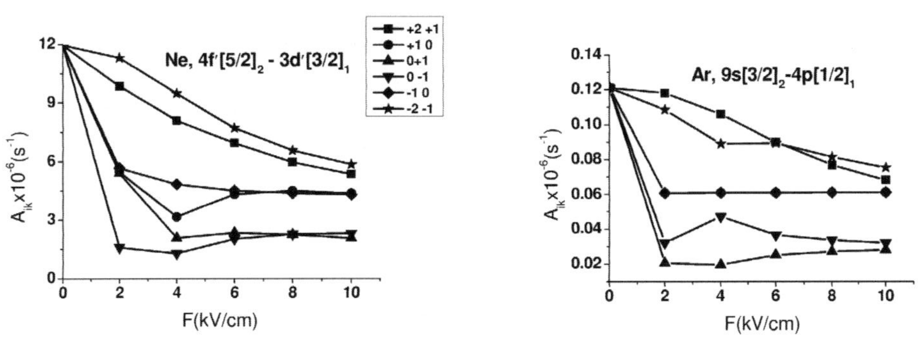

Figure 2. Dependence of transition probabilities between the Stark states of rare gas atoms on the electric field strength at $J > J'$ ($\omega = 10^2$ MHz).

The M-ordering of transition probabilities $A(JM \rightarrow J'M')$ is observed not only for $J > J'$ transitions. When $J < J'$, we also have the M-ordering of transition probabilities, but now probabilities of the

$\pm M \rightarrow \max|M'|$ transitions have maximal values, and these probabilities diminish with M'. Figure 3 is an obvious illustration of this fact.

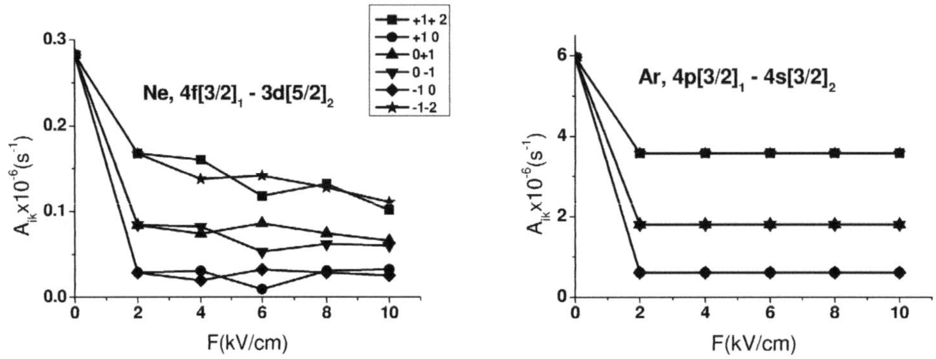

Figure 3. Dependence of transition probabilities between the Stark states of rare gas atoms on the electric field strength for $J < J'$ ($\omega = 10^2$ MHz).

Finally, our calculation results have shown that the M-ordering of transition probabilities $A(JM \rightarrow J'M')$ is observed for all rare gas atoms in alternating circularly polarized electric fields with frequencies changing from 10^2 to 10^8 MHz. As an illustration, figure 4 shows an evolution of probabilities of the $6d[5/2]_{2M} - 5p[3/2]_{1M'}$ transitions with increasing electric field frequency for the Kr atom. The examined frequencies are the electric field frequencies of different lasers, a namely: $\omega = 3.4352 \cdot 10^6$ MHz (an HCN laser), $\omega = 2.4181 \cdot 10^7$ MHz (a CO_2 laser), and $\omega = 2.8301 \cdot 10^8$ MHz (a Nd laser).

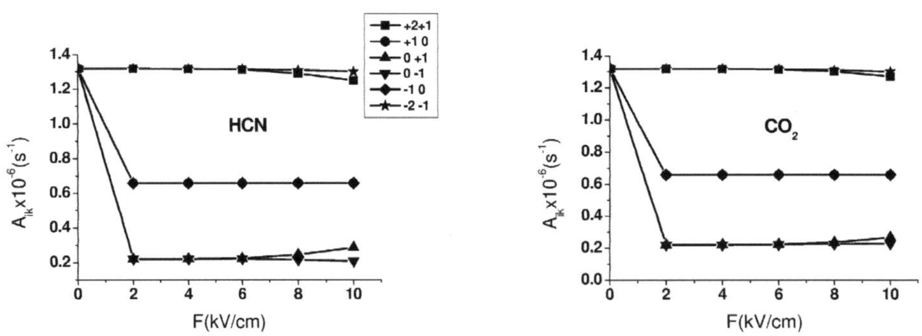

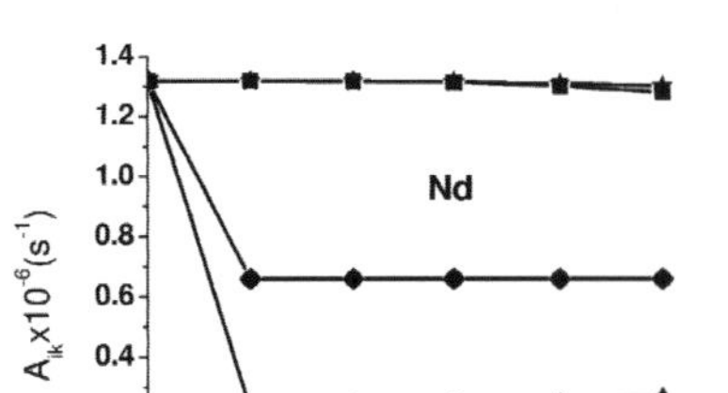

Figure 4. Dependence of transition probabilities between the Stark states on the electric field frequency for the $6d[5/2]_2 - 5p[3/2]_1$ transition of the Kr atom:
HCN: $\omega = 3.4352 \cdot 10^6$ MHz,
CO_2: $\omega = 2.4181 \cdot 10^7$ MHz,
Nd: $\omega = 2.8301 \cdot 10^8$ MHz.

It is seen from our calculation results that the M-ordering of transition probabilities has a global character, because it is detected for all rare gas atoms influenced by electric fields.

Owing to a global character of the M-ordering of transition probabilities $A(JM \rightarrow J'M')$, it is possible to assume that the above-described effect will take place in atomic emission spectra in strong laser fields. Our calculation results listed in table 4 confirm this assumption. As follows from this table, in strong laser fields the bound-bound transition probabilities are also M-ordered, and the probabilities of transitions between the Stark states with $M > 0$ are greater than those with $M < 0$ by analogy with the ionization rates.

Table 4. Probabilities of the $np[1/2]_1$–$n's[3/2]_2$ transitions (in 10^6 s^{-1}) for rare gas atoms in the electric field with parameters $F = 0.06$ a.u. and $\omega = 0.057$ a.u.

Transition $M \rightarrow M'$	$1\rightarrow2$	$1\rightarrow0$	$0\rightarrow1$	$-1\rightarrow-2$	$-1\rightarrow0$	$0\rightarrow-1$
Ne, $3p[1/2]_1$–$3s[3/2]_2$	2.58	2.5	1.17	2.03	0.21	0.35
Ar, $5p[1/2]_1$–$5s[3/2]_2$	7.00	5.25	1.59	3.52	3.50	1.55

Thereby, based on our calculation results, we can conclude that the influence of strong circularly polarized laser fields leads to the M-ordering of both the bound-bound transition probabilities and the ionization rates of the Stark states of rare gas atoms, and this ordering has the same character.

4. Conclusions

The suggested theoretical approach can be useful for calculations of the excitation and ionization processes of atoms in strong laser fields when perturbation theory is inapplicable. Atomic spectra in the electric field obtained within the framework of our method has allowed us to reveal dynamic resonances arising in strong laser fields owing to the non-resonant Stark effect and to define the ground-state energy shifts in the electric fields needed for refinement of the formulas for calculating the ionization rates of the ground states of rare gas atoms. Finally, the calculations of the bound-bound transition probabilities are of interest for the study of the influence of electric fields on atoms when the bound-bound and bound-free transition probabilities are of the same order.

Acknowledgements

The author thanks Ingo Barth for useful discussions. This work was supported by the Programme of Improving TSU's International Competitiveness.

References

[1] Rapoport L B, Zon B A, and Manakov N L *Theory of Many-Photon Processes in Atoms* (Moscow: Atompress)
[2] Delone N B and Krainov V P 1999 *Progr. of Phys. Sciences* **169** 753
[3] Keldysh L V 1964 *Zh. Eksp. Teor. Fiz.* **47** 1945
[4] Delone N B and Fedorov M V 1989 *Progr. of Phys. Sciences* **158** 215
[5] Goreslavski S P and Popruzhenko S V 1996 *Laser Phys.* **6** 780
[6] Popov V S 2004 *Progr. of Phys. Sciences* **174** 921
[7] Delone N B 2000 *Soros Educational J.* **6** 66
[8] Barth I and Smirnova O 2011 *Phys. Rev.* A **84** 063415
[9] Barth I and Smirnova O 2013 *Phys. Rev.* A **87** 013433
[10] Chernov V E, Dorofeev D I, Kretinin I Yu and Zon B A 2005 *Phys. Rev.* A **71** 022505
[11] Koryukina E V 2005 *Rus. Phys. J.* **48** 891
[12] Koryukina E V 2005 *J. Phys. D: Appl. Phys.* **38** 3296
[13] Koryukina E V 2006 *Proc. SPIE* **6263** 175
[14] Koryukina E V and Koryukin V I 2008 *Atmos. Oceanic Opt.* **21** 616

[15] Koryukina E V 2014 *Izv. Vyssh. Uchebn. Zaved., Fiz.* **57** 259
[16] Frank A G, Gavrilenko V P, Kirie N P and Oks E 2006 *J. Phys. B: Atom., Mol. and Opt. Phys.* **39** 5119
[17] Bunkin F V and Prokhorov A M *Zh. Eksp. Teor. Fiz.* 1964 **46** 1091
[18] Sambe H 1973 *Phys. Rev.* A **7** 2203
[19] Delone N B and Krainov V P 1978 *An Atom in a Strong Light Field* (Moscow: Energoatomizdat)
[20] Jucys A P and Savukynas A J 1973 *Mathematical Foundations of the Atomic Theory* (Vilnius: Mokslas)
[21] Bates D R and Damgaard A 1949 *Philos. Trans. Roy. Soc.* A **242** 101
[22] Saxon R P 1973 *J. Chem. Phys.* **59** 1539
[23] Larsen H, Olsen J, Hattig C, Jørgensen P, Christiansen O and Gauss J 1999 *J. Chem. Phys.* **111** 1917

Prototyping of Dental Structures Using Laser Milling

A O Andreev[1], M S Kosenko[1], V N Petrovskiy[1] and V D Mironov[1]
[1]National Research Nuclear University "MEPhI", Kashirskoe sh., 31, Moscow, Russia
E-mail.: andreev07@mail.ru

Abstract. The results of experimental studies of the effect of an ytterbium fiber laser radiation parameters on processing efficiency and quality of ZrO_2 ceramics widely used in stomatology are presented. Laser operating conditions with optimum characteristics for obtaining high quality final surfaces and rapid material removal of dental structures are determined. The ability of forming thin-walled ceramic structures by laser milling technology (a minimum wall thickness of 50 μm) is demonstrated. The examples of three-dimensional dental structures created in computer 3D-models of human teeth using laser milling are shown.

1. Introduction

Zirconium dioxide is used for making electronic equipment, dielectric and piezoelectric ceramic, solid electrolytes, fuel elements and heat insulators. A variety of applications for ZrO_2 is caused by a unique combination of its thermal, optical, electric and other properties [1-3].

The prosthetics of teeth is dynamically developing medicine area. The market of dental structures which are based on zirconium dioxide is growing every year. Nowadays the computer system of laser scanning of a tooth imprint made by the doctor is used to create a 3D-model. Further, this model is used to create individual dental crowns or dental bridge structures from partially stabilized zirconium dioxide.

Manufacturing constructions is carried out using a device with CAD/CAM-systems by mechanical milling [4]. For ceramic blocks milling diamond bur with water-cooling is required. Process of yttria tetragonal zirconia polycrystal (Y-TZP) mechanical milling is very expensive, it demands a lot of time and it leads to significant wear of diamond burs. Moreover the production of dental structures with thickness of walls less than 400 microns is impossible, and the formation of microcracks in the material structure is possible by mechanical milling.

In this paper we propose an alternative precision laser milling technology of dental structures made of zirconium dioxide. Possibility of decrease in consumption of material on production of zirconium structures, increase of blank processing speed, increase in accuracy of processing and decrease of the ceramic structures minimum thickness to an optimum is considered.

2. Experimental part

Nowadays dental structures are prototyped widely using ZrO_2 ceramics with added yttrium from 1 to 7% (so called partially stabilized zirconia). In the production of a dental ceramic structure the most important factors are the speed of the final material production and roughness of the surface. Since the treated dental blank subsequently is processed with dye solutions and enamels which requires a surface roughness for better adhesion. The optimum value of dental crown surface roughness should be at a level of 10 - 30 microns.

For the experimental determining of the effect of the repetition rate and pulse duration of the laser radiation on the laser milling productivity the fiber laser (λ = 1,068 mkm) with an average power P_{av} =

20 W output radiation was directed onto a flat surface of ceramic blanks with the help of scanning mirrors. The diameter of the laser spot was d = 50 mm, and scanning speed of the ceramic blank surface was v = 1 m / s and was not changed. During the experiment the parameters of laser pulse duration (from 4 to 200 ns) and the pulse repetition frequency (from 1 to 20 kHz) were varied.
The ceramic sample surface was processed by scanning the laser beam over square areas 5×5 mm in size. The scanning square area was progressively filled with laser passes or "paths". Laser "paths" with various degrees of overlap (0, 25, 50 and 75%) were applied to ceramics to determine the effect of this parameter on the material removal rate. The average ZrO_2 ceramics removal rate was estimated at the repetition rate f = 20 kHz and average radiation power P_{av}=20 W. The results are shown in Table 1.

Table 1. Material removal rate (cm^3/h) depending on the pulse duration and the degree of overlap of the elementary "paths".

Degree of "path" overlap, %	***Pulse duration, ns***			
	30	**50**	**100**	**200**
0	0.16	0.70	2.04	3.93
25	0.21	0.80	2.27	4.60
50	0.17	0.54	1.73	4.70
75	0.17	1.26	2.26	3.58

The pictures of surface after laser treatment were taken with an optical microscope at magnifications of 2x and 20x. Typical surface topography is shown in Fig. 1.

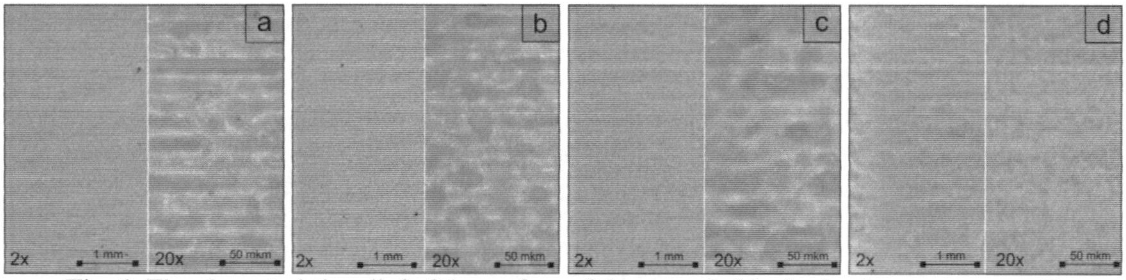

Fig. 1. The characteristic relief of the surface of the ceramic blank after laser treatment. Degrees of overlap: a – 0%, b – 25%, c – 50%, d – 75%.

As one may see from the above data, the degree of overlap slightly affects the material removal rate, while with increasing pulse duration the productivity of the laser milling process increases significantly.
The data obtained were used to construct the dependences of the material removal rate on the laser pulse duration for various degrees of "path" overlap (Fig. 2).

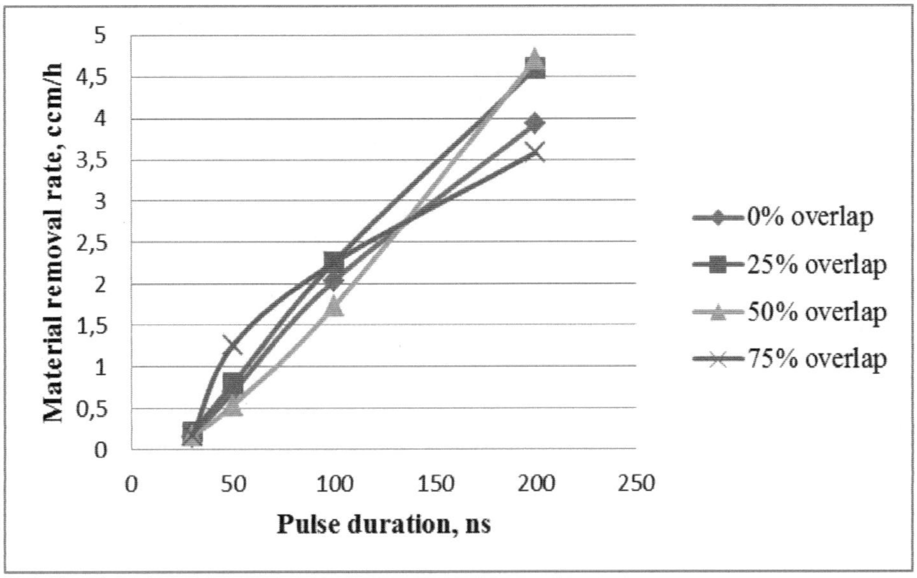

Fig. 2. Dependences of the material removal rate on the laser pulse duration (for various degrees of "path" overlap)

To investigate the influence of the pulse repetition rate and the pulse duration on the surface quality after laser processing, the roughness of processed areas was measured using the mechanical profilometer Dektak 150. During this experiment the parameters of pulse duration (from 30 to 200 ns) and repetition rate (from 2 to 20 Hz) were varied, while the average radiation power (P_{av} = 20 W), the scanning speed (v=1 m/s) and degree of "path" overlap (equal to zero) remained unchanged. The measurement results are shown in Table 2.

Table 2. Measurements of the processed surface roughness (μm) versus the pulse duration and the pulse repetition rate (Pav=20W, v=1 m/s, no overlap)

Repetition rate, Hz	***Pulse duration, ns***			
	30	**50**	**100**	**200**
2	2.31	3.70	4.49	10.36
5	3.65	5.45	9.02	9.01
10	4.61	6.41	8.22	9.35
15	4.90	7.72	7.15	6.49
20	5.38	7.89	7.10	9.51

As seen from the above data, the factor of pulse duration effects greater on roughness of the laser treated surface than the factor of repetition rate. Thus the entire frequency range with increasing pulse duration increases surface roughness.

The data obtained were used to find the dependence of the surface roughness on the repetition rate and pulse duration of the laser radiation (Fig. 3, 4).

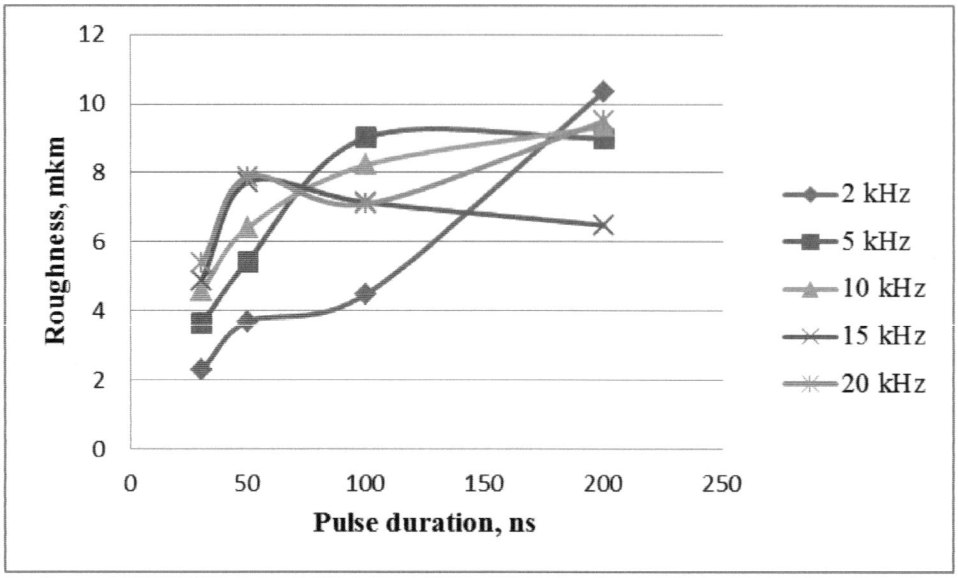

Fig. 3. The dependence of the surface roughness on the pulse duration (different repetition rate).

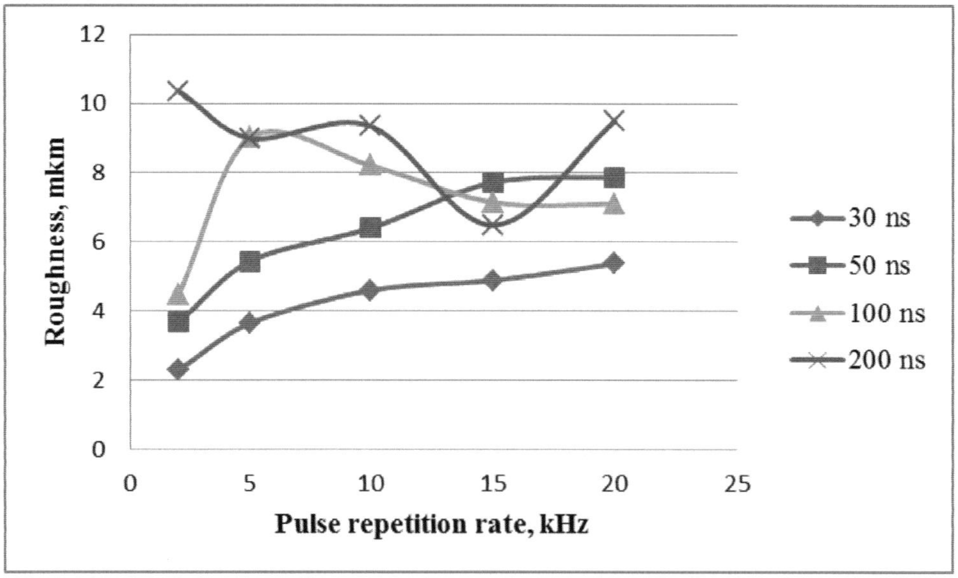

Fig. 4. The dependence of the surface roughness on the pulse repetition rate (different pulse duration).

3. Discussion

For the procedure of forming three-dimensional ZrO_2 ceramics dental structure prototypes the process productivity (material removal rate) and the final surface quality (total roughness) are the determining factors.

The material removal rate increases with the increase of pulse duration (Fig. 2). The dependence is almost linear in the chosen region of pulse durations. Thus, the use of laser pulses with a longer duration increases the material removal rate. The productivity of laser milling technology (at average power P_{av}=20 W) is comparable with the material removal rate using conventional mechanical milling.

The factor of laser pulse duration has a greater effect on the surface quality (roughness) than the factor of the pulse repetition rate. At shorter pulses, the surface remains smoother than at long pulses. However, as noted above, dental structures require a roughness of 10-30 μm, which means the choice of the mode of operation with pulse durations from 100 to 200 ns. The pulse repetition rate affects the surface quality to a lesser extent, and can be chosen in the range of 15-20 kHz for corresponding optimum pulse durations.

The combination of optimum parameters for prototyping of ZrO_2 ceramics dental structures using laser milling technology is as follows (for λ=1,068 μm, P_{av}=20 W, v=1 m/s):

- Pulse duration is 200 ns
- Pulse repetition rate is 20 kHz
- Degree of "path" overlap is 25%

The cross section of the cylindrical structure of ZrO_2 ceramics (thickness of the wall is 50 μm) produced by laser milling technology at optimum parameters given above is shown on Fig. 5. This technology allows the creation of thin-walled dental structures. Walls of dental frameworks with such thicknesses have light transmittance similar to that of human tooth tissue, which is an important esthetic requirement in stomatology.

Fig. 5. Half cut section of the cylindrical structure with a wall thickness of 50 μm (side and top views).

Three-dimensional dental structure obtained by laser milling technology are shown on Fig. 6. Processing ZrO_2 ceramic blanks was carried out from pre-prepared 3D-model of a human tooth.

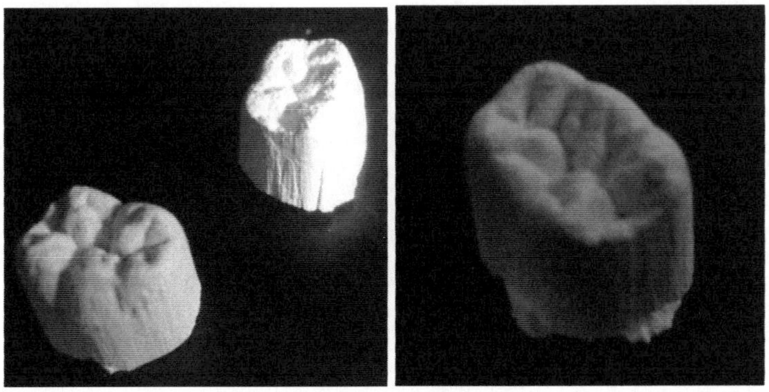

Fig. 6. Three-dimensional dental structure obtained from 3D-model of a human tooth by technology of laser milling.

It must be noted that the technology of laser milling of ceramic articles can be used to produce three-dimensional structures which can find application in other science and engineering fields.

This work was supported by the Russian Foundation for Basic Research (grant № 14-02-00369-a).

Journal of Physics: Conference Series **691** (2016) 012007 doi:10.1088/1742-6596/691/1/012007

References

[1] Pugachevskiy M A, Zavodinskiy V G and Kuz'menko A P 2011 Zirconium dioxide dispergating by impulse laser radiation *Technical Physics* **81** pp 98-102

[2] Ryabochkina P A, Sidorova N V, Ushakov S N and Lomonova E E 2014 Spectroscopic properties of erbium-doped yttria-stabilised zirconia crystals *Quantum Electron.* **44(2)** pp 135–137

[3] Ganin D V, Mikolutskiy S I and Tokarev V N 2014 Formation of micron and submicron structures on a zirconium oxide surface exposed to nanosecond laser radiation *Quantum Electron.* **44** pp 317-322

[4] Vaphin S V and Chvan V I 2011 Ceramics on the basis of zirconium dioxide *Dental-practice* **1** pp 20-27

Analysis of microstructure and properties of multilayer coatings produced by laser cladding

D P Bykovskiy, V N Petrovskiy, P S Dzhumaev, V I Polskiy, V M Yermachenko

National Research Nuclear University "MEPhI", Moscow, Russia

E-mail: d.bykofsky@gmail.com

Abstract. Purpose of the work is to prepare multilayer coatings corresponding to specified requirements to recovery and improvement of surface details. Requirements to coatings: providing durable and reliable adhesion base and filler materials, absence of pores, cracks, delaminations, reducing mixing metal base and cladding. We used iron-based PR-10R6M5 and tungsten carbide Hoganas 44712 powders. Experimental determination of the optimal technological mode of application of the single track, the coefficient of overlapping tracks to create a full layer, the angle of the second cladding layer, relative to the first one and, finally, the determination of the optimal additive tungsten carbide to achieve increased durability were produced to fulfill these requirements.

1. Introduction

The technology of the recovery process and the application of protective coatings using the fiber laser is one of the priority technologies in mechanical engineering, with great prospects and require more extensive research [1]. During application of cladding coatings it is necessary not only to take into account the whole set of parameters that affect the quality of the coating, but also the economic feasibility when choosing one of them [2]. Processes affecting the quality of education of the deposited coatings are not fully known. The study of these processes allows more efficient to select process conditions to ensure the specified quality of the applied coating. Used powder PR-10R6M5 was made by spraying molten gas. It is usually used for the restoration and consolidation of punching tools, the creation of the working surfaces of cutting tools, hardening roll cold rolling welding and other purposes [3]. Powder Hoganas 44712 is agglomerated powder WC-Co and has spherical particles given more uniform distribution of carbide in the matrix [4].

2. Experimental equipment

Photo of experimental setup is shown in figure 1a. The equipment consists of a filler unit (1) (view inside is shown in figure 1b), the controler (2), a powder feeder SulzerMetco Twin 10-C (3) for feeding a metal powder with a particle size of 5 to 200 μm. We used a fiber laser LS-3.5 (4) with a power of up to 3.5 kW, the wavelength $\lambda = 1069$ nm, the fiber diameter of 100 μm. Cladding head in figure 1b consists of a collimating and focusing optical system, water cooling system optics and nozzle, delivery system of filler powder and shielding gas.

The metal powder is provided in the treatment zone via a coaxial nozzle symmetrically from all the sides, under laser radiation it is heated and melted (figure 2). Simultaneously, the substrate

(a)

(b)

Figure 1. Experimental equipment (a) and view inside of cladding module (b).

surface is melted. When the relative motion of the nozzle relative to the substrate after cooling is formed on the surface of the track. The sequential application of tracks with some overlapping formed a new layer.

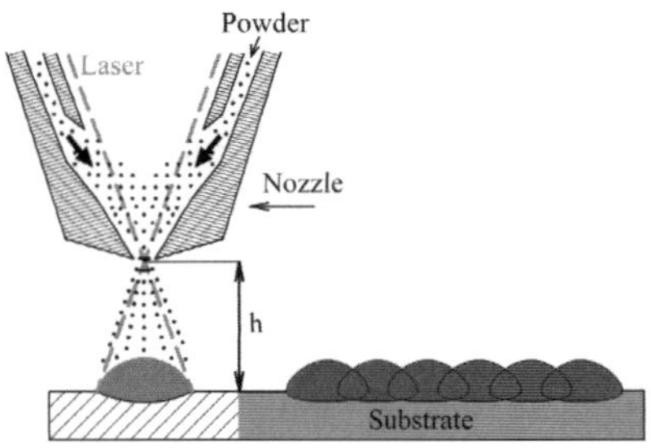

Figure 2. Chart of coaxial laser cladding.

3. Experimental results

3.1. Forming one track

It was necessary to prepare the filler material before cladding. Grain size of PR-10R6M5 is in the range up to 140 μm. Sieving through a flat sieve was performed for improving the properties of the cladding. The cell size varied from 63 to 100 μm. The powder became more homogeneous, without large teardrop-shaped beads after screening (fig. 3).

It is different forming tracks during cladding with different modes of laser power, speed,

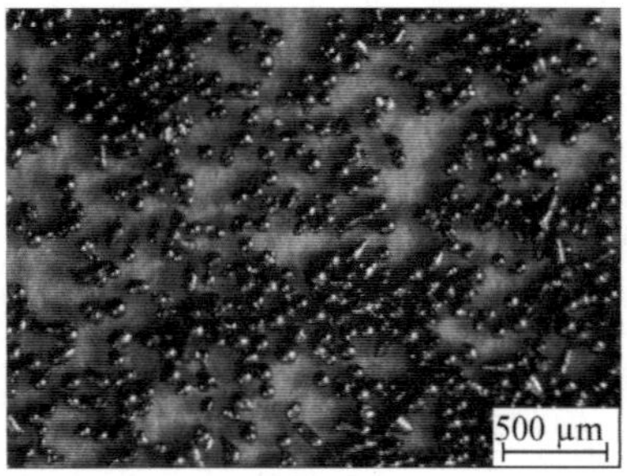

(a)

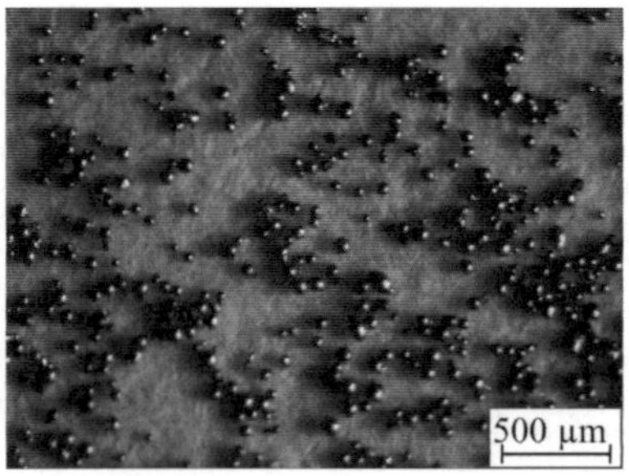

(b)

Figure 3. Photo of powder before (a) and after (b) sieving through a flat sieve.

powder flow, gap between nozzle and substrate surface. Under cladding with different modes of laser power, speed, powder flow, gap between nozzle and substrate surface different tracks are forming. Deep penetration of the filler material into the substrate can be seen in a technological window of these parameters, this results in a significant mixing of materials and the deterioration of coating properties. Coating, produced under the other process parameters window with a small penetration of filler material into the substrate, will have insufficient adhesion. One track geometrical parameters were measured for proper selection of cladding technological window: width, height and depth of penetration of the substrate, depending on adjusting of the following parameters: laser power, speed, powder flow, gap between the nozzle and the substrate. The measurement results are shown in figure 4.

The measurements evidenced that a change of the laser power has the most crucial impact on

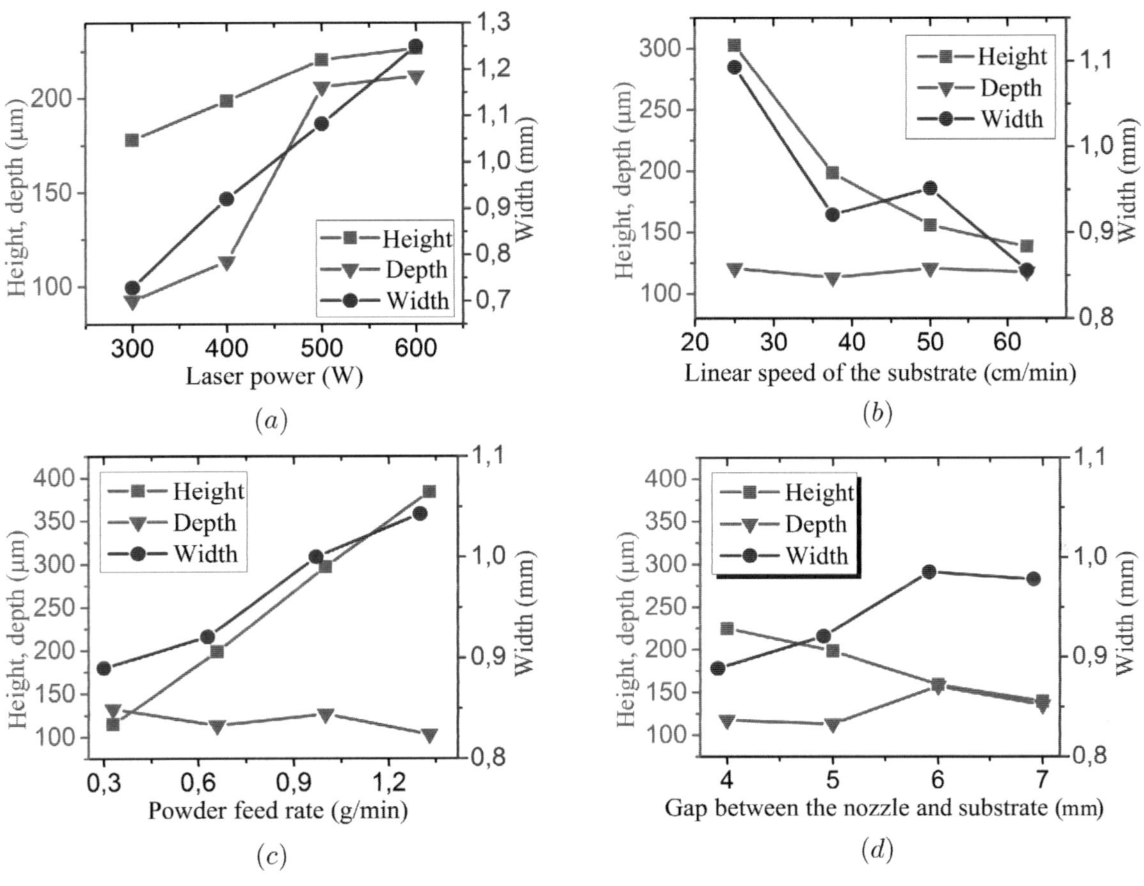

Figure 4. Dependences of geometric parameters of cladded tracks on laser power (a), powder feed rate (b), linear speed (c) and gap between the nozzle and substrate (d).

the width of the cladding track and depth of penetration of filler material in the substrate and the powder feed rate - on the height of the tracks. The most appropriate mode was selected for further research on the following parameters: uniformity of track form, absence of pores and cracks, the least gauge of mixing substrate materials and powder, and a minimal depth of penetration of filler material in the substrate with good adhesion of the coating to the substrate without delamination from it.

A study of topography and microstructure of the surface was carried out under the electron microscope (SEM) Carl Zeiss EVO 50 XVP, determined the elemental composition of the samples by the method Vickers on the setup HVS-1000 with automatic loading of the indenter to the

Journal of Physics: Conference Series **691** (2016) 012008 doi:10.1088/1742-6596/691/1/012008

load at 1 N. The dwell time at load has been chosen to 20 s. The results showed (figure 5a) that the HAZ is 300 μm, cladding structure - equiaxed mesh, near the boundary with the substrate - dendritic structure, grain boundaries are enriched in heavy elements (Mo, W). The structure of the HAZ - milled grain ferrite-pearlite with traces of partial martensitic transformation. The structure of the substrate - polyhedral (equiaxed) grains of ferrite and pearlite. The boundary of the substrate and facing a clear, there is no peeling, cracking and porosity. Microhardness (figure 5b) changes at the boundary of the substrate and cladding (points 3 and 4) from 5700 MPa (cladding) to 2000 MPa (substrate material).

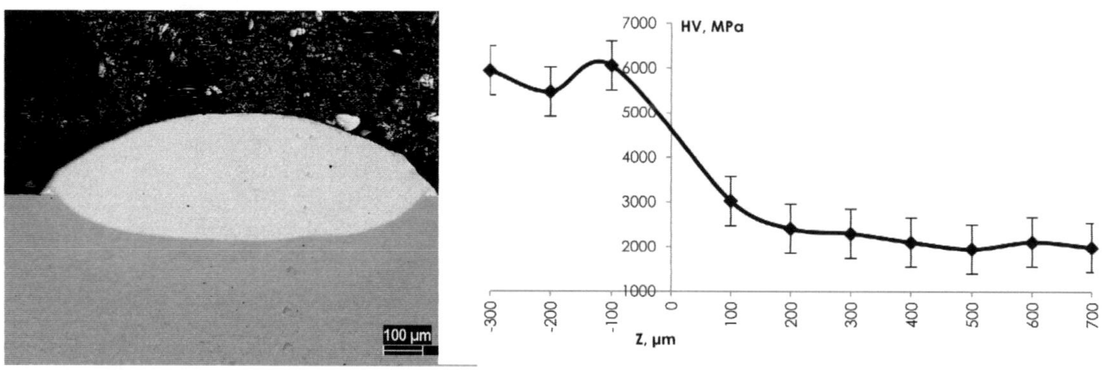

(a) (b)

Figure 5. One track producing by optimal regime (a) and change of microhardness over the cross section (b).

3.2. Forming one layer

Single-layer coatings were formed by depositing sequential tracks with different coefficients of overlapping by width: 0.33, 0.5, 0.6. The temperature distribution of the substrate was also measured during the deposition. The measured distribution determined the distance in the transverse direction, which is a significant decrease in temperature. Measurements demonstrated that the change in the surfacing (laser power, deposition rate, flow rate of the powder, gap between the nozzle and substrate) influences the maximum temperature in the vicinity of the melting of the metal. The distance in a transverse direction, at which crucial attenuation of temperature in all cases happens, may be taken as 5 mm.

Next, a single-layer coating was performed considering the temperature distribution of the substrate (figure 6).After application of the track 1 (figure 7) substrate material was displaced at the distance D where thermal field of one track falls significantly. On the basis of the analysis devoted to the measured distribution of substrate material temperature within welding the value of D = 5 mm was chosen. Then track 2 was applied, and once again shift at the distance D happened, track 3 was applied. Further substrate material was turned back to the track 1 and was moved in direction of the track 1 at the pace L = 0.33W, and after that the layers 4, 5 and 6 were applied. The procedure was repeating, until the whole layer was completed.

Figure 8 show images of produced coatings obtained due to electron microscope, and figure 9 - the measurement results of their microhardness. For all coatings structure - equiaxed mesh, near the boundary with the substrate - dendritic structure, grain boundaries are enriched in heavy elements (Mo, W). The structure of the HAZ - milled grain ferritepearlite with traces of partial martensitic transformation. The boundary of the substrate and facing a clear, there is no peeling, cracking and porosity. Microhardness changes at the boundary with the cladding

Figure 6. The temperature distribution during the cladding (thermal camera picture).

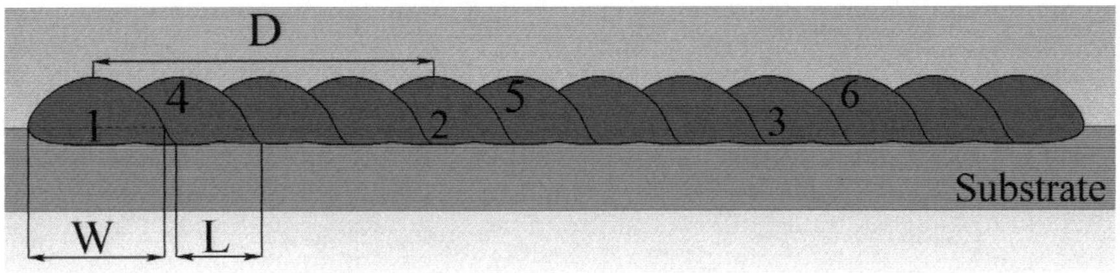

Figure 7. Strategies for creating layer considering thermal field.

and the substrate 6000 MPa (cladding) to 2000 MPa (substrate material). The Fe content in the powder is 80% and about 98% in the substrate. The resulting coating with overlapping 0.33 contents 86% Fe, with an overlap of 0.5 - 85%, with overlapping 0.66 - 83%, with overlapping 0.66 and considering the temperature distribution - 81.5%. Cover with a large overlap ratio has minimal mixing of substrate materials and powder that indicates improved quality. Coatings by strategy based on the temperature distribution, have the best characteristics concerning the mixing of substrate materials and the powder, the highest microhardness, but penetration of substrate and coating thickness are uneven at the site of the last series of tracks.

3.3. Forming two layers

The bilayer coatings were produced under different conditions. Initially the first layer was created by sequential application of tracks with overlaping 0.66. Next, a second layer was formed with the same mode, but with different directions of the tracks: along the first layer, at an angle of 45° and 90°. Structure of coatings is presented in figure 10 and change of microhardness over the cross section this bilayer coatings in figure 10d.

Coatings have equiaxed mesh, near the boundary with the substrate - dendritic structure, grain boundaries are enriched in heavy elements (Mo, W). The width of the HAZ 300 m, its structure - the crushed grain ferrite-pearlite with traces of partial martensitic transformation. The iron content in the three cases was 85%. The microhardness increased to 7,000 MPa (figure 10d), its dependence is more uniform in the case with the direction of the clad layers of 0°. Since measured the remaining coating characteristics are the same, one can conclude the best quality in the event of cladding layers at an angle of 0°.

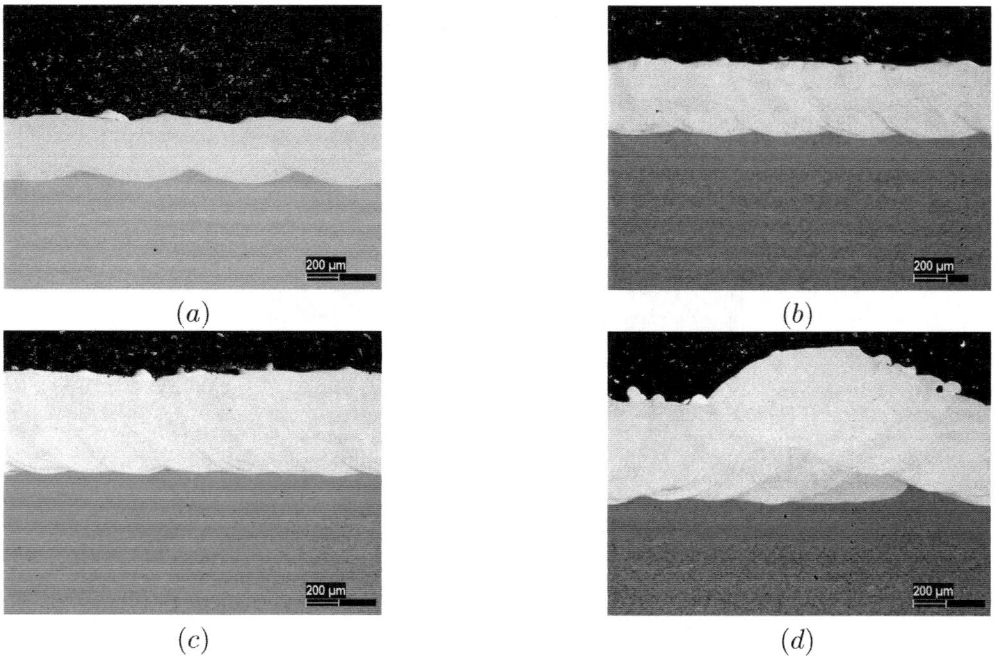

Figure 8. . Cross-sections of one layer with overlapping 0.33 (a), 0.5 (b), 0.66 (c), 0.66 and considering thermal field (d).

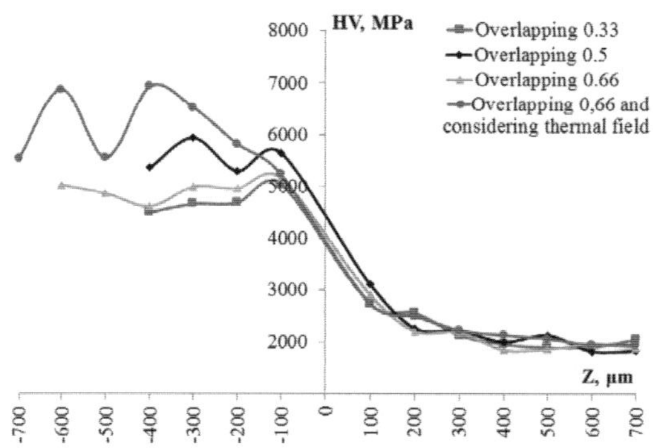

Figure 9. Change of microhardness over the cross section of one layer.

3.4. Forming high wear resistance coatings

Coatings with additives of tungsten carbide- based 44712 Hoganas powder were created under obtained technological regime. Additives ranged from 3 to 9% by weight. The coating, containing 9% Hoganas 44712 powder cracked, so did not participate in further tests. Another samples were tested for wear resistance by standard Brinell-Haworth with a rubber disc and sand. The experiment is based on the current method of comparative evaluation of steel and alloys tribological characteristics (weight loss) by dry friction. Steel disk with a rubber rim is rotating at 200 rpm and is pressed against the fixed sample with a force of 15 N during 10 min. Between the disk and the sample is fed from the bunker by gravity silica sand. Addition of 7% of the powder based on tungsten carbide resulted in an increase in wear resistance of the alloy layer 2 times in comparison with cladding iron-based powder (figure 11).

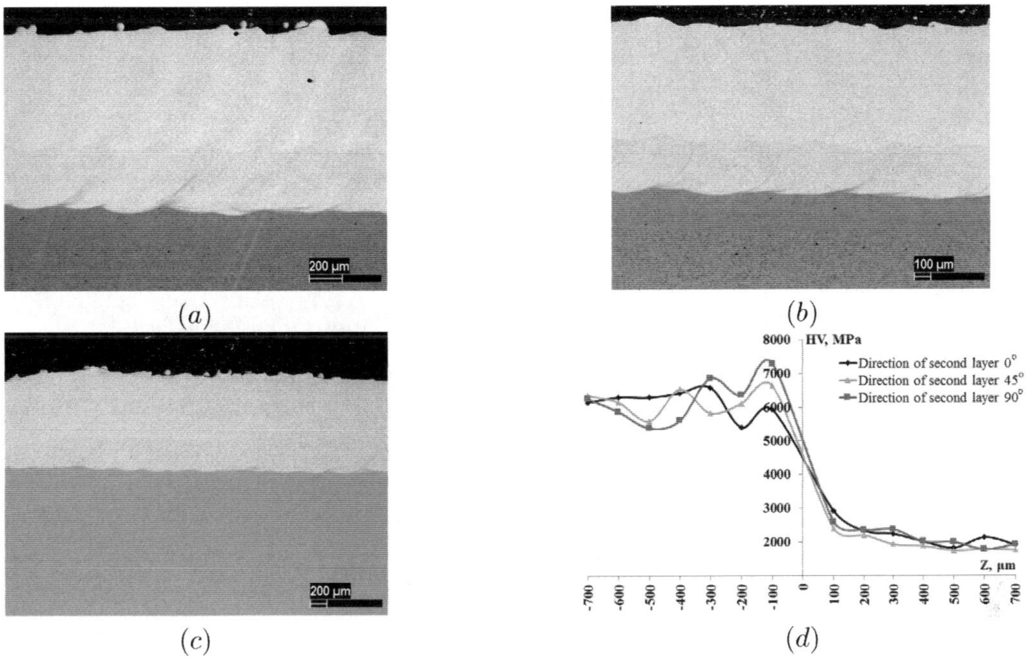

Figure 10. . Bilayer coatings with direction of second layer 0° (a), 45° (b), 90° (c) and change of microhardness over the cross section (d).

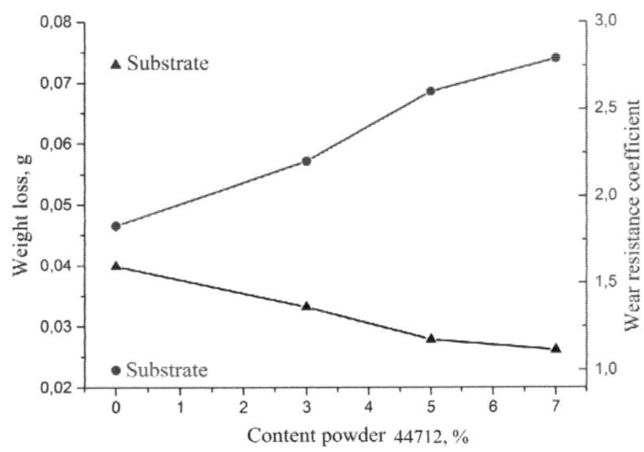

Figure 11. Dependence of wear resistance in content powder Hoganas 44712.

4. Conclusion

The greatest influence on the height of cladding track has the powder feed rate and on the width of the cladding track and depth of penetration of filler material in the substrate of the tracks - change of the laser power. Monolayer coatings with a coefficient of 0.66 overlapping tracks have lower gauge of mixing powder material and the substrate than coatings with a coefficient of 0.33 and 0.5. This is a consequence of smaller area thermal effects on the substrate. Bilayer coatings with the direction of the clad layers 0° are more uniform than in the case when a direction angle of 45° and 90°. Differences in the ratio of mixing for the three different directions are not observed. Addition of 7% powder Hoganas 44712 improved wear resistance 2 times in comparison with cladding iron-based powder. The coatings can be used to restore worn-out and manufacture of new machine parts and mechanisms: the body parts of various internal combustion engines, electrical and crank shafts, valves, pulleys, flywheels, wheel hubs, etc.

Acknowledgments
This work was financial supported by the Ministry of Education and Science of the Russian Federation (unique identier PNIER RFMEFI58214X0004).

References

[1] Nenadl O, Ocelk V, Palavra A and Hosson J T D 2014 *Physics Procedia* **56** 220 – 227
[2] Grigoryants G, Shiganov I and Misyurov I 2006 *Technological processes of laser processing* (MGTU)
[3] Oliveira U, Ocelk V and Hosson J D 2005 *Surface and Coatings Technology* **197** 127 – 136
[4] Murzakov M, Petrovskiy V, Birukov V, Dzhumaev P, Polski V, Markushov Y and Bykovskiy D 2015 *Physics Procedia* **71** 202 – 206

Receiving Wear-Resistance Coverings Additives of Nanoparticles of Refractory Metals at a Laser Cladding

M A Murzakov[1], V N Petrovskiy[1], D P Bykovskiy[1], A O Andreev[1], V P Birukov[2], Y V Markushov[1]

[1]National Research Nuclear University MEPhI, Moscow, Russia
[2]IMASH RAN, Moscow, Russia

E-mail: clericfull@mail.ru

Abstract. Laser cladding technology was used to conduct experiments on production of wear-resistant coatings with additive nanoparticles of refractory metals (WC, TaC). Mechanical testing of coating abrasion was made using Brinell-Howarth method. The obtained data was compared with wear-resistance of commercial powder containing WC. It was found that at a concentration 10-15% coating with nanopowder additives shows a dramatic increase in wear-resistance by 4-6 times as compared to carbon steel substrate. There were conducted metallurgical studies of coatings on inverse electron reflection. There was determined elemental composition of deposited coating and substrate, and microhardness measured. It was found that structure of deposited coating with nanoparticles is fine.

1. Introduction

Laser cladding is widely used in production of protective coatings in various industries: chemical, mining, oil and gas, and others [1]. Durability is an important parameter of protective coatings that operate under abrasive wear. There are studies on laser cladding of wear-resistant coatings with tungsten carbide additives (grain size - 100 microns) [2-4]. The main problems of these coatings are cracks and pores in the body of cladding. One of the ways to solve the problem of cracking is substrate preheating [5, 6].

In recent years interest to nanoparticle additives as a mean to improve mechanical and tribological properties of coatings began to grow. In a number of references smaller grain size was marked in the structure of coating [7, 8]. Nanoparticles prevent the spread of defects in coating, thereby improving its quality. Besides the advantages, nanopowders still have some disadvantages, in particular, there is no possibility to feed nanopowder directly to weld deposition zone since there is no special feeding device. A solution to this problem is a preliminary substrate covering with powder and subsequent laser radiation processing.

This study describes the work on the production of wear-resistant coatings with different concentrations of TaC and WC nanoparticle additives, and tests made using Brinell Howarth method. The results obtained were compared with results of tests of commercial powders Ni + 60% WC. Previously, there were obtained investigation results on reflection of inverse electrons of coating and substrate, defined mixing zone and measured micro-hardness of coatings and nano-powders without additives [9]. It was found that coating microstructure is fine-grained, and flaw-free [10].

2. Experimental setup

Experimental work was carried out using a universal laser system consisting of the following elements:

1. Robotic arm of ABB production;
2. Fiber laser of NTO "IRE-Polus" production;
3. Scanning system with processing field150x150 mm;
4. SULZER powder feeder.

Microhardness measurements were performed using Vickers method. The load during the measuring process was 100 g, the exposure time - 10 seconds. Abrasion testing was carried out using Brinell-Howarth method. The test procedure is as shown on figure 1: a rotating rubber disk 3 is pressed against the sample 2 and abrasive (river sand) pours through the feed magazine 1. The load is 15N, exposure time is10 minutes.

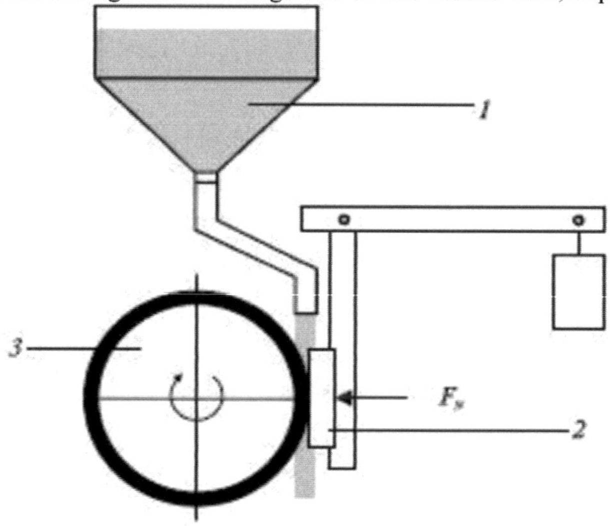

Figure 1. Overall schematic of wear testing according to Brinell-Howarth method

3. Technique of the experiment

The experiments were carried out with different types of HOGANAS powders: 1535-40, 1360. As a powder of tungsten carbide powder TP40/40 by TECHNOGENIA production was chosen. Carbon steel AISI C1030 was chosen as a substrate. Chemical composition of the substrate and HOGANAS powders is shown in Table 1.

Table 1. Chemical composition of powders 1360 and 1535-40 by HOGANAS production, and substrate.

	Ni	C	Fe	Cr	B	Si
Powder 1360	base	0.45	2.9	11.0	2.3	3.6
Powder 1535-40	base	0.03	0.14	-	2.75	2.86
Substrate AISI C1030	0.25	0.22	98	0.15	-	0.041

Mixtures of powders with different concentrations (5, 10, 15, 20% respectively) of TaC and WC nanopowders with conventional HOGANAS powders have been preliminary placed on steel substrates measured 70x15x10 mm. The layer thickness was about 1 mm.

Pre-process parameters were selected for laser cladding of nanopowders mixture and ordinary powders. Laser power was 1.5 kW, processing speed - 10 mm / sec, blooming - 20 mm, overlap factor - 35-40%. The main criterion in the selection of process parameters were as follows:

$$H \approx \frac{1}{4}L \tag{1}$$

where H is roller height, L is roller width as shown in Figure 1.

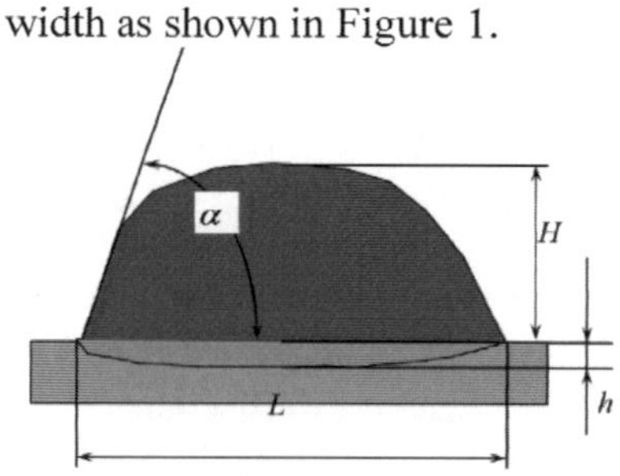

Figure 2. General view of roller and its basic geometrical parameters.

During the selection of process parameters the following values were changed:
- Radiation power;
- Linear speed;
- Size of laser radiation spot on surface of built-up sample;
- Overlap degree %.

4. Results and discussion
Samples with abrasion resistant coating containing nanoparticles of tungsten carbides and tantalum were obtained. Abrasive material for the test was river sand, which was pre-screened. Also, wear-resistance ratio was calculated. Figure 3 shows the results of wear measurement of coating with nanoparticles WC and TaC, as well as wear resistance of AISI C1030 steel substrate and ordinary commercial powder TP40/40 produced by TECHNOGENIA.

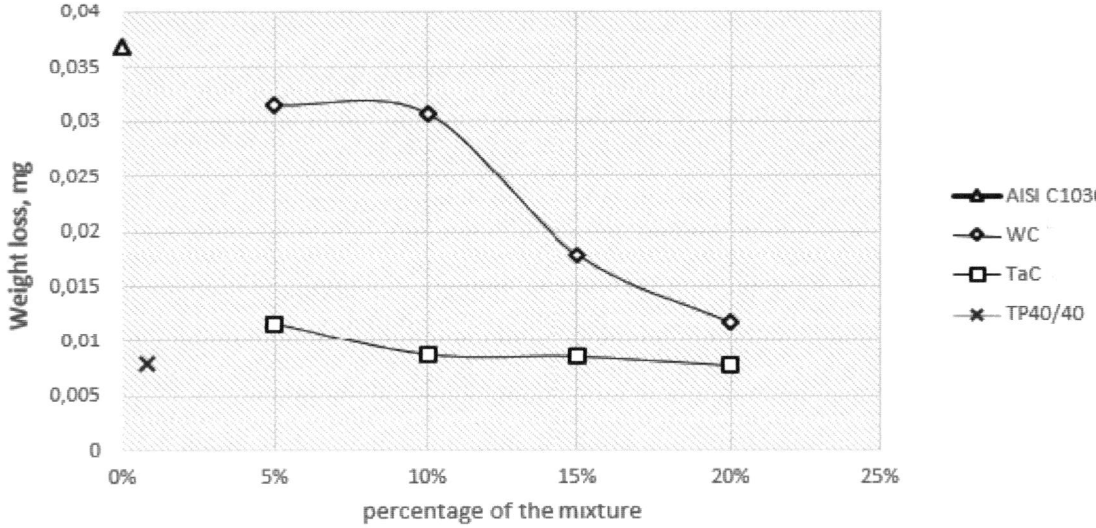

Figure 3. Measurement of wear-resistance of powders with nano-WC and nano-TaC and powder TP40/40.

As one can see on figure 3, wear of coating with nano-WC additives is lower than of TECHNOGENIA commercial powder. This may be due to the fact that during the overlaying of coating with nanoparticles, tungsten carbide begins to decompose into secondary phases, which do not increase wear resistance of the coating. Coatings with tantalum carbide nanoparticles, at 10% concentration of carbides, possess the same wear as coating with TP40/40 powder. This can be explained by the fact that during the process of cladding tantalum carbide hardfacing remains unchanged and do not break up because the melting point of TaC particles is ≈3800 °C. In future, other process parameters will be selected to obtain stable carbide phases of tungsten, which will improve wear resistance of coating.

It was observed that coating with nanoparticles of tungsten and tantalum carbides has no defects such as cracks or porosity, in contrast to coatings produced by cladding with powder TP40/40. This may be due to the fact that nanoparticles prevent propagation of defect in coating. Nanoparticles of tungsten and tantalum carbides are evenly distributed throughout the microstructure of the coating that has been defined earlier in the work [10].

5. Conclusion
This article describes an experiment on laser surface coating using nickel based powders with different concentrations of nanoparticles of tungsten and tantalum carbide as additives. There were selected technological parameters for laser surface coating: P - laser radiation power, v - processing speed and f - blooming. Power density in this experiment was 0.68-0.98 MW / cm2.

Comparative endurance tests were conducted with the same parameters for coatings containing nano-WC and nano-TaC. It was found that coatings, which contain particles of TaC, were highly durable, unlike coatings with nano-WC. This is because tantalum carbide has a high melting point of 3800 °C, while the tungsten carbide has a melting point of about 2800 °C. Thus, tungsten carbide during surface coating breaks into either tungsten and carbon, or into secondary W2C carbides. Wear resistance of coating with TaC additives increases

4 times as compared with the coating without a nanopowder additive. This is a marked influence on the wear resistance of nano-powder coating. There were conducted wear tests of coatings with conventional powders TP40/40 by TECHOGENIA production, the results were compared with test results of nanopowders. It was found that wear resistance of coatings with nano-WC is worse than of coatings with powder TP40/40, while wear of coating with nano TaC is on a level with these powders. The concentration of the tantalum carbide is 10%, while the concentration of tungsten carbides in TP40/40 is 40%.

Acknowledgments
The study was financially supported by the Ministry of the Education and Science (unique identifier PNIER RFMEFI58214X0004).

References

[1] Torims T 2013 *DAAAM International Scientific Book*, p. 587-608
[2] Yong-Chwang Chen, Kwo-An Chiang 2006 *Journal of the Chines Institute of Engineers* **29** **3**, p. 423-431
[3] Hu Cheng, Jian Yi, Zhigang Fang, Sheng Dai, Xianrui Zhao 2013 *Material Transactions* **1**, p. 50-55
[4] Qianlin Wu, Wenge Li 2011 *Material Transactions* 52 3, p. 560-563
[5] M.R. Fernández, A.García, J.M.Cuetos, R.González, A.Noriega, M.Cadenas 2015 *Wear* **324-325**, p. 80-89
[6] Parisa Farahmand, Radovan Kovacevic 2015 *Journal of Materials Processing Technology* **222**, p. 244-258
[7] Parisa Farahmand, ShuangLiu, ZheZhang, RadovanKovacevic 2014 *Ceramics International* **40**
[8] Zhang Wei, Yao Jian Hua, Zhang Qun Li, Kong Fan Zhi 2006 *Solid State Phenomena* **118**, p. 579-584
[9] M A Murzakov, V N Petrovskiy, V I Polski, V D Mironov, N M Prokopova, E V Tret`yakov 2015 *23rd International Laser Physics Workshop (LPHYS`14), Journal of Physics Conference Series* **594**
[10] M. Murzakov, V. Petrovskiy, V. Birukov, P. Dzhumaev,V. Polski, Y. Markushov, D. Bykovskiy 2015 *Physics Procedia* **71** p.202-206

SINGLE BACTERIUM DETECTION USING SERS

[1]S.A. Gonchukov, [1]T.V. Baikova, [1]M.V. Alushin, [2]T.S. Svistunova, [3]S.A. Minaeva, [4]A.A. Ionin, [1,4]S.I. Kudryashov, [4]I.N. Saraeva, [4]D.A. Zayarny

[1]*National Research Nuclear University MEPhI, Moscow, Russia*
[2]*Infectious Clinical Hospital No 2, Moscow, Russia*
[3]*Institute of Laser and Information Technologies RAS, Troitsk, Russia*
[4]*P.N. Lebedev Physical Institute RAS, Moscow, Russia*
gonchukov@mephi.ru

Abstract

This work is devoted to the study of a single *Staphylococcus aureus* bacterium detection using surface-enhanced Raman spectroscopy (SERS) and resonant Raman spectroscopy (RS). It was shown that SERS allows increasing sensitivity of predominantly low frequency lines connected with the vibrations of Amide, Proteins and DNA. At the same time the lines of carotenoids inherent to this kind of bacterium are well-detected due to the resonance Raman scattering mechanism. The reproducibility and stability of Raman spectra strongly depend on the characteristics of nanostructured substrate, and molecular structure and size of the tested biological object.

1. Introduction

Pathological bacteria are the cause of many serious diseases. Their early diagnostics is the indispensable condition of the successful fight against these diseases. Many methods have been developed for detection of bacteria which are the causative agents of human illnesses. Bacteria invasion into the cerebral spinal fluids (CSF) is a serious problem leading to the development of meningitis – a dangerous bacterial form of the central nervous system illness with a high level of mortality. First bacteria identification is very important for the early diagnosis and treatment of this illness. The current tools available for detecting the bacteria penetration into CSF are cytological, bacteriological, and serological analysis techniques. They are time-consuming and not always reliable, simple, and inexpensive. At present the optical methods of CFS analysis unfortunately are poorly developed.

In view of its potential for the fast and noninvasive analysis at a molecular level the Raman spectroscopy has been the method of precise research of biological objects for the last years. Raman spectroscopy undoubtedly holds promise for CSF analysis, because it is a prompt and noninvasive technique capable of providing reliable information about the kind of microorganisms in a few seconds [1-5]. An increase in measurement sensitivity becomes actual if the detection of a single bacterium is the goal.

This work is devoted to the study of bacterium detection using surface-enhanced Raman spectroscopy (SERS), which combines the molecular fingerprint specificity with an increased sensitivity due to an enhancement of the Raman signal intensity for molecules of interests in close proximity to a plasmonic nanostructure. The theory of this phenomenon has been still poorly developed. In spite of this, experimental study showed a very high gain of spectral lines using the plasmonic effect. This fact was experimentally demonstrated for the cases of simple molecular substances detection. Concerning the bacteria as the objects of the study the spectroscopic results are more modest [6-9].

Pure bacterial cultures of *Staphylococcus aureus* were chosen for the study in vitro. This kind of bacteria is a causative agent of meningitis. As known, this bacterium can be effectively detected using resonant version of Raman spectroscopy (RS) due to carotinoids content [10,11]. As a result, the detection efficiency of resonance version of Raman spectroscopy and surface-enhanced Raman spectroscopy could be compared in the frames of this work.

2. Materials and Methods

2.1. Nanoparticles

Gold sphere nanoparticles (Au-NPs) of $\approx$ 40 nm in diameter were used for SERS. Au-NPs were fabricated by pulsed laser radiation (1030 nm, 300 fs) focused on a gold plate. In so doing, the gold plate was placed under a water or isopropyl alcohol layer two millimeters thick. Au-NPs were put on the substrate of silver coated glass plate. The silver film was deposited on the glass plate by magnetron unit SC7620 Mini Sputter Coater. Colloidal Au-NPs were dropped onto the substrate surface (1-22 drops) that was rotated in a centrifuge at a speed up to 400 rpm. As a result, the substrates with different Au-NPs concentration and the substrate without NPs were prepared for the bacteria deposition.

2.2. Bacterium samples

Cultivation and preparation of in vitro cultured bacterium of *Staphylococcus aureus* were performed by the bacteriological laboratory of the Moscow Infectious Clinical Hospital No 2. The bacterial species were cultured in nutrition medium and their colonies suspended afterwards in phosphate buffered saline solution. About 1 ml of the suspension were spotted onto the clean surface of a substrate and dried at room temperature. 70% ethanol solution was used for bacteria inactivation. As was shown earlier [2], a bacterium sample in the form of a droplet dried at room temperature keeps the main characteristic features of its Raman spectrum.

2.3. Spectroscopic instrumentation

The spectral measurements were fulfilled with the help of Nicolet Almega XR spectrometer supplied by a 532-nm laser excitation source (Nd:YAG CW laser producing 20-mW second-harmonic radiation). The spectrometer was equipped with a microscope capable of X10, X50 and X100 magnification. The diameter of the probe laser beam spot was 1 μm in the microscopic research. The Stokes shift components in the range of 400–3100 cm^{-1} at the spectral resolution of 2 cm^{-1} were used in our analysis. Wave number calibration was performed by using the characteristic silicon reference peak at 521 cm^{-1}. The phenomenon of photobleaching for 30 seconds prior to the recording of the Raman spectra was used to reduce the fluorescence influence. The experimental spectra were computer processed using the OMNIK software.

3. Experimental Results

The measurements showed that maximum absorption of Au-NPs in water solution just after their preparation takes place at the plasmonic resonance wavelength of 525 nm (Fig. 1). Because of NPs aggregation, the absorption is decreasing and plasmonic resonance is broadening in time. So, the absorption coefficient is degraded as much as twice in two days.

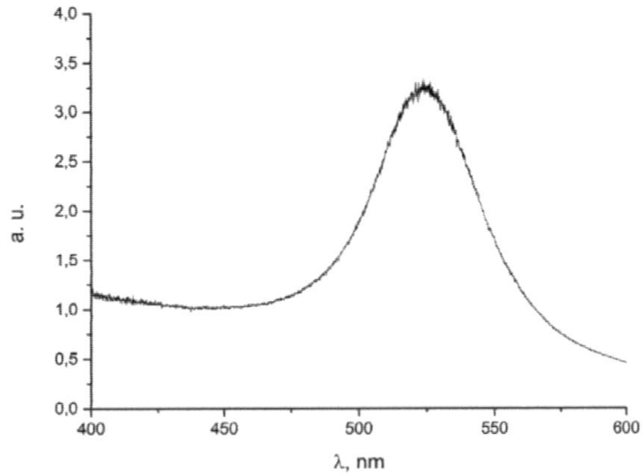

Figure 1. Absorption spectrum for gold sphere nanoparticles (≈ 40 nm in diameter) in water

Raman spectra of *Staphylococcus aureus* dried samples registered using resonant RS (without NPs) and SERS (with Au-NPs) are demonstrated in Fig. 2. It can be seen that the advantage of plasmonic effect using is equal to one order and more for the lines, connected with the vibrations of Amide, Proteins and DNA (1673.39, 1547.97, 1498.89, 1449.82, 1321.67 и 730 cm^{-1}) [12]. At the same time the lines of carotenoids (in the range of 1152 cm^{-1} and 1528 cm^{-1}) typical of *Staphylococcus aureus* are amplified insignificantly because of the plasmonic effect. However, they are well-detected due to the resonance Raman scattering mechanism because the excitation wavelength (532 nm) is within the carotenoids absorption band [13,14].

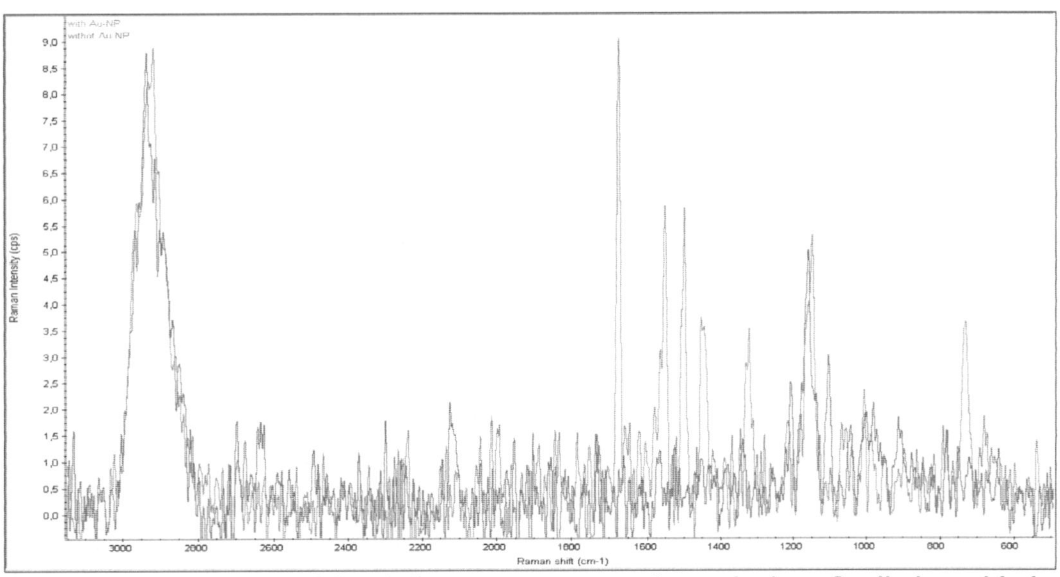

Figure 2. Raman spectra of *Staphylococcus aureus* under excitation of radiation with the wavelength of 532 nm for resonant RS (red) and SERS (green)

It was also observed that some spectral lines intensities are changing from one measurement to another. This fact of the spectra registration non-reproducibility and specific difference between the resonant RS and SERS spectra mentioned above is obviously the result of joint action of substrate parameters, bacteria orientation on substrate surface and the distance between the surface and bacteria. The influence of these factors was discussed during spectral analysis of single molecules [8,15]. In the case of bacteria as the test object these factors can be manifested more specific and complicated ones because of the large difference between characteristics of nanostructured substrate surface and molecular structure, and the size of bacterium.

4. Conclusion

The potentiality of surface-enhanced Raman spectroscopy (SERS) for a single bacterium detection as compared to the resonant Raman spectroscopy (RS) is demonstrated in this work. Specific difference between the resonant RS and SERS spectra and their non-reproducibility can be explained by large difference between the parameters of nanostructured substrate, and the size and molecular structure of bacterium and their spatial orientation in the vicinity of substrate surface.

This work was supported by the Russian Foundation for Basic Research (project No 15-02-08400, 13-02-00971) and RAS Presidium program.

References

[1] Harz M, Kiehntopf M, Stöckel S, Rösch P, Straube E, Deufel T and Popp J 2009 *J. Biophot.* **2** 70
[2] Baikova TV, Minaeva SA, Sundukov AV, Svistunova TS, Bagratashvili NV, Alushin MV, and Gonchukov SA 2015 *J. Phys.: Conf. Series* **594** 012029
[3] Sathyavathi R, Dingari NC, I. Barman, Prasad P, Prabhakar S, Rao DN, Dasari RR and Undamatla J 2012 *J. Biophoton,* **1,** 6
[4] Gonchukov SA, Lonkina TV, Bagratashvili VN, Minaeva SA, Sundukov AV, Migmanov TE, Lademann J and Darvin ME 2013 *Meditsinskaya Fizika* **2** 97 in Russian
[5] Gonchukov SA, Lonkina TV, Minaeva SA, Sundukov AV, Migmanov TE, Lademann J, Darvin ME and Bagratashvili VN 2014 *Laser Physics Letters* **11** 015602
[6] Fan X, Zheng W and Singh DJ 2014 *Light: Science & Applications* **3** e179
[7] Lezec HJ, Degiron A, Devaux E, Linke RA, Martin-Moreno L, Garcia-Vidal FJ and Ebbesen TW 2002 *Science* **297** 820
[8] Le Ru E and Etchegon P 2009 *Principles of Surface Enhanced Raman Spectroscopy and related plasmonic effects* (Elsevier)
[9] Howes PD, Rana S and Stevens MM 2014 *Chem. Soc. Rev.* **43** 3835
[10] Darvin ME, Zastrov L, Gonchukov SA and Lademann J 2009 *Opt. Spectrosc.* **107** 917
[11] Gonchukov S, Sukhinina A, Bakhmutov D, Biryukova T, Tsvetkov M and Bagratashvily V 2013 *Laser Physics Letters* **10** 075610
[12] Movasaghi Z, Rehman S and Rehman IU 2007 *Applied Spectroscopy Reviews* **42** 493
[13] Darvin ME, Gersonde I, Ey S, Brandt NN, Albrecht H, Gonchukov SA, Sterry W and Lademann J 2004 *Laser Phys.* **14** 231
[14] Darvin ME, Zastrov L, Gonchukov SA and Lademann J. 2009 *Opt. Spectrosc.* **107** 917
[15] Camden JP, Dieringer JA, Zhao J and Van Duyne RP 2008 *Accounts of Chemical Research* **41** 1653

PHOTODYNAMIC INACTIVATION OF BACTERIA AND BIOFILMS USING CATIONIC BACTERIOCHLORINS

[1]Meerovich G.A., [2]Tiganova I.G., [3]Makarova E.A., [4]Meerovich I.G., [2]Romanova Ju.M., [2]Tolordova E.R., [2]Alekseeva N.V., [2]Stepanova T.V., [2]Koloskova Yu., [3]Luk'anets E.A., [5]Krivospitskaya N.V., [5]Sipailo I.P., [5]Baikova T.V. [1]Loschenov V.B., [5]Gonchukov S.A.

[1]*A.M. Prokhorov General Physics Institute, Moscow, Russia*
[2]*N.F. Gamaleya Federal Research Center for Epidemiology and Microbiology, Moscow, Russia*
[3]*Organic Intermediates and Dyes Institute, Moscow, Russia*
[4]*Federal Research Centre "Fundamentals of Biotechnology" of the Russian Academy of Science, Moscow, Russia*
[5]*National Research Nuclear University MEPhI, Moscow, Russia*
gonchukov@mephi.ru

Abstract
This work is devoted to the study of two new synthetic bacteriochlorins with four and eight cationic substitutes as the photosensitizers in the photodynamic process. The spectral and antibacterial properties of these photosensitizers in saline solution were investigated. It is shown, that the aggregation ability decreases and the antibacterial efficiency grows as the cationic substitute number increases.

1. Introduction

As it is known, the photodynamic process (PDP) is widely used in the treatment of different illnesses. Photodynamic therapy (PDT) is based on the cytotoxic mechanism of PDP that leads to the loss of cells vital function. A key element of this process is a photosensitizer (PS), the properties of which determine the treatment effectiveness. A number of PS are used in medical practice. However, the quest for more effective PS continues. Strong absorption in the transparency spectral region of biological tissue, a high quantum yield of the triplet state for the singlet oxygen formation, a good solubility in water, as well as the available synthesis are the most important requirements to PS. Cancer treatment is traditionally given a lot of attention [1,2]. PDP also has a great potential in the fight against pathogenic bacteria [3,4], because PS can possess the property of selectively interaction with a bacterium. Cationic bacteriochlorins can meet these requirements [5-7].

Unlike antibiotics, all of which act on a specific target in a microbial cell, the photodynamic antibacterial process is a nonspecific damage of cellular components potentially subjected to oxidative reactions. In so doing the bacteria do not develop resistance to PD treatment, bactericidal effect has an addressable character and has no systemic effects on the body flora.

Antibacterial efficiency should increase with PS concentration increasing. But, at the same time, it may be accompanied by an aggregation of PS, that can significantly reduce the treatment effectiveness. The phenomenon of PS aggregation plays an essential negative role in the PDP [8-11]. However, it should be expected that the Coulomb repulsion will help reduce of this factor influence in the case of cationic bacteriochlorins using [12].

The study of PDP antibacterial effect on the bacterial biofilms, which are formed during severe infectious diseases development, is of considerable interest. Normally biofilms are surrounded by a matrix which protects the bacteria from exposure to antibiotics and immune protection factors [13,14]. Biofilms formed by gram-negative bacteria *Pseudomonas aeruginosa*, peculiar to the nosocomial infections, possess a very high resistance [15]. These biofilms can be inactivated only by cationic PS using [5-7]. The inactivate efficiency of gram-negative bacteria essentially depends on the number of positive charges in the molecular structure of cationic PS [16].

Antibacterial efficiency of both gram-positive and gram-negative bacteria using synthetic cationic bacteriochlorins was demonstrated in the works [17-19]. However the bacteriochlorin derivatives, used in these studies, have no more than four cationic substitutes.

Two original synthetic cationic bacteriochlorins: 5,10,15,20-Tetrakis [1-(4'-bromobutyl)-3-pyridyl] bacteriochlorin tetrabromide [(3-PyBuBr)4BCBr4] and 5,10,15,20-Tetrakis [1-(4'-pyridiniobutyl)-3-pyridyl] bacteriochlorin octabromide [(3-PyBuPy)4BCBr8)], which have four and eight cationic groups (BCl-4 and BCl-8) are under consideration in this work. These BCls have been previously synthesized for the cancer therapy [20,21]. In this work the possibilities of these BCls for inactivating of plankton (free floating) and biofilm cultures of gram-negative *Pseudomonas aeruginosa* are studied.

2. Materials and Methods

2.1. Sample preparation

BCl-4 and BCl-8 photosensitizers were synthesized according [20,21].

2.2. Spectroscopic instrumentation

The absorbance spectra were measured with the help of Hitachi-S3410 and Cary-300 UV-Vis spectrometers. Fiber-optic spectrometer LESA-5 was used for the spectral analysis of fluorescence, excited by Nd:YAG CW laser at the wavelength of 532 nm.

2.3. Antibacterial efficiency measurements

Biological studies were done on bacterial cultures of *Pseudomonas aeruginosa* in plankton and biofilm forms. Bacterial biofilms were grown on the glass slides (8x12 mm) in a Petri dish filled with the LB nutrient broth for 20 h at 37°C. The slides were put in the wells of 24-well plate, washed and exposed to PS in saline for 1 h at 37°C. After incubation samples were washed with fresh saline and irradiated using LPhD-03-Biospec arc lamp source with narrow-band filter (670-840 nm) or LED matrix source with the wavelength of 761 nm. Power density in spectral range of PS absorption was 10-20 mW/cm^2. Then serial dilutions were plated to determine viable cells using colony-forming unit (CFU) microbiological method.

Journal of Physics: Conference Series **691** (2016) 012011 doi:10.1088/1742-6596/691/1/012011

Effectiveness of photodynamic antibacterial chemotherapy (PACT) was demonstrated by fluorescence microscopy using Nikon H600L microscope. Biofilms of bacteria were grown on cover slips for 3 h, incubated with PS, washed and irradiated. Three cover slips were stained with Live/Dead Biofilm Viability Kit ("Invitrogen"), and pictures were taken with 800× magnification, with subsequent overlay of images acquired with green and red filters.

3. Results and discussion

As shown on figures 1 and 2, the absorption spectra for both bacteriochlorins are similar with maxima at 763-765 nm. The analysis of experimental data showed that shapes of spectra and positions of maxima did not significantly change with concentration. The concentration dependence of absorption intensity for BCl-8 is very close to predicted by Bouguer's law, while for BCl-4 it is possible to observe some sublinearity at concentrations higher than 0.12 mM.

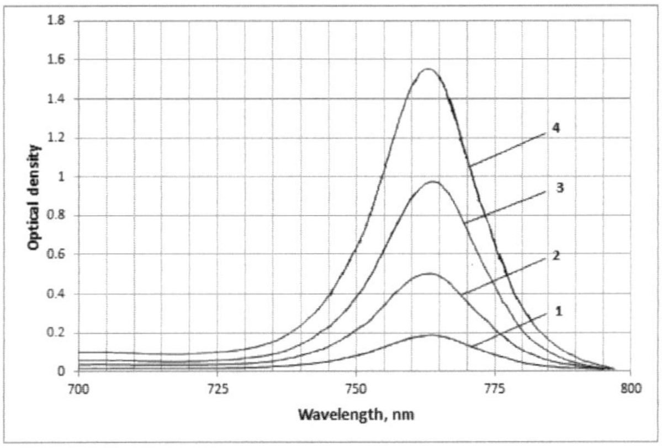

Figure 1. Absorption spectra of BCl-4 in saline solution for concentrations: 0.02 мM (1), 0.06 мM (2), 0.12 мM (3) and 0.2 мM (4).

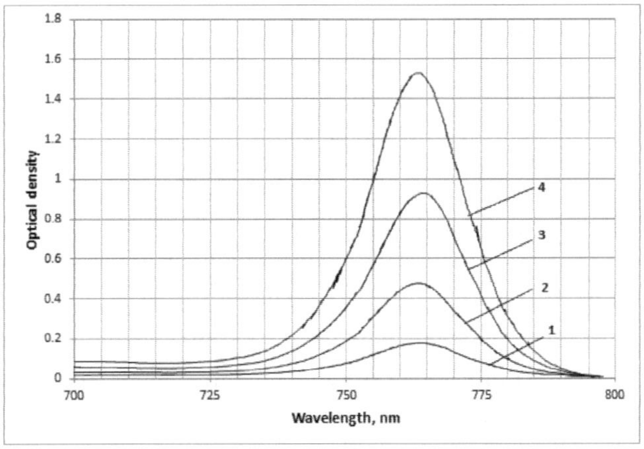

Figure 2. Absorption spectra of BCl-8 in saline solution for concentrations: 0.02 мM (1), 0.06 мM (2), 0.12 мM (3) and 0.2 мM (4).

The fluorescence spectra of bacteriochlorins at different concentrations are presented in figures 3 and 4. As can be seen, the main spectral lines take place in the region of 770-780 nm. The comparison of the absorption and fluorescence spectra shows that they significantly overlap. Stokes shifts do not exceed a few nanometers and slightly increase with the concentration: 8.4-13.2 nm for BCl-4 and 9.1-15.3 nm for BCl-8.

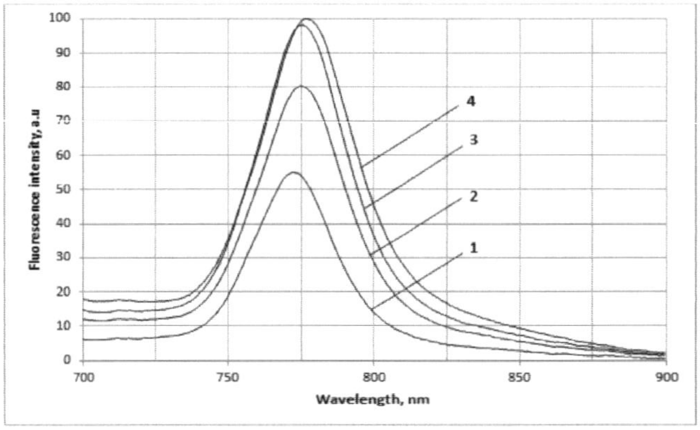

Figure 3. Fluorescence spectra of BCl-4 in saline solution for concentrations: 0.02 mM (1), 0.06 mM (2), 0.12 mM (3) and 0.2 mM (4). Excitation wavelength is equal to 532 nm.

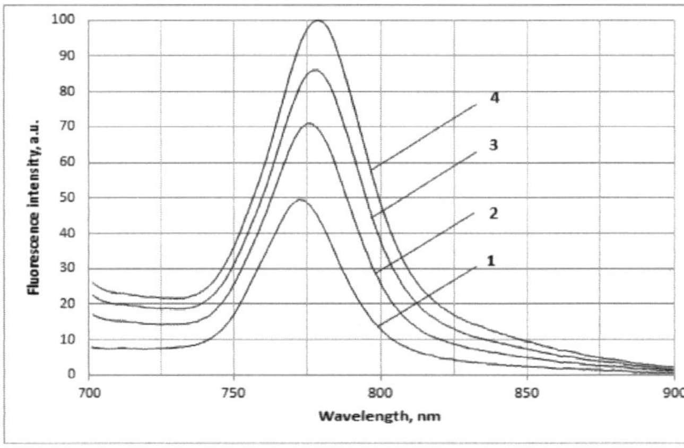

Figure 4. Fluorescence spectra of BCl-8 in saline solution for concentrations: 0.02 mM (1), 0.06 mM (2), 0.12 mM (3) and 0.2 mM (4). Excitation wavelength is equal to 532 nm.

The intrensity of BCl-8 fluorescence at high concentrations (>0.1 mM) shows the sublinear growth (Fig.5), with its spectral maximum shifting to longer wavelengths, while the halfwidth of fluorescence band does not noticeably change. Together with linear concentration growth of absorption, this behaviour is closer to the case of re-absorption of fluorescence in concentrated solutions than behavior connected to aggregation of PS molecules [11]. On the other hand, in BCl-4 solutions the sublinearity of concentration dependence on fluorescence starts much earlier, at the concentrations above 0.05 mM, coming by the level of 0.14 mM to saturation. Combined with behaviour of absorption, this allows us to assume that BCl-4 does express some (not significant) aggregation of molecules at the upper limit of the concentration range used in this study. In general, it can be concluded that the aggregation process in solutions of both bacteriochlorins is minimized due to the Coulomb repulsion, allowing us to expect high antibacterial photodynamic efficiency of both photosensitizers.

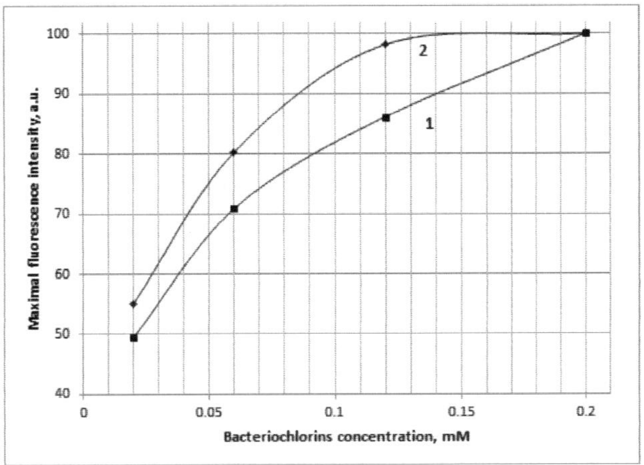

Figure 5. The dependences of fluorescence intensity on the bacteriochlorin concentration for BCl-8 (1) and BCl-4 (2)

The results of PDP investigation against gram-negative bacteria in the plankton state and in the biofilms are demonstrated in fig. 6 and 7. As can be seen in fig.6, the photodynamic inactivation is efficient even at photosensitizers dose of 0.005 mM. Under these conditions, irradiation up to the light dose of 8 J/cm^2 is sufficient for decrease of viability bacteria with BCl-4 and BCl-8 by 4 and 5 decimal orders respectively. Fig.7 shows that antibacterial photodynamic treatment with PS under study is also efficient against biofilms of *Pseudomonas aeruginosa*. The antibacterial efficiency of BCl-8 is higher than for BCl-4.

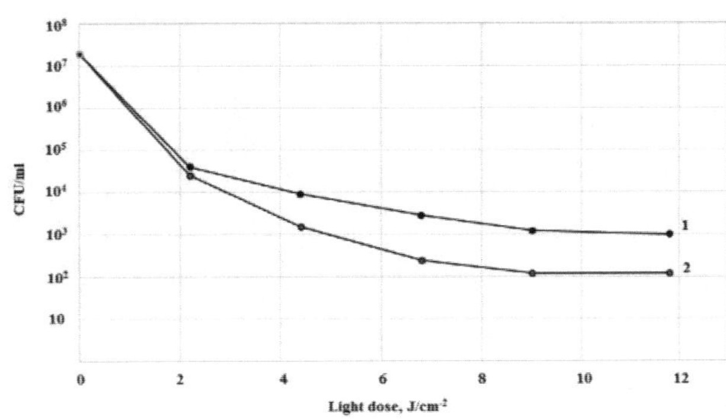

Figure 6. The dependences of CFU on the light dose for *Pseudomonas aeruginosa* plankton bacteria at 0.005 mM solutions using BCl-4 (1) and BCl-8 (2) photosensitizers.

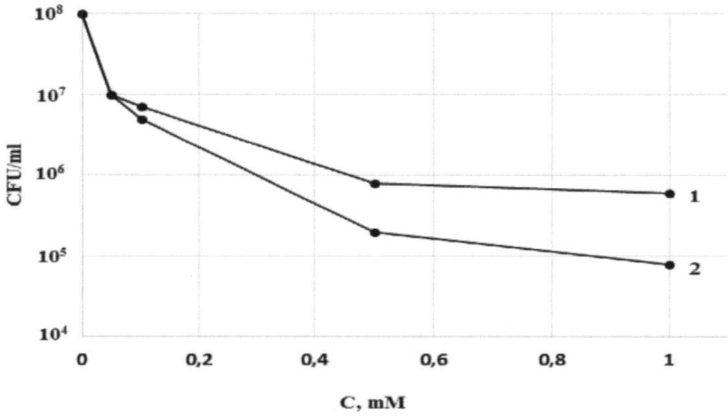

Figure 7. The dependences of CFU on the concentration for *Pseudomonas aeruginosa* in the biofilms after irradiation (light dose 50 J/cm^2) for BCl-4 (1) and BCl-8 (2) photosensitizers

The photodamage of bacteria in biofilms after PDP was demonstrated by fluorescent microscopy using Live/Dead Biofilm Tracer, as shown on fig.8. Live/Dead Tracer includes two components: SYTO 9 traces the DNA with the green fluorescence in all bacteria of biofilm, while propidium iodide provides intense green fluorescent marking of DNA only in cells with damaged membranes. Overlay of images of red and green fluorescence images of cells stained with this techniques allows the undamaged bacteria (with green fluorescence) from the damaged ones (with red sum fluorescence) to be distinguished.

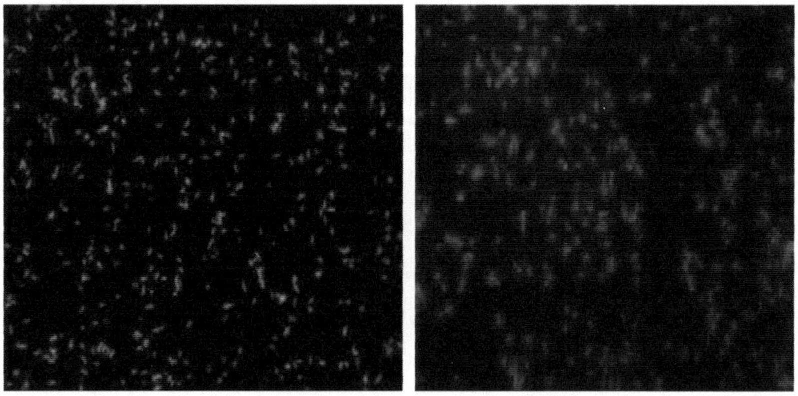

Figure 8. The microphotos of *Pseudomonas aeruginosa* biofilm before (left) and after (right) PDP

4. Conclusion

It was shown that the photodynamic treatment with new synthetic cationic photosensitizers 5,10,15,20-Tetrakis [1-(4'-bromobutyl)-3-pyridyl] bacteriochlorin tetrabromide and 5,10,15,20-Tetrakis [1-(4'-bromobutyl)-3-pyridyl] bacteriochlorin octabromide with high cationic substitute number provides efficient inactivation of gram-negative *Pseudomonas aeruginosa* bacteria both in plankton and biofilm forms. The high positive charge of molecules of these photosensitizers can realize the efficiency of the photodynamic antibacterial therapy.

References

[1] Gelfond M L 2007 *Practical oncology* **8** 204 in Russian
[2] Chissov VI Sokolov VV and Filonenko EV 1998 *Russian Chemical Journal* **XLII** 5
[3] Jori G and Brown S 2004 *Photochem. Photobiol. Sci.* **3** 403
[4] Hamblin MR and Hasan T 2004 *Photochem. Photobiol. Sci.* **3** 436
[5] Malik Z Ladan H and Nitzan Y 1992 *J. Photochem. Photobiol.* **14** 262
[6] Minnock A et al 1996 *J. Photochem. Photobiol.* **32** 159
[7] Brusov SS et al 2014 *Russian Journal of Biotherapy* **4** 59
[8] Meerovich GA et al 2013 *Russian Journal of Biotherapy* **3** 45
[9] Ma L Moan J and Berg K 1994 *Intern. Journ. of Cancer.* **57** 883
[10] Edrei R et al 1998 *Journ. of Porphyrins and Phthalocyanines* **2** 191
[11] Dhami S et al 1995 *Photochemistry and Photobiology* **61** 341
[12] Makarov DA et al 2009 *Russian Journal of Physical Chemistry A* **83** 1044
[13] Donlan RM and Costerton JW 2002 *Clin. Mic. Rev.* **15** 167
[14] Pace JL et al 2005 *CRC journal* **7** 109
[15] Hassett DJ Sutton MD and Schurr MJ 2009 *Trends Microbiol.* **17** 130
[16] Strakhovskaya MG et al 2009 *Biochemistry* **74** 1305 in Russian
[17] Huang L et al 2010 *Antimicrobial Agents and Chemotherapy* **54** 3834
[18] Schastak S et al 2010 *PLOS ONE* **5** 1
[19] Schastak S et al 2008 *Methods Find Exp Clin.* **30** 129
[20] Dudkin SV et al. 2013 *Russian patent* № 2476218
[21] Dudkin SV et al. 2013 *Russian patent* № 2479585

Soliton Formation and Superluminality Effect due to Nonlinear Absorption of Femtosecond Laser Pulse Energy by the Medium Containing Nanorods

Vyacheslav A. Trofimov, Tatiana M. Lysak

Lomonosov Moscow State University, Moscow 119992, Russia

E-mail vatro@cs.msu.ru

Abstract. We investigate a femtosecond pulse propagation in a medium, containing nanorods, with taking into account the dependence of multi-photon absorption from the aspect ratio of nanorods. Nanorods melting due to the laser energy absorption leads to the non-stationary interaction of laser pulse with the medium and time-dependent nanorod aspect ratio changing. Under certain conditions, we found out the soliton-like mode of a laser pulse propagation and the superluminality effect: acceleration of light (fast light) in comparison with light propagation in a linear medium. We discuss a physical mechanism of superluminality effect for considering laser pulse propagation. Using spatio-temporal analogy, one can see the similarity between the pulse centre evolution along longitudinal coordinate and the beam centre evolution under the infrared optical radiation propagation in a cloud, or fog, which moves across the beam, with taking into account its thermal blooming.

1. Introduction

In recent years, thin films, doped with noble (gold or silver) metal nanoparticles, have attracted attention as recording media due to their strong nonlinear response resonance to the frequency of incident optical radiation [1-4]. Detailed understanding of laser pulse interaction with metal nanoparticles is of great interest for various fields of application. For example, nanoparticles aspect ratio changing because of photo-thermal melting provides the ability for five-dimensional recording [1,5]. A number of investigations are devoted to the measurements on a single-particle level, focusing on the dependence of nanoparticle optical response on the nanoparticle aspect ratio, orientation and local environment [6-11].

Among a number of problems of laser radiation interaction with the medium containing nanorods, self-similar mode of propagation is of great importance. This is due to the fact that the laser radiation spectrum distortions, because of nonlinear refraction, can cause false recording of information in all-optical data storage devices. Indeed, nanoparticles melting due to the laser radiation absorption leads to the laser pulse spectrum changing caused by the pulse chirping. This results in the condition violation for optimal information recording, as well as (that is more essential) in the false information recording and reading. That is why the investigation of the laser pulse interaction dynamic with the medium, containing nanoparticles, is an actual problem.

In our previous papers [12-14] we have investigated a femtosecond pulse propagation in a medium with nanorods under the conditions of both changing the ellipticity (aspect ratio) of nanorods and dependence of TPA from nanoparticle aspect ratio. We analyzed the influence of the relation between the nanorods absorption spectrum bandwidth and laser pulse spectrum bandwidth on laser pulse spectrum transformation under the pulse propagation in a medium containing nanorods, and found that, under certain conditions, the laser pulse spectrum distortion is really absent despite the action of various nonlinearities.

Our attention in this paper (see also [14]) is attracted by the effects of superliminality [15-19], as well as a soliton-like formation of a laser pulse at its propagation in a medium with nanorods. Such kind of the light propagation occurs if weak laser energy absorption takes place on the laser pulse dispersion length. We also take into account one- or multi-photon of laser radiation by nanorods and time-dependent nanorod aspect ratio changing due to their melting. Due to the superluminality effect, the total time interval of light pulse interaction with nanoparticles decreases thus decreasing the nonlinear response. In this paper, we compare the features of the superluminality effect in the media with one-photon and multi-photon (three- and four-photon) absorption with the similar effect in the media with two-photon absorption, discussed in [14].

The light acceleration effect takes place for a chirped incident pulse, and we confirmed this fact by analytical consideration in [14]. Using spatio-temporal analogy, one can see the similarity between the pulse centre evolution along longitudinal coordinate and the beam centre evolution at the infrared optical radiation propagation in a cloud, or fog, which moves across the beam, under the condition of thermal blooming [20-23]. As it is well known, the beam centre changing appears due to the refraction of the beam on the thermal blooming grating and amplitude grating induced by nonlinear absorption of laser energy by the cloud particles.

2. Problem statement

We consider a femtosecond laser pulse propagation in a medium with nanorods, taking into account the aspect ratio (ellipticity) of nanorods changing due to their melting because of multi-photon absorption (MPA) of optical energy. In the framework of slowly varying envelope of wave packet, this process can be described by the following dimensionless nonlinear:

$$\frac{\partial A}{\partial z} + iD\frac{\partial^2 A}{\partial t^2} + (\varepsilon - 1)(\delta_0 + i\xi)|A|^{2(k-1)} A = 0 \quad , \tag{1}$$

$$\frac{\partial \varepsilon}{\partial t} = -\tilde{\delta}(\varepsilon - 1)|A|^{2k} \tag{2}$$

with initial and boundary conditions for complex amplitude and aspect ratio of nanorods

$$A(z,0) = A(z,L_t) = 0, \quad 0 \le z \le L_z, \; A(0,t) = A_0(t), \quad 0 \le t \le L_t, \; \varepsilon(z,t=0) = \varepsilon_0, \; 0 \le z \le L_z .$$

Above A is dimensionless slowly varying envelope of electric field pulse, normalized on the maximum value of square root from the incident pulse intensity ($z = 0$). The nanorods are melting under the action of laser light. As a consequence, their aspect ratio ($\varepsilon = a/b$) is varied, where a and b are major and minor axes of nanorods, ε_0 is its initial value. Parameter k is the number of photons, involved in the absorption. E.g., $k = 1$ is one-photon absorption, $k = 2$ is two-photon absorption (TPA) and so on. Variable z is a dimensionless longitudinal coordinate, along which the optical radiation propagates; t is a dimensionless time in the system of coordinates moving with a pulse, time is measured in the units of τ_{pulse} - duration of the incident pulse; the dimensionless duration of laser pulse is denoted as τ_p; L_t is a dimensionless time interval, during which the laser pulse interaction with nanorods is analyzed, L_z is a dimensionless length of nonlinear medium. Parameter D characterizes the group velocity dispersion (GVD). Parameters δ_0 and $\tilde{\delta}$ characterize the absorption of laser light. Coefficient ξ characterizes self-action of laser pulse due to detuning of the carrier frequency of wave packet from the central frequency of the nanorod absorption spectrum. The case of $\xi = 0$ corresponds to an optical pulse propagation in a medium with pure amplitude grating. It means an influence only of absorption on the laser pulse propagation. In the opposite case ($\xi \neq 0$), the phase grating is also induced by the laser radiation. It should be mentioned, that the positive sign of the parameter ξ (this case is named by us as positive grating) corresponds to pulse compression and the laser pulse decompression occurs at negative sign of this parameter (this case is named by us as negative grating).

We specify the following complex amplitude of incident pulse

$$A(z=0,t) = A_0(t) = \exp\left(-((t-L_t/2)/\tau)^{m_t}\right), 0 \le t \le L_t$$

τ is a dimensionless pulse duration, m_t=2 describes the Gaussian pulse shape.
We also follow the laser pulse centre position

$$\tau_c(z) = \int_0^{L_t} (t-L_t/2)|A(z,t)|^2\, dt \bigg/ \int_0^{L_t} |A(z,t)|^2\, dt$$

3. Analytical analysis

Below we develop the solution of equations (1)-(2) in the framework of nonlinear geometrical optics approximation [24] for one-photon absorption medium and a chirped incident pulse. This helps us to understand and explain the features of laser pulse propagation in a medium, containing nanorods, with a phase grating induced by laser radiation as well.

Let us represent the complex amplitude as $A = \sqrt{I}\exp(-iS)$, S is a real function describing a pulse phase. In the new variables, these equations are transformed to

$$\begin{gathered}
\frac{\partial I}{\partial z} + 2D\frac{\partial I}{\partial t}\frac{\partial S}{\partial t} + 2DI\frac{\partial^2 S}{\partial t^2} + 2\delta_0 fI = 0, \\
\frac{\partial S}{\partial z} + D\left(\frac{\partial S}{\partial t}\right)^2 - D\frac{1}{\sqrt{I}}\frac{\partial^2 \sqrt{I}}{\partial t^2} - \xi f = 0, \\
\frac{\partial f}{\partial t} = -\tilde{\delta} fI \ .
\end{gathered} \tag{3}$$

Introducing the instantaneous frequency $\Omega(z,t) = \dfrac{\partial S(z,t)}{\partial t}$ of the optical pulse and omitting the third term in equation (3) we get the following set of equations

$$\frac{\partial I}{\partial z} + 2D\frac{\partial (I\Omega)}{\partial t} - 2\frac{\delta_0}{\tilde{\delta}}\frac{\partial f}{\partial t} = 0 \ , \tag{4}$$

$$\frac{\partial \Omega}{\partial z} + 2D\theta\frac{\partial \Omega}{\partial t} - \xi\frac{\partial f}{\partial t} = 0,$$

$$\frac{\partial f}{\partial t} = -\tilde{\delta} fI \ .$$

Integrating equation (4) with respect to t and taking into account the zero-value pulse intensity till $t \le 0$, we get

$$\begin{gathered}
\frac{\partial P}{\partial z} + 2D\Omega\frac{\partial P}{\partial t} - 2\frac{\delta_0}{\tilde{\delta}} f_0\left(\exp(-\tilde{\delta}P) - 1\right) = 0 \ , \\
\frac{\partial \Omega}{\partial z} + 2D\Omega\frac{\partial \Omega}{\partial t} - \xi f_0\left(\exp(-\tilde{\delta}P) - 1\right) = 0
\end{gathered} \tag{5}$$

where $P(z,t)=\int_0^t I\ (z,\tau)d\tau$, $I(z,t)=\left|A(z,t)\right|^2$, is the laser pulse energy, passed through a medium section z till time moment t. We can solve these equations for a pure amplitude grating ($\xi=0$). In this case, the solution of the Cauchy problem for these equations can be constructed [14] by the characteristics method

$$\Omega(z,t)=\tilde{\Omega}_0\left(t-2D\Omega z\right),\ w(z,t)=\tilde{w}_0\left(t-2D\Omega z\right)$$

where $\tilde{\Omega}_0(t)=\Omega(0,t)=\dfrac{\partial S\left(z=0,t\right)}{\partial t}$ is an instantaneous frequency of the incident wave packet, and

$$w=\ln\left(2\delta_0 f_0\left(\exp\left(\tilde{\delta}P\right)-1\right)\right)+2\delta_0 f_0 z\,.$$

If the incident pulse is a chirped one

$$S\left(z=0,t\right)\equiv\Omega_0(t+b)^2, \tag{6}$$

an instantaneous frequency of incident laser pulse is $\tilde{\Omega}_0(t)\equiv 2\Omega_0(t+b)$ and the solution of equation (5) is

$$\Omega(z,t)\equiv\frac{2\Omega_0(t+b)}{1+4D\Omega_0 z}\,.$$

Therefore, the solution of equation (4) with respect to the beam intensity is

$$I(z,t)=\frac{I_0\left(\zeta\right)}{1+\exp\left(-\tilde{\delta}P_0(\zeta)\right)\left(\exp(2\delta_0 f_0 z)-1\right)}\cdot\frac{1}{1+4D\Omega_0 z},\quad \zeta=\frac{t-4D\Omega_0 zb}{1+4D\Omega_0 zb}\,. \tag{7}$$

Let us note, that formula (7) is invalid for z-sections satisfying inequality: $1+4D\Omega_0 z<<1$, because we derived (7) using the geometrical optics approximation.

A laser pulse shift in time coordinate depends on the sign of parameter Ω_0. Let us remember that the initial pulse is up-chirped (has positive chirp) for positive value of Ω_0 and down-chirped (has negative chirp) for negative value of Ω_0. In figure 1 we compare the computer simulation results (solid lines) with the results obtained on the basis of analytical formula (7) derived for incident Gaussian pulse shape and quadratic law of the frequency distribution (6) (dashed lines). As we can clearly see, the solid and dashed lines are aligned at a distance of no more than 0.05 dimensionless units despite the negligible GVD under the analytical consideration.

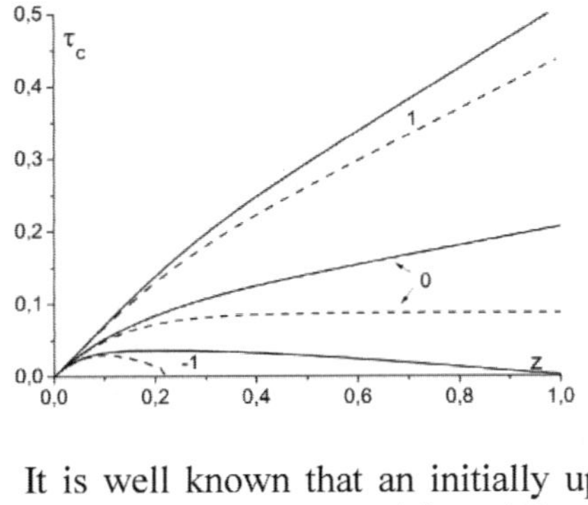

Figure 1. Pulse center shifting along z-coordinate in dependence of the chirping parameter Ω_0 (shown in Figure) for $D=1$, $\delta_0=5$, $\tilde{\delta}=0.5$ and one-photon absorption (k=1) of light energy. Dashed lines correspond to analytical solution, solid lines correspond to computer simulation results

It is well known that an initially up-chirped pulse, propagating in a linear medium with normal GVD, is broaden more rapidly, while a down-chirped pulse first is compressed and then it is broaden [25], and, of course, for both signs of Ω_0 no pulse center shift is observed. In the medium, containing nanorods, the pulse can shift towards the area of time increasing or time decreasing. The nanorods melting leads to the slowing down of the laser pulse propagation. On the other hand, the affect of nanorods melting on the incident down-chirped pulse can reverse the pulse center shifting direction and the laser pulse starts accelerating (figure 1). As it will be shown in the next Section 4, this effect is similar to a light propagation in a medium with a positive phase-amplitude grating (compare curves -1 in figure 1 and solid line 1 in figure 2a).

4. Computer simulation results

Below we present computer simulation results for the propagation of an incident Gaussian pulse ($m_t = 2, \tau = 1$) at $D = 0.1$ and positive phase-amplitude grating ($\xi = 5$), and sufficiently small strength of melting ($\tilde{\delta} = 5$), and small enough depletion ($\delta_0 = 0.005$) of laser energy under propagation on a short distance ($z \approx 1$). We compare the process of soliton formation and light acceleration discussed in [14] for TPA of light energy with the similar superluminality effects for one-photon ($k = 1$) and multi-photon ($k = 3, 4$) absorption of light energy.

For all considered number of photons, involved in the process of absorption, the effect of pulse centre shifting in the direction of time decreasing takes place (figure 2a). Nevertheless, soliton formation occurs only for two- and three photon absorption (compare figures 3,4,8,9).

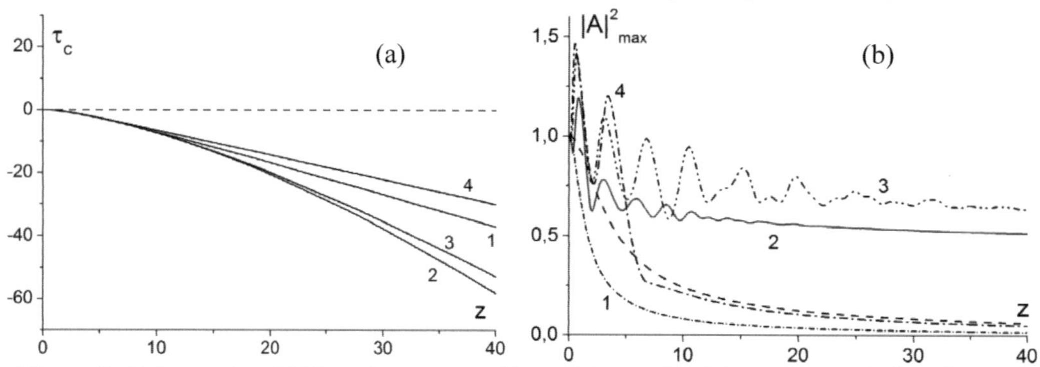

Figure 2. Pulse centre shifting along z-coordinate (a), maximal intensity evolution (b) for the positive phase amplitude grating $\xi = 5$, $\delta_0 = 0.005$, $\tilde{\delta} = 5$. Dashed lines correspond to the pulse propagation in a linear medium. Number of photons, involved in the absorption process, is shown by figures.

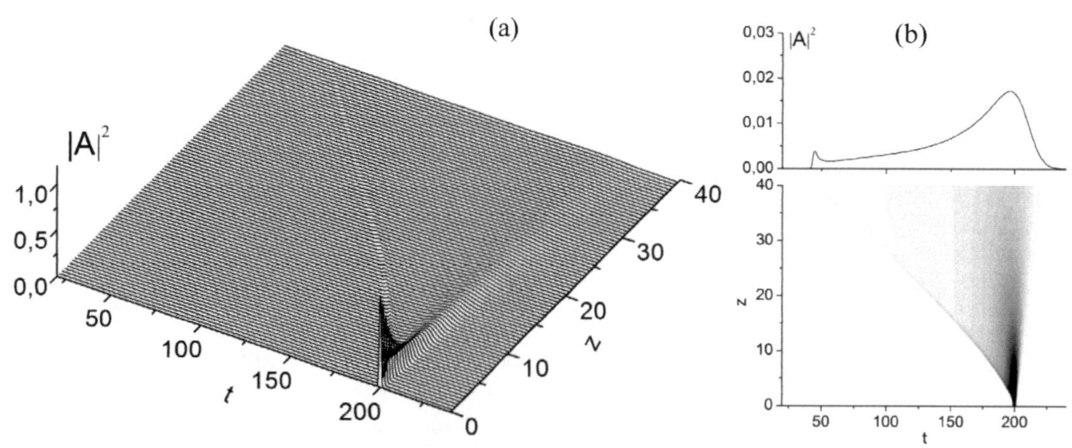

Figure 3. Soliton formation (a), maximal intensity position for both sub-pulses (b) at the incident Gaussian pulse propagation for one-photon absorption ($k = 1$) of light energy. The top plane in Figure b shows sub-pulse shapes at section z=40.

In the case of one-photon absorption, the broadening of the initial pulse in the direction of time decreasing occurs with a well pronounced peak at the pulse front (figure 3). This broadening is accompanied by significant decreasing of maximal pulse intensity, which is much stronger than even the intensity decreasing for the pulse propagation in a linear medium (compare curve 1 and dashed line in Figure 2b).

If two photons are involved in the process of absorption, two solitons are formed as a result of the incident pulse splitting at the initial stage of propagation (figures 4,5). Maximal intensity growth occurs at the initial stage of propagation as a result of a nonlinear absorption. After splitting, the left sub-pulse transforms into a soliton and propagates with high velocity without its shape changing for at least 30 dimensionless units (figure 4b). The right sub-pulse also gradually transforms into a soliton. However, its velocity corresponds to the dispersion spreading of the pulse in a linear medium, while the left sub-pulse velocity exceeds it many times. Figure 5 clearly shows that the velocity of the right sub-pulse is close to the velocity of the pulse front motion, propagating in a linear medium.

As it was already mentioned, the positive phase grating action is similar to the incident chirped pulse propagation (see (6)) with $\Omega_0 < 0$. Moreover, in this case, the processes of higher frequency waves absorption due to nanorods melting and lower frequency waves generation due to the positive phase grating are not limited by rapid energy absorption. Therefore, the higher velocity sub-pulse (left sub-pulse in Figures 4,5) transforms into a soliton at its further propagation.

To clarify the soliton formation mechanism and the pulse acceleration we combined in figure 6 the pulse shape and frequency chirp $\frac{ds}{dt}$ ($s(t)$ is the pulse phase) with the aspect ratio distribution. Mention that the negative slope of a frequency chirp distribution in figure 6 corresponds to the positive chirp (up-chirp) in generally accepted terminology and is the result of our notations. At the initial stage of propagation, the pulse is splitting into two sub-pulses. A positive phase grating, which is formed at the pulse propagation, produces a negative chirp (see Section 3 above). This negative chirp compensates a positive chirp caused by the normal dispersion action. As a result, waves with lower frequencies go to the trailing edge of the pulse, and waves with higher frequencies – to its leading edge. This leads to the pulse compression, the wave packet power grows and the nanorods melting increases due to the high intensity waves absorption. So, the pulse splitting into two sub-pulses occurs. The waves with lower frequencies are concentrated in the left sub-pulse, while the waves with higher frequencies – in the right sub-pulse. So, the left sub-pulse moves with higher velocity than the right sub-pulse does. The further compression of the sub-pulses due to the action of positive phase-amplitude grating is restricted by the normal dispersion and energy absorption. So, after a certain distance of propagation, each sub-pulse is transformed into a soliton and becomes trapped at the boundary between domains with high and low nanorod aspect ratio. The soliton form of the self sub-pulse is also proved by figure 7, where the pulse spectrum is also split into two parts, the one with lower frequencies corresponds to the higher velocity soliton. It should be mentioned that, in figure 7, there are the frequencies with zero value spectral amplitude between the two parts of the spectrum. Up to the section $z = 30$, the right sub-pulse takes a soliton form as well (figures 4,5), being trapped by the melting front. Constant values of frequency chirp for both sub-pulses are well seen in figure 6b.

The influence of the positive phase grating results in a permanent shifting of higher frequency waves to the sub-pulse front, where they are absorbed by nanorods. So, the spectrum of both sub-pulses is permanently shifting into the range of lower frequencies, which is well seen in figure 6, especially for the high velocity left sub-pulse. As a result, the velocity of each sub-pulse is growing. In figure 4 the acceleration of the higher velocity left sub-pulse is also well seen. As for the right sub-pulse, its velocity is close to the velocity of the pulse front in a linear medium. Indeed, lower nanorod aspect ratio decreases the self-action in the area of the right sub-pulse. As a result, the sub-pulse compression decreases and the right sub-pulse trapping occurs at a lower peak intensity compared to the left sub-pulse. As it was already mentioned, the right sub-pulse consists of higher frequency waves compared to the left sub-pulse. That's why its velocity is much smaller than the velocity of the left sub-pulse, though it is bigger than the velocity of the pulse, propagating in a linear medium. As a result of both sub-pulses accelerating, the pulse center is shifting in the direction of time decreasing (fast light, line 2 in figure 2a).

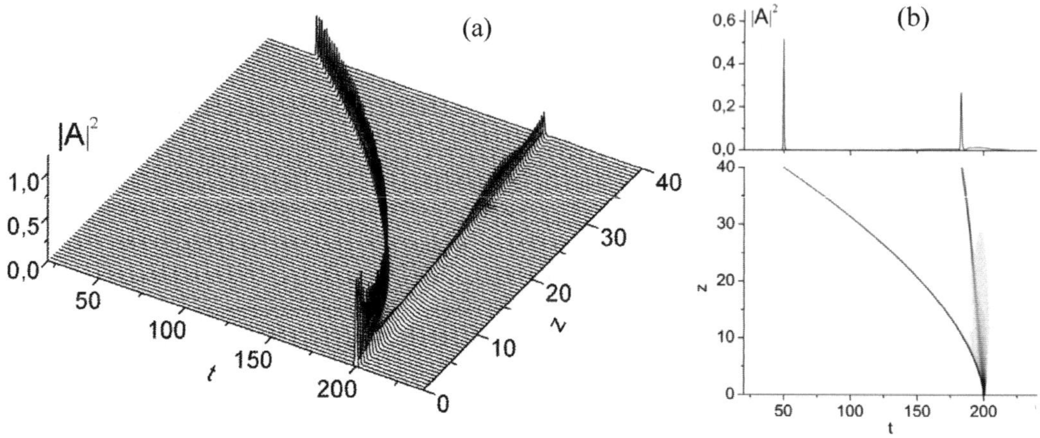

Figure 4. Soliton formation (a), maximal intensity position for both sub-pulses (b) at the incident Gaussian pulse propagation for TPA ($k = 2$) of light energy. The top plane in Figure b shows sub-pulse shapes at section z=40.

The similar process of soliton formation and light acceleration also takes place for laser pulse propagation in a medium with three-photon absorption (figure 8). Nevertheless, in comparison with TPA, pulse centre shifting is smaller, but the left sub-pulse maximal intensity is greater (figure 2). And the left sub-pulse does not transform into a soliton up to section z=40. This is obviously due to the smaller effective strength of phase grating and laser energy depletion because of the fourth and the sixth power of the amplitude modulus in Equations (1) and (2), correspondingly.

Further photon number increase results in decreasing of the phase grating effective strength and laser energy depletion. So, the optimum ratio of GVD and strength of phase grating, necessary for soliton formation, is violated. As a result, after the pulse splitting, both sub-pulses start spreading and the decreasing of maximal intensity is similar to the one for the pulse propagation in a linear medium (figure 2b, compare curve 4 and dashed line). Nevertheless, even for k=4, the pulse centre shift in the direction of time increasing occurs because of slow waves absorption for nanorods melting (figure 2a).

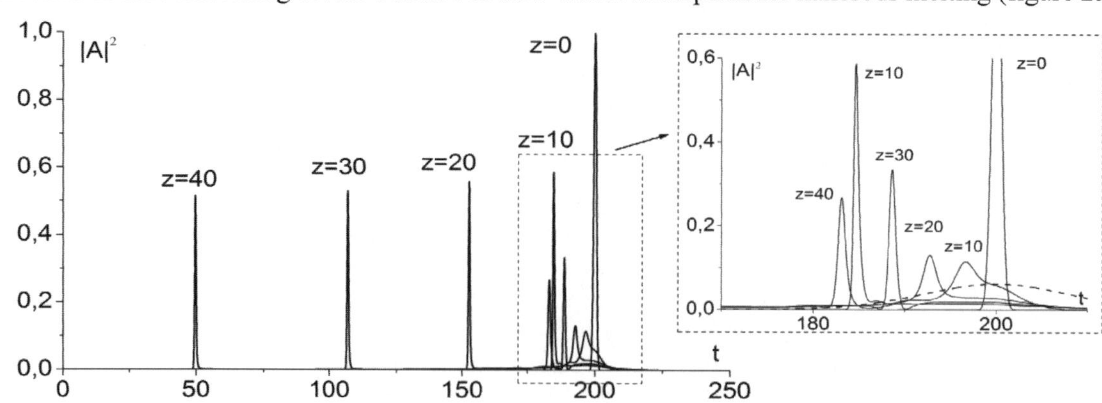

Figure 5. Shapes of both sub-pulses combined at various sections for TPA ($k = 2$) of light energy. Dashed lines correspond to section $z = 40$ for the pulse, propagating in a linear medium.

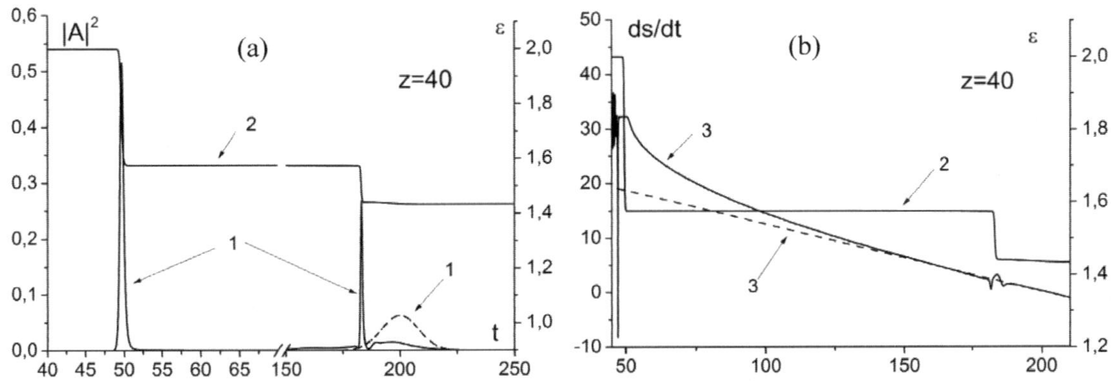

Figure 6. Pulse shape (curve 1) and aspect ratio distribution (curve 2) (a); frequency chirp (curve 3) and aspect ratio distribution (curve 2) (b) at z=40 for TPA ($k = 2$) of light energy. Dashed lines show corresponding distributions for the pulse, propagating in a linear medium.

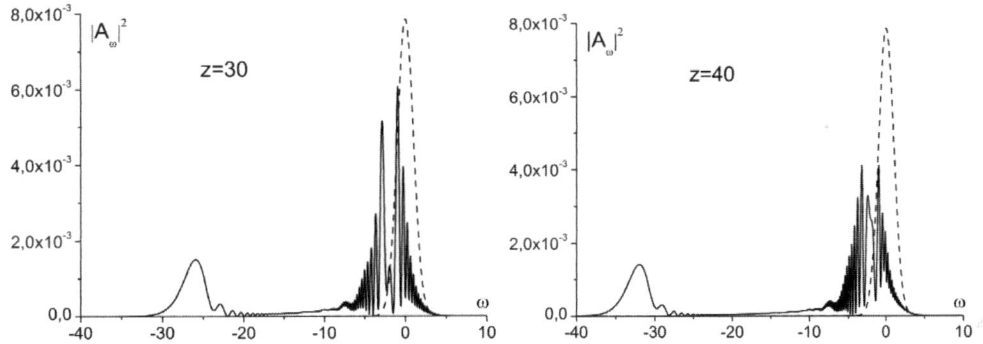

Figure 7. Spectrum of laser pulse propagating in a medium with TPA ($k = 2$) of light energy. Dashed lines correspond to the incident light spectrum.

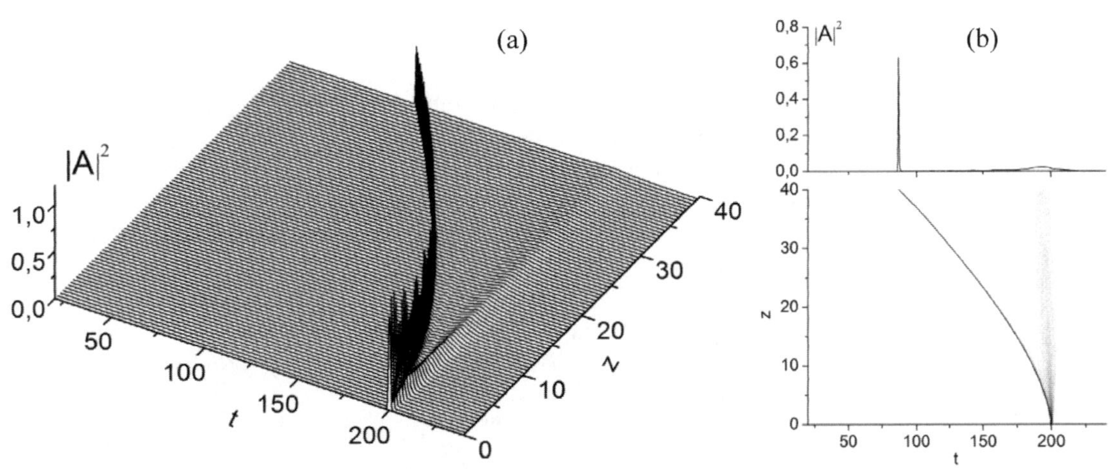

Figure 8. Soliton formation (a), maximal intensity position for both sub-pulses (b) at the incident Gaussian pulse propagation for three-photon absorption ($k = 3$) of light energy. The top plane in Figure b shows sub-pulse shapes at section z=40.

5. Conclusions

Under the weak absorption on the dispersion length, we found out acceleration of light (fast light) in comparison with light propagation in a linear medium with one-photon and multi-photon (three- or four-photon) absorption.

The physical reason for laser pulse acceleration is the pulse chirping due to the induced positive phase-amplitude grating and absorption of light energy for nanorods melting which leads to non-stationary interaction of laser radiation with the medium and time-dependent changing of aspect ratio of nanorods. Acceleration of light is accompanied by the soliton formation due to the trapping of laser radiation by nanorods melting fronts for TPA and three-photon absorption. As a result of soliton formation, very slow decrease of maximal intensity occurs at the pulse propagation in the medium with TPA or three-photon absorption, especially in comparison with the pulse propagation in a linear medium.

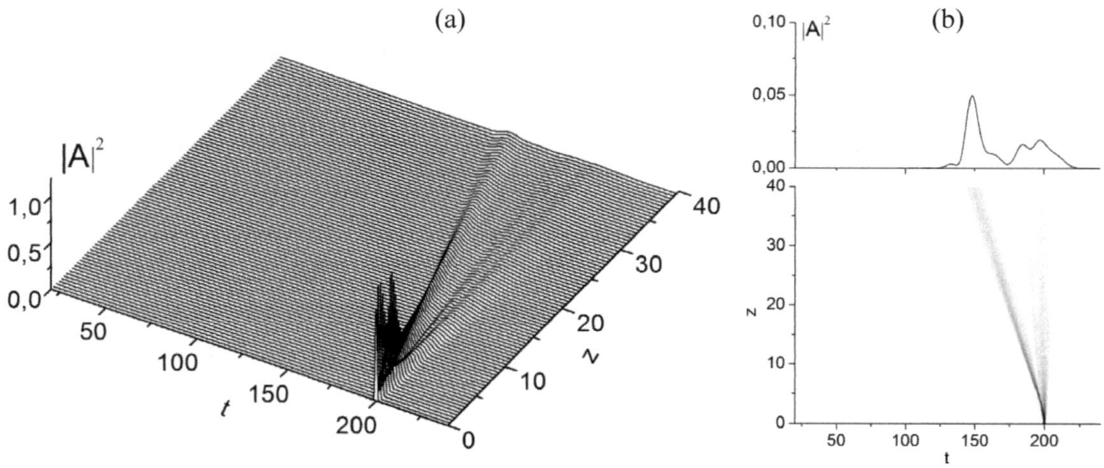

Figure 9. Soliton formation (a), maximal intensity position for both sub-pulses (b) at the incident Gaussian pulse propagation for four-photon absorption ($k = 4$) of light energy. The top plane in Figure b shows sub-pulse shapes at section z=40.

Acknowledgments

The investigation was made using support of the Russian Science Foundation (Grant № 14-21-00081).

References

[1] P. Zijlstra, J. W. M. Chon, and M. Gu, "Five-dimensional optical recording mediated by surface plasmons in gold nanorods," *Nature* **459**, 410-413 (2009).

[2] H. Ditlbacher, J. R. Krenn, B. Lamprecht, A. Leitner, and F. R. Aussenegg, "Spectrally coded optical data storage by metal nanoparticles," *Opt. Lett.* **25**, 563–565 (2000).

[3] O.Wilson, G. J. Wilson, and P. Mulvaney, "Laser writing in polarized silver nanorod films," *Adv. Mater.* **14**, 1000–1004 (2002).

[4] J. Pérez-Juste, B. Rodríguez-González, P. Mulvaney, and L. M. Liz-Marzán, "Optical control and patterning of gold-nanorod-poly(vinyl alcohol) nanocomposite films," *Adv. Funct. Mater.* **15**, 1065–1071 (2005).

[5] A. B. Taylor, J. Kim, and J. W. M. Chon, "Detuned surface plasmon resonance scattering of gold nanorods for continuous wave multilayered optical recording and readout" *Opt. Express* **20** (5), 5069–5081 (2012).

[6] N. Del Fatti, D. Christofilos, and F. Vallée, "Optical response of a single gold nanoparticle," *Gold Bulletin* **41**(2), 147-157 (2008).

[7] C. M. J. Wijers, J.-H. Chu, J. L. Liu, and O. Voskoboynikov, "Optical response of layers of

embedded semiconductor quantum dots," *Phys. Rev. B* **74**, 035323, (2006).

[8] A. Mohammadi, V. Sandoghdar, and M. Agio, "Gold nanorods and nanospheroids for enhancing spontaneous emission," *New Journal of Physics* **10**, 105015 (2008).

[9] R. Jin, J. E. Jureller, H. Y. Kim, and N. F. Scherer, "Correlating second harmonic optical responses of single Ag nanoparticles with morphology," *J. Am. Chem. Soc.* **127**, 12482-12483 (2005).

[10] V. P. Drachev, A. K. Buin, H. Nakotte, and V. M. Shalaev, "Size dependent χ^3 for conduction electrons in Ag nanoparticles," *Nano Lett.* **4** (8), 1535-1539 (2004).

[11] S. Eustis, and M.A. El-Sayed, "Why gold nanoparticles are more precious than pretty gold: Noble metal surface plasmon resonance and its enhancement of the radiative and nonradiative properties of nanocrystals of different shapes," *Chemical Society Reviews* **35**, 209-217 (2006).

[12] V. A. Trofimov, T. M. Lysak, and Sh. Lan, "Influence of ellipticity of nanorods on both TPA of femtosecond laser pulse and transformation of pulse spectrum," *Proc. of SPIE* **8424**, 842435 (2012).

[13] V. A. Trofimov, T. M. Lysak, and Sh. Lan, "Influence of absorption spectrum width of nanorods on TPA of femtosecond laser pulse," *Proc. of SPIE* **8463**, 84630J (2012).

[14] V. A. Trofimov and T. M. Lysak, " Acceleration of light and soliton formation due to nonlinear absorption of femtosecond laser pulse energy by the medium containing nanorods," *J. Opt. Soc. Am. B* **33** (1), 00-00 (2016).

[15] A. Hache´, L. Poirier, "Anomalous dispersion and superluminal group velocity in a coaxial photonic crystal: Theory and experiment," *Phys. Rev. E.* **65** (3), 036608-1:6 (2002)

[16] G. M. Shore, **"**Superluminality and UV completion," *Nuclear Physics B* 778 (3), 219-258 (2007).

[17] G. Nimtz, **"**Superluminal signaling by photonic tunneling,"*IEEE Journal of selected topics in quantum electronics* **9** (1), 79-85 (2003).

[18] G. Nimtz, "On superluminal tunneling," *Progress in Quantum Electronics* **27** (6), 417- 450 (2003).

[19] I. Blonskyi, V. Kadan, O. Shpotyuk, and I. Dmitruk, "Manifestations of sub- and superluminality in filamented femtosecond laser pulse in fused silica," *Opt. Commun.* **282** (9), 1913–1917 (2009).

[20] J.R. Penano, P. Sprangle and B. Hafizi , "Propagation of high energy laser beams through atmospheric stagnation zones", *J. Directed Energy* **2**, 107-129 (2006).

[21] R. Ruquist, "Variable atmosphere effects on high energy laser propagation," *J. Phys. Colloques* **41**, *C9-121-C9-128* (1980).

[22] S. A. Akhmanov , D. P. Kindach , A. V. Migulin , A. P. Sukhorukov, and R. V. Khokhlov "Thermal self-action of laser beams", *IEEE J. Quantum Electron.* **QE-4** (10), (1968).

[23] D. C. Smith, "High-power laser propagation: Thermal blooming," *Proceedings of the IEEE* 65 (12), 1679-1714 (1977).

[24] S. A. Akhmanov , A. P. Sukhorukov , and R. V. Khokhlov , "Self-focusing and diffraction of light in a Nonlinear medium," *Sov. Phys. Usp.* **10,** 609–636 (1968).

[25] G.P. Agrawal, *Nonlinear Fiber Optics,* Boston, MA: Academic, 1989.

2D Self-Similar Profile for Laser Beam Propagation in Medium with Saturating Multi-Photon Absorption

Vyacheslav A. Trofimov, Tatiana M. Lysak, Irina G. Zakharova

Lomonosov Moscow State University, Moscow 119992, Russia

E-mail vatro@cs.msu.ru

Abstract. We study a self-similar mode of 2D laser beam propagation in media with multi-photon absorption (MA) taking into account a resonant nonlinearity and nonlinear absorption saturating. An analytical solution of the corresponding equations describing the problems under consideration is derived using an eigenvalue problem method generalization for soliton-like solution finding. The developed solution is used as incident beam profile and phase front for computer simulation of the 2D laser beam propagation. In particular, we demonstrate numerically that the laser beam propagation in a self-similar mode occurs within a certain distance, which depends on medium properties. Under certain relations between the nonlinear absorption and resonant nonlinearity, and cubic nonlinear response, we observe the super long distance of the beam propagation without any beam profile distributions.

1. Introduction

Soliton (or soliton-like) profile of a laser beam (shape of a laser pulse) is very attractive problem among other nonlinear optics problems (see, for example [1-20]). In most researches, the soliton mode of laser pulse or beam propagation is investigated for transparent media or media with linear absorption. This can be explained by wide practical applications. Nevertheless, in some cases for soliton formation understanding, a study of laser pulse soliton appearance due to a nonlinear absorption of a medium is needed. Thus, in [21] it was emphasized that the two-photon absorption (TPA) and free-carrier effects impact on the soliton behaviour in silicon. The paper authors reported the first experimental demonstration of soliton-effect pulse compression in silicon despite TPA and free carrier generation at the picosecond pulse propagation. Analytical and numerical study of multi-photon absorption (MA) influence and free carrier influence, as well as disorder-induced linear scattering influence on laser pulse propagation in slow-light photonic crystal waveguides was discussed in [22]. In [23] a soliton evolution was demonstrated experimentally and numerically for the plasma multi-photon generation regime in highly-nonlinear and highly-dispersive photonic crystal. Numerical simulation of femtosecond pulse propagation in a silicon photonic nano-wire waveguide was conducted using the generalized nonlinear Schrödinger equation taking into account a TPA. The soliton propagation in optical fibers with non-Kerr nonlinearities was investigated analytically and numerically in [24-25].

Journal of Physics: Conference Series **691** (2016) 012013 doi:10.1088/1742-6596/691/1/012013

Among a number of methods, developed for soliton finding, an approach based on the appropriate nonlinear eigenvalue problem solution, should be emphasized. Let us note, that each this problem requires a unique iterative method for soliton finding because the corresponding equation is a nonlinear one. Therefore, it is also true for the medium with nonlinear absorption. However, in this case, the soliton (self-similar profile of intensity distribution) finding is a more difficult task. Such kind of laser pulse propagation modes was discussed in [26-28] for both homogeneous and layered medium. In these papers we investigated also a cubic nonlinear response influence on the self-similar mode realization for the laser pulse propagation. In [29-30] we demonstrated the possibility of laser pulse propagation with the self-similar profile in a medium, possessing a resonant or cubic nonlinearity (self-focusing or defocusing) and MA of optical energy. Below we generalize our approach for 2D problem solution and carry out the corresponding computer simulation.

2. Formulation of the problem

2D laser beam propagation in a medium with MA is described by the dimensionless Schrödinger equation

$$\frac{\partial A}{\partial z}+iD_x\frac{\partial^2 A}{\partial x^2}+iD_y\frac{\partial^2 A}{\partial y^2}+i\beta A+\beta(\delta(1+i\Delta)\frac{|A|^k}{1+|A|^{k+2}/I_{sat}^{k/2+1}}+i\alpha|A|^2)A=0,\ \ 0<z\le L_z,\ \ 0<x<L_x,\ \ 0<y<L_y \quad (1)$$

with the following initial and boundary conditions

$$A(z,0,y)=A(z,L_x,y)=A(z,x,0)=A(z,x,L_y)=0,\ 0\le z\le L_z,\ A(0,x,y)=A_0(x,y),\ 0\le x\le L_x, 0\le y\le L_y. \quad (2)$$

In (1)-(2) $A(z,x,y)$ is the dimensionless complex amplitude; z is the dimensionless coordinate, along which a laser beam propagates, x and y are normalized coordinates which are transverse to the propagation direction, L_x, L_y are their maximal values. Parameters D_x, D_y characterize the laser beam diffraction. Parameter β describes a constant phase shift at a laser beam propagation, in particular. The integer constant k determines the MA type. The value k=2 corresponds to two-photon absorption (TPA), and k=4 means three-photon absorption (ThPA). The constant δ characterizes an amplitude of the medium nonlinear absorption (or amplification). The sign of this constant depends on the medium type: $\delta>0$ means a medium with absorption, and vice versa, if $\delta<0$, we deal with an amplifying medium. The constant α describes a cubic nonlinear response of the medium. The parameter Δ relates to the detuning between the carrier frequency of a wave packet and the energy transition frequency for an atom or a molecule. Parameter $I_{sat}^{k/2+1}$ describes absorption saturation.

3. Beam propagation in a medium with nonlinear absorption only

Firstly, we consider a beam propagating in a medium without Kerr and resonant nonlinear responses ($\alpha,\Delta=0$) without an absorption saturation ($I_{sat}^{k/2+1}\to\infty$). Let us represent the complex amplitude in the form

$$A(z,x,y)=B(x,y)e^{-i\lambda z}. \quad (3)$$

Substituting (3) in (1) and solving the resulting eigenvalue problem, we get the following soliton-like solution

$$B(x,y)=ch^{-n}(c_1x+c_2y)e^{if(x,y)},n=\frac{2}{k},\ D_xc_1^2+D_yc_2^2=\frac{\beta\delta k^2}{3\sqrt{2(k+2)(k+4)}},$$

$$f(x,y)=\sqrt{\frac{k}{2}(\frac{k}{2}+1)}\ln(ch(c_1x+c_2y))+const,\cdot\lambda=(\beta-\frac{2k\beta\delta}{\sqrt{2(k+2)(k+4)}};\ -\ \frac{4\beta\delta}{(k+4)}). \quad (4)$$

In (4) c_1 and c_2 are constants, $f(x,y)$ is the phase distribution, and λ is the eigenvalue, determined by the MA type. Function B(x, y) is the corresponding eigenfunction. It is important that the self-similar solution (4) is in a perfect accordance with 1D solution derived in [27]. Let us note that the equalities (4) determine a variety of the eigenfunctions since for each value of the parameter k we obtain different combinations of c_1 and c_2 values, satisfying the third expression in (4).

Obviously, beam radii as well as its maximal intensity change at the beam propagation in a medium with nonlinear absorption. To take into account both these changes and eigenfunction (4) one can search a self-similar beam profile evolution in the following form

$$A(z,x,y)=\rho(z,x,y)e^{is(z,x,y)},\ \rho=\rho_0(z)ch^{-n}(c_1(z)x+c_2(z)y),\ s=\varphi(z)+\sqrt{n^2+n}\ln(ch(c_1(z)x+c_2(z)y)).\quad (5)$$

Substituting the functions (5) into equation (1), and neglecting the derivatives of c_1 and c_2 with respect to the longitudinal coordinate, after some transformations we get

$$n=\frac{2}{k},\ \varphi(z)=\frac{1}{2\sqrt{2(2+k)}}\ln(z+z_0)-\beta z,\ z_0=\frac{k+4}{4k\beta\delta}.$$
$$(D_x c_1^2+D_y c_2^2)=\frac{k}{4\sqrt{2(2+k)}}(z+z_0)^{-1},\ \rho_0(z)=\left(\frac{k+4}{4k\beta\delta}\right)^{1/k}(z+z_0)^{-1/k}.\quad (6)$$

From (6) it is easy to see that the expression for $D_x c_1^2+D_y c_2^2$ turns into (4) in section $z=0$. Obviously, disregarding of derivatives of c_1 and c_2 is in a contradiction with the assumption (5). Nevertheless, it is necessary to keep in mind two circumstances. First, the maximal intensity of the beam connects with its radii via the power integral and we can see this dependence from (6). Second, these derivatives are less with respect to the derivative from other functions on z-coordinate. Third, if we consider the laser beam propagation in a medium with TPA (k=2), for simplicity, and neglect the diffraction influence, the amplitude $\rho_0(z)$ evolution is similar to a solution of the simplified equation (see (1))

$$\frac{dA}{dz}+\beta(i+\delta|A|^2)A=0,\quad 0<z\le L_z,\quad 0<x<L_x,\quad 0<y<L_y.\quad (7)$$

Multiplying both parts of (7) by A^* and supposing a equality $|A(z=0)|=1$, we come to the following formula which will be referenced below

$$|A(z)|^2=(2\beta\delta z+1)^{-1}.\quad (8)$$

This intensity evolution law is similar to the corresponding expression for the function $\rho_0(z)$ from (6). Therefore, it is to be expected that at least for weak diffraction case, our analytical consideration will be valid. As will be shown below, the application field of the expressions (6) is much wider.

4. Self-similar mode of beam propagation in a medium with TPA

In addition to a soliton-like profile developing it is necessary to demonstrate that this profile is rather stable at beam propagation. In this section, we present the computer simulation results of the problem (1)-(2) for $k=2,\ \alpha=0,\ \Delta=0,\ I_{sat}^{k/2+1}\to\infty, c_1=c_2$. The last equality corresponds to the symmetrical beam propagation.

As initial complex amplitude distribution $A_0(x,y)$, the derived solution for the eigenvalue problem in the form (4) is used. For definitely, in our computations the parameter β is equal to $\beta=1$, if other its value is not specified. For practice, the preservation of the incident profile of a beam, propagating

Journal of Physics: Conference Series **691** (2016) 012013 doi:10.1088/1742-6596/691/1/012013

at significant diffraction influence ($D_x = D_y = 1$), is of interest. It is a crucial key for developed solution. For example, computer simulation results, obtained at $\delta = 5$, is depicted in Fig. 1. We compare the analytical dependences (6) with the numerically computed maximum intensity $I_{max} = \max_{x,y} |A(x,y,z)|^2$ in Fig. 1a. It is evident that two curves are in a concordance, although we see a certain discrepancy between them due to the laser beam diffraction influence. Almost the same maximal intensity changing occurs if we change only the diffraction coefficients down to $D_x = D_y = 0.1$. This fact can be explained by the prevailed influence of TPA. Absorption coefficient decreasing to $\delta = 2.5$ results in almost perfect coincidence between I_{max} and $\rho_0^2(z)$ in the first stage of the laser beam propagation up to z=0.2 (Fig. 1b). It is also important a correlation between $\rho_0^2(z)$ and a solution of (8), written in the frame-work of negligible diffraction assumption. In Fig. 1c we can see a good agreement between these two curves.

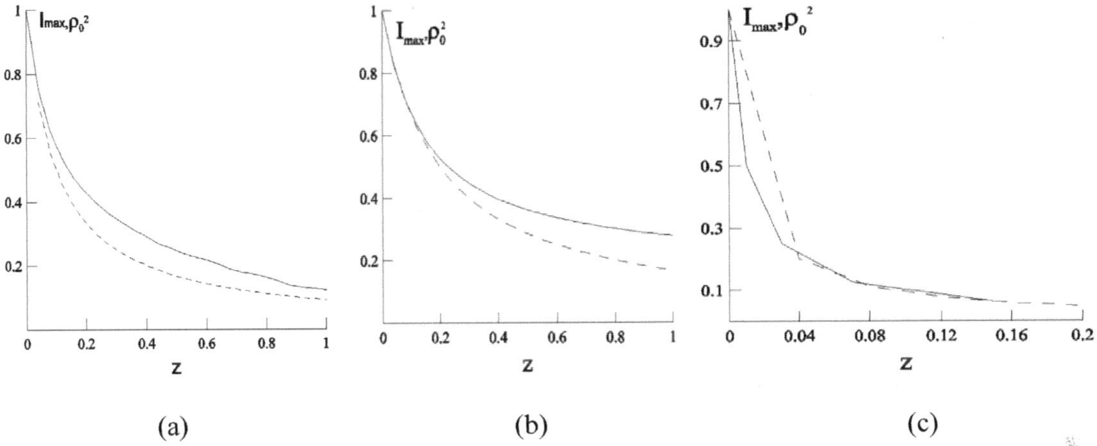

Fig.1. Maximum intensity computed using (1) (a, b) or (8) (c) (solid line) in comparison with $\rho_0^2(z)$ (dashed line) for $D_x = D_y = 1$, $\delta = 5$ (a), 50 (c); $D_x = D_y = 0.1$, $\delta = 2.5$ (b).

5. Self-similar profile of a beam propagating in a medium with TPA and cubic nonlinearity

In this section we consider simultaneous action of nonlinear cubic response and TPA i.e. $k = 2, \alpha \neq 0, \Delta = 0$, $I_{sat}^{k/2+1} \to \infty$. Using the representation (3), substituting it into (1), and analyzing the arising eigenvalue problem, we write the following expressions

$$B(x,y) = ch^{-n}(c_1 x + c_2 y) e^{if(x,y)}, f(x,y) = a \ln ch(c_1 x + c_2 y)) + const,$$
$$(D_x c_1^2 + D_y c_2^2)(a^2 - 2) = -\alpha\beta, 3a(D_x c_1^2 + D_y c_2^2) = \beta\delta, \lambda = (\beta - \beta(\frac{\delta}{3a} - \alpha); -\frac{2\beta\delta}{3}). \tag{9}$$

Let us note that the parameter a in (9) is positive and can be defined as follows

$$a = -\frac{3\alpha}{2\delta} + \sqrt{\frac{9\alpha^2}{4\delta^2} + 2}.$$

It is possible to take into account the dependence of the derived solution on z if we include in (9) the factor $|e^{2\operatorname{Im}\lambda}|$ in the following way:

$$(D_x c_1^2 + D_y c_2^2)(a^2 - 2) = -\alpha\beta e^{2\operatorname{Im}\lambda}, 3a(D_x c_1^2 + D_y c_2^2) = \beta\delta e^{2\operatorname{Im}\lambda}.$$

Then, for simplicity, assuming an equality validity $D_x = D_y = D$ and $c_1 = c_2 = c$, we finally get

$$a = -\frac{3\alpha}{2\delta} + \sqrt{\frac{9\alpha^2}{4\delta^2} + 2},\ c^2 = \frac{\beta\delta}{6D(-\frac{3\alpha}{2\delta} + \sqrt{\frac{9\alpha^2}{4\delta^2}} + 2)}(1 + \frac{4\beta\delta}{3}z)^{-1}, \mathrm{Im}\,\lambda = -\frac{1}{2}\ln(1 + \frac{4\beta\delta}{3}z),$$
$$\mathrm{Re}\,\lambda = \beta z + \frac{3}{12}\left(\frac{1}{-\frac{3\alpha}{2\delta} + \sqrt{\frac{9\alpha^2}{4\delta^2}} + 2} - 3\frac{\alpha}{\delta}\right)\ln(1 + \frac{4\beta\delta}{3}z). \tag{10}$$

6. Computer simulation of a self-similar profile beam propagation in a medium with TPA and Kerr nonlinearity

Below we discuss computer simulations results, obtained for a medium with TPA and Kerr nonlinearity: $k = 2, \alpha \neq 0, \Delta = 0, c_1 = c_2, I_{sat}^{k/2} \to \infty$. Incident beam with the self-similar profile (9) decays at its propagation in a medium with both focusing and defocusing nonlinearity. However, defocusing nonlinearity (strongly speaking, this nonlinear response can be realized due to resonant nonlinearity) leads to slower intensity attenuation. Moreover, with nonlinearity increasing a plateau on the curve of the maximum intensity evolution can be well seen (Fig. 2a).

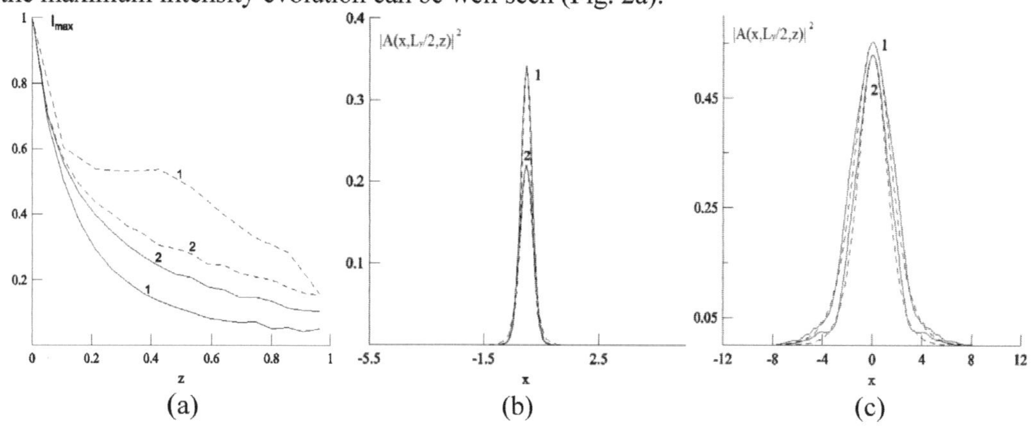

Fig. 2. Maximum intensity evolution for $\alpha = 7.5, 1$ (solid lines 1, 2) and -7.5, -1 (dashed lines 1,2) (a); Intensity profile, obtained from computer simulation (solid line 1 – z=0.18, 2 – z=0.28), and from (9) with the computed radius (dashed lines) for $\alpha = 7.5$ (b), -7.5 (c). Other parameters are: $D_x = D_y = 1$, $\delta = 5$.

For rather weak focusing/defocusing nonlinearities ($\alpha = \pm 1$), a beam with the initial profile (9) attenuates at a certain distance (z=0.3) without visible profile distortions. Then, the beam profile becomes distorted near the beam edges. With increasing the nonlinearity strength, a beam profile disturbs at smaller propagation. This regularity is the most pronounced for defocusing nonlinearity. In Fig. 2c, we see the considerably deformed profile of a beam ($\alpha = -7.5$) while, for the focusing nonlinearity, the profile is still self-similar one excepting its edges (Fig. 7b). Let us note, for this focusing nonlinearity value, the beam profile keeps its central part even after its intensity has been decreased twenty times from the initial value.

To clarify the beam profile perturbation scenario, we follow the beam phase distribution dynamics (Fig. 3 a - c). It is seen that the beam profile edge changes are accompanied by the phase distribution changes. At the initial stage of laser beam propagation the beam phase curvature corresponds to the beam focusing (Fig. 3 a). Then, phase distribution disturbs near the beam edges and, as consequence of this, the beam profile violates from its initial distribution (Fig. 3 b). Further laser beam propagation leads to a wave front curvature changing out the central part of laser beam. Thus, two sub-beams appear (Fig. 3 c) and they move from the central sub-beam which has a self-similar profile.

It should be emphasized that the phase distribution of a beam, propagating in a focusing medium, preserves its initial form.

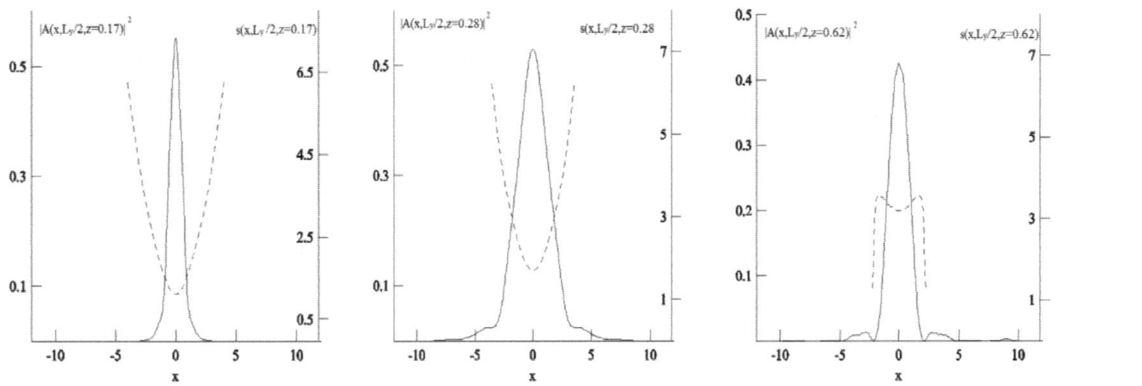

(a) (b) (c)

Fig. 3. Intensity profile (solid lines) and phase distribution (dashed lines) depicted in consecutive sections z and computed for the parameters: $D_x = D_y = 1$, $\delta = 5$, $\alpha = -7.5$.

Our study of 2D beam propagation in a medium with nonlinearity absorption would not be properly completed without considering self-similar beams with different initial radii on transverse coordinates: parameters c_1 and c_2. We analyzed the propagation of the beams with various relations between these parameters i.e. $c_1 > c_2$, $c_1 < c_2$, with the aim of their influence showing up. Typical regularities are shown in Fig. 4. It is evident that the beam propagation scenario remains the same for all relations between these parameters. Each picture of Fig. 4 demonstrates the beam profile, obtained

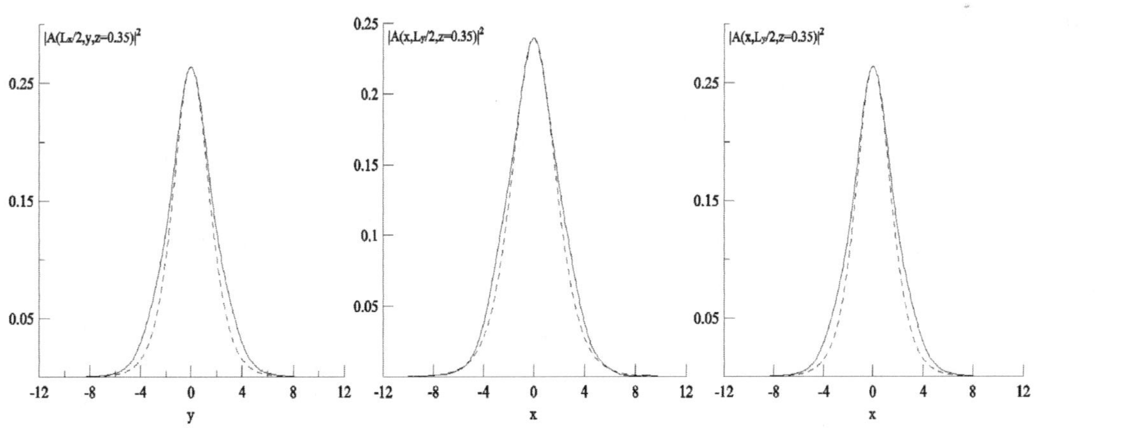

(a) (b) (c)

Fig.4. Intensity profile, obtained from computer simulation (solid lines) and from (9) with the computed radius (dashed lines) for $\alpha = 1$, $D_x = D_y = 1.0$, $\delta = 5$ and $c_1 = 0.7, c_2 = 0.5$ (a); $c_1 = 0.5, c_2 = 0.5$ (b); $c_1 = 0.5, c_2 = 0.7$ (c).

using computer simulation, and the beam profile given by (9). If $c_1 > c_2$ (Fig. 4a), we show the beam profile with respect to y coordinate at the centre of the domain along x coordinate. If $c_1 < c_2$ (Fig. 4c) we change coordinates. Comparing these pictures with the middle one (Fig. 4b), which represents the case of equal radii, we see no essential difference between them. In all cases, the beam central part coincides with the theoretical prediction. It is obvious that the beam loses its profile starting from the edges.

The last but not the least of our results, which we discuss in this section, concerns an important finding described briefly in [29]. Computer simulation of a 1D beam propagation in a medium with a resonant nonlinear response revealed that the beam propagates without any profile distortions of its profile up to the distance of 8 diffraction lengths for certain values of parameters. This scenario is especially apparent for the following set of nonlinear absorption, resonant nonlinearity, and cubic nonlinear response with parameters $\beta = 0.1$, $\alpha = 1$, $k = 4$, $\delta = 5$, $\Delta = 0.7$ (Fig. 5a). Let us remind that the following equality takes place k=4. It means three-photon absorption occurring.

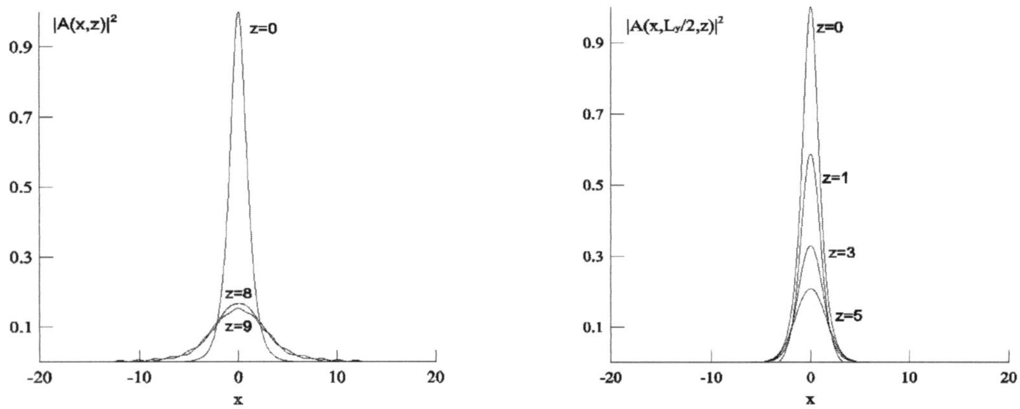

(a) (b)

Fig. 5. Iintensity profile evolution, obtained using computer simulation for the parameters $\alpha = 1$, $D_x = D_y = 0.4$, $\delta = 5$, $\Delta = 0.7$ in 1D case (a) or in 2D case (b).

Note that we have not yet derived an analytical solution for a 1D beam, nor have done it in 2D case. For the numerical simulation in [29] we used a self-similar profile, computed by numerical solving of the eigenfunction problem, corresponding to (1). This technique was developed in [26-30] for 1D case. 2D case is considerably more complicated and has not yet been realized for the nonlinearities mentioned above. In order to simulate a 2D beam propagation in self-similar mode we constructed the following approximation for the initial beam profile. The 1D beam profile, say discrete complex function $F(x_j)e^{is(x_j)}$, where index j relates to a mesh nodes, computed from the eigenvalue problem, is used as a base for the intensity profile approximation along both axes x and y. Complex amplitude coincides with this function at the beam centre along y-coordinate. Then, to construct the dependence of the initial distribution of complex amplitude on y-coordinate, we multiply the function $F(x_j)e^{is(x_j)}$ by $F(y_l)e^{is(y_l)}$, which coincides with value of $F(x_j)e^{is(x_j)}$ if values of the discrete coordinates are equal: $y_l = x_j$.. In fact, our approximation is given by the formula $F(x_j, y_l) = F(x_j)F(x_l)e^{is(x_j)+is(x_l)}$, where indices j and l relate to the grid nodes in a 2D domain.

Fig. 5 sums up our computer simulation results. It is well seen that the 2D beam attenuates at a greater rate than the 1D beam. In addition, the distance of beam travelling at the self-similar profile is shorter than in 1D case. After the distance, which is equal to 5 diffraction lengths, the 2D beam starts to disturb. Its profile is completely broken down and the amplitude decreases more than ten times.

Probably, the achieved length of the 2D beam propagation in a self-similar mode could be explained by a rather rough approximation of the initial profile.

7. Propagation of laser beam in a medium with resonant nonlinearity and saturation

In this section we report the computer simulation results, obtained for the parameters: $I_{sat}^{k/2+1}=1.7$, $k=4, \alpha \neq 0, \beta = 0.1, \Delta \neq 0, \delta \neq 0$. It means the joint action of MA, and resonant nonlinearity, and Kerr nonlinearity, as well as absorption saturation. In our knowledge, there are no papers, in which the analytical solution for this problem is developed. However, in [30] we reported about self-similar profile of a 1D beam, propagating the medium with resonant nonlinear response and its saturation. This result was obtained using a computer simulation. Below, we use this beam profile as initial beam profile for the problem (1)-(2). Particularly, remarkable result, got for 1D case, is shown in Fig. 6 for both signs of laser beam frequency detuning.

The beam attenuates greatly for positive sign of Δ, but the central part of its self-similar profile preserves at a longer distance than at the laser beam propagation in the medium with negative resonant response (Fig. 6a,b). The beam phase differs essentially for both signs of Δ. The latter circumstance is illustrated by Fig. 6 c, d.

Similar regularities are revealed for 2D beam propagation. Note that in the process of computations we used a beam with the self-similar profile, described in the last part of the previous section, as initial condition. In Fig. 7 the significant self-similar profile deformation is clearly seen for negative resonant medium response.

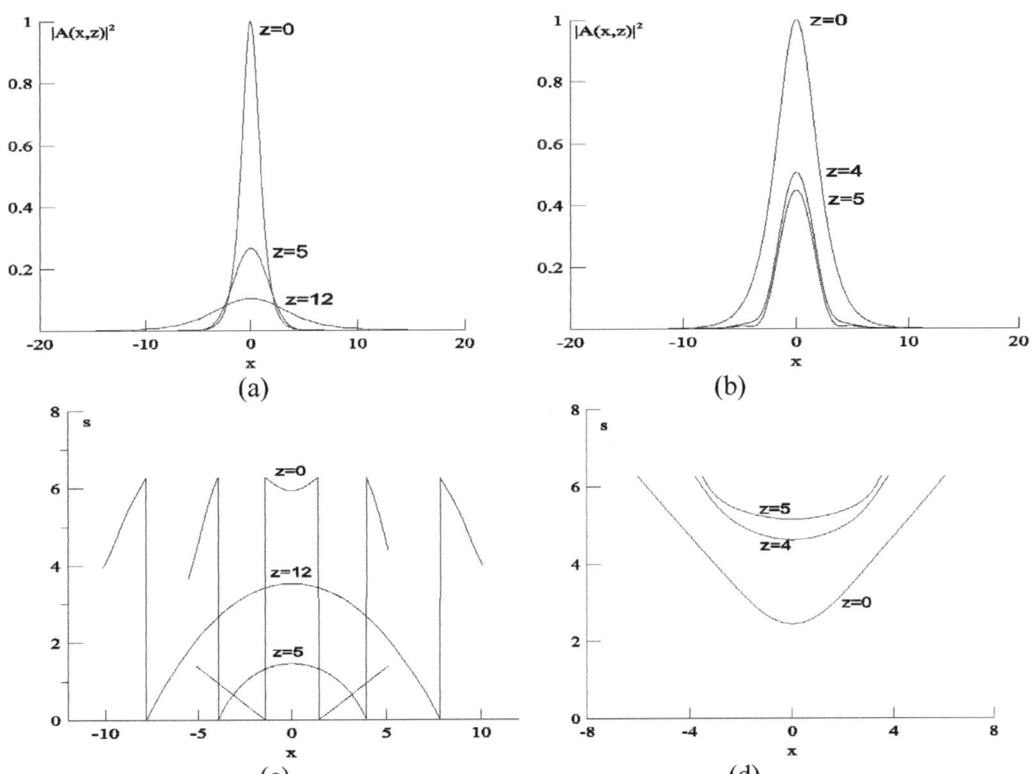

Fig. 6. Intensity profile (a, b) and phase distribution (c, d) at various distances (shown in the pictures), computed for 1D laser beam propagation with parameters: $D_x = 0.4$, $\delta = 5$, $\alpha = 2$, $\Delta = 0.7$ (a, c), - 0.7 (b, d).

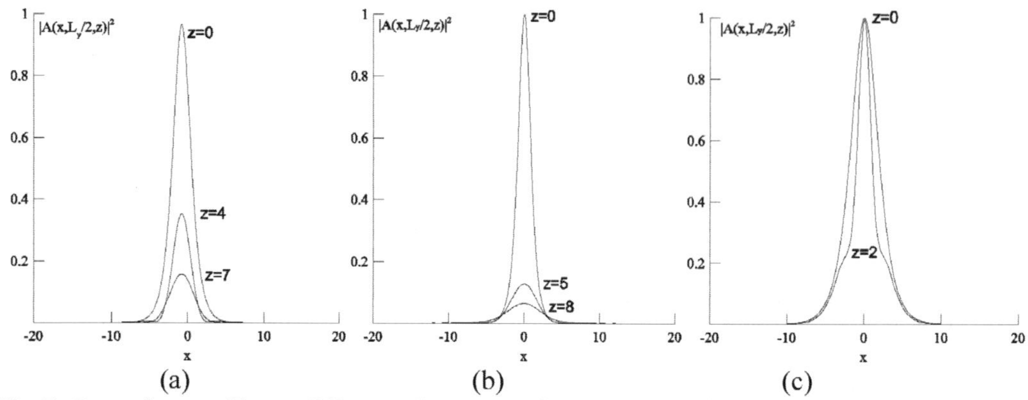

Fig.7. Intensity profile at different distances (shown in the pictures), computed for 2D laser beam propagation with parameters: $D_x = D_y = 0.4$, $\delta = 5$, $\alpha = 2$, $\Delta = 0$(a), 0.7 (b), -0.7 (c).

8. Conclusion

We derived analytically the 2D self-similar profile of a laser beam propagating in a media with both various types of MA and various nonlinear phase grating. Our computer simulation confirmed the results of analytical consideration. Under certain nonlinear absorption, and resonant nonlinearity, and cubic nonlinear response, the self-similar beam propagates along super long distance without any disturbance of its profile.

9. Acknowledgments

The investigation was made using support of the Russian Science Foundation (Grant № 14-21-00081).

References

[1] Akhmanov S A, Vysloukh V A and Chirkin A S 1992 *Optics of femtosecond laser pulses* (NY: American Institute of Physics)
[2] Shen Y R 1984 *The Principles of Nonlinear Optics* (NY: J. Wiley Press)
[3] Agrawal G P 1995 *Nonlinear fiber optics* (NY: Academic Press)
[4] Sukhorukov A P 1988 *Nonlinear Wave Interactions in Optics and Radiophysics* (Moscow: Nauka)
[5] Kivshar Y S and Agrawal G P 2003 *Optical Solitons: From Fibers to Photonic Crystals* (San Diego: Academic Press)
[6] Akhmediev N N and Ankiewicz A 1997 *Solitons: Nonlinear Pulses and beams* (London: Chapman and Hall Press)
[7] Brull L and Lange H 1986 Stationary, oscillatory and solitary wave type solution of singular nonlinear Schrödinger equations *Math. Meth. in the Appl. Sci.* **8** (4) 559–75
[8] Malomed B A 2006 *Soliton management in periodic systems* (NY: Springer)
[9] Berge L, Mezentsev V K, Rasmussen J J et al 2000 Self-guiding light in layered nonlinear media *Opt. Lett.* **25** (14) 1037–39
[10] Tower I and Malomed B A 2002 Stable (2+1)-dimensional solitons in a layered medium with sign-alternating Kerr nonlinearity *JOSA* B **19** (3) 537–43
[11] Yannopapas V, Modinos A and Stefonou N 2003 Anderson localization of light in inverted opals *Phys. Rev.* B **68** 193205
[12] Goodman R H, Holmes Ph J and Weinstein M I 2004 Strong NLS soliton-defect interactions *Physica* D **192** 215–42
[13] Driben R, Oz Y, Malomed B A et al. 2007 Mismatch management for optical and matter-wave quadratic solitons *Phys. Rev.* E **75** 026612
[14] Serkin V N and Hasegawa A 2002 Exactly integrable nonlinear Schrödinger equation models

with varying dispersion, nonlinearity and gain: application for soliton dispersion *IEEE Journal of Selected Topics in Quantum Electronics* **8** (3) 418

[15] Serkin V N, Hasegawa A and Belyaeva T L 2007 Nonautonomous solitons in external potentials *Physics Review Letters* **98** 074102

[16] Baizakov B B, Malomed B A and Salerno M 2003 Multidimensional solitons in periodic potentials *Europhysics Letters* **63** (5) 642–48

[17] Al Khawaja U 2009 Integrability of a general Gross–Pitaevskii equation and exact solitonic solutions of a Bose–Einstein condensate in a periodic potential *Phys. Letters* A **373** 2710–16

[18] Xu Q Q, Yao D Z, Liu X N,Zhou Q, Xiong G G 2012 Solitary propagation effect of a well-defined chirped femtosecond laser pulse in a resonance-absorbing medium Physical *Review A* **86** (2) 86.023853

[19] Mani Rajan M S, Hakkim J., Mahalingam A, and Uthayakumar A 2013 Dispersion management and cascade compression of femtosecond nonautonomous soliton in birefringent fiber *Eur. Phys. J. D* **67** 150

[20] Loomba Sh, Gupta R, Kaur H, Mani Rajan M S 2014 Self-similar rogue waves in an inhomogeneous generalized nonlinearSchrödinger equation *Physics Letters A* 378 2137

[21] Blanco-Redondo A., Husko C., Eades D., Zhang Y., Li J., Krauss T.F., and Eggleton B.J. 2014 Observation of soliton compression in silicon photonic crystals *Nature Communications* **5** 3160

[22] Husko C.A., Colman P., Combrié S., De Rossi A., and Wong C.W. 2011 Effect of multiphoton absorption and free carriers in slow-light photonic crystal waveguides *Optics Letters* **36**(12) 2239-41

[23] Husko C.A., Combrié S.,Colman P., Zheng J., De Rossi A., and Wong C.W. 2013 Soliton dynamics in the multiphoton plasma regime *Scientific Reports* **3** 1100

[24] Kohl R., Biswas A., Milivic D., and Zerrad E. 2008 Optical soliton perturbation in a non-Ker law media *Optics & Laser Technology* 40, 647-62

[25] Biswas A., Fessak M., Johnson S., Beatrice S., Milivic D., Jovanoski Z, Kohl R., and Majid F. 2012 Optical soliton perturbation in a non-Ker law media: traveling wave soliton *Optics & Laser Technology* 44, 263-68

[26] Trofimov V A, Matysevich O V and Smotrov D A 2011 Mode of propagation of optical radiation with self-similar pulse shape in layered medium with nonlinear absorption *Proc. of SPIE* **8095** 80951K

[27] Trofimov V A, Zakharova I G, Smotrov D A, Lan Sh 2013 Self-similar pulse shape mode for femtosecond pulse propagation in optical fiber with multi-photon absorption and nonlinear refraction *Proc. of SPIE* **8785** 87750K

[28] Trofimov V, Zakharova I, Fedotov M 2014 Self-similar Shape Mode of Optical Pulse Propagation in de-focusing Medium with two-photon Absorption *Journal of Physics: Conference Series (22nd International Laser Physics Worksop (LPHYS 13))* **497** 012023.

[29] Trofimov V A, Zakharova I G, Konar S 2014 Self-similar pulse-shape mode for femtosecond pulse propagation in medium with resonant nonlinearity *Proc. of SPIE* **9136** 91360Z

[30] Trofimov V A, Zakharova I G. 2015 Propagation of femtosecond pulse with self-similar shape in medium with nonlinear absorption *Proc. of SPIE* **9503** 9503-27

24th International Laser Physics Workshop (LPHYS'15) IOP Publishing
Journal of Physics: Conference Series **691** (2016) 012014 doi:10.1088/1742-6596/691/1/012014

Phonon instability of insulating states in optical lattices

V.I. Yukalov[1] **and K. Ziegler**[2]

[1]Bogolubov Laboratory of Theoretical Physics, Joint Institute for Nuclear Research, Dubna 141980, Russia
[2]Institut für Physik, Universität Augsburg, Augsburg D-86135, Germany

E-mail: yukalov@theor.jinr.ru

Abstract. The influence of collective phonon excitations, due to intersite atomic interactions, on the stability of optical lattices is analyzed. These phonon excitations are shown to essentially reduce the ability of atoms to be localized. The states that seem to be insulating in the absence of the phonon excitations can become delocalized when the latter are present. The delocalization effect exists for both long-range as well as local atomic interactions.

1. Introduction

Physics of cold atoms in optical lattices is an intensive field of research, as can be inferred from the review articles [1–5]. Depending on the depth of the wells at the lattice sites, temperature, and interaction strength, atoms can be in an insulating localized state or in delocalized itinerant state. Here we consider insulating states.

In an insulating state, an atom, localized at a lattice-site well, being in the ground state, is described by a well localized atomic wave function, as in Fig. 1a. This function can exhibit two types of variations. First, the wave function of the ground state can be transformed into an excited state, e.g., as is pictured in Fig. 1b, which corresponds to a single-particle excitation. Second, because of interactions with other atoms, the wave packet can oscillate around the lattice site, as is shown in Fig. 1c, which manifests collective phonon excitations.

Thus, in physical reality, there always exist two types of atomic motion, the transfer between different energy levels, accompanied by an essential wave function deformation, and small oscillations around a lattice site, preserving the wave-function shape.

If one postulates that the system is described by a Hubbard Hamiltonian, this implies that small oscillations are disregarded and only transitions between energy bands are taken into account. The Hubbard model is formulated in terms of single-particle states and does not contain collective atomic fluctuations. Clearly, there are no phonon excitations in the standard Hubbard model. However, one should not confuse a model and the physical reality. It is well known that in any real physical system, whether it is gas, liquid, or solid, there always exist sound waves caused by particle interactions. And this does not depend on the presence or absence of any external fields. In nature, any realistic physical system of interacting atoms does exhibit the presence of density, or sound, waves characterizing collective phonon excitations. Any model is a cartoon of reality, and may take or not take into account the existence of phonons. But one should not confuse cartoon models and physics.

In the present paper, we show how it is possible to modify the Hubbard model in order to take into account collective phonon excitations. And we analyze the consequence of the phonon existence on the stability of insulating states. It turns out [6] that phonon excitations can strongly influence the region of stability of insulating states in optical lattices. In those cases, where the phonon instability occurs, the lattice can be metastable, and living sufficiently long time for being experimentally observed.

2. Phonon-dressed lattice model

Let us start with the general form of the Hamiltonian in the second-quantization representation,

$$\hat{H} = \int \psi^\dagger(\mathbf{r}) \left[-\frac{\nabla^2}{2m} + V_L(\mathbf{r}) \right] \psi(\mathbf{r})\, d\mathbf{r} + \frac{1}{2} \int \psi^\dagger(\mathbf{r})\psi^\dagger(\mathbf{r}')\Phi(\mathbf{r}-\mathbf{r}')\psi(\mathbf{r}')\psi(\mathbf{r})\, d\mathbf{r}d\mathbf{r}' \,, \tag{1}$$

in which the field operators can satisfy either Bose or Fermi statistics. The optical-lattice potential

$$V_L(\mathbf{r}) = \sum_{\alpha=1}^{d} V_\alpha \sin^2(k_0^\alpha r_\alpha) \tag{2}$$

is formed by laser beams with the laser wave vector

$$\mathbf{k}_0 = \left\{ k_0^\alpha = \frac{2\pi}{\lambda_\alpha} = \frac{\pi}{a_\alpha} \right\} \tag{3}$$

prescribing the lattice vector

$$\mathbf{a} = \left\{ a_\alpha = \frac{\lambda_\alpha}{2} = \frac{\pi}{k_0^\alpha} \right\} . \tag{4}$$

The real-space dimensionality is $d = 1, 2, 3$. The characteristic kinetic energy of an atom, caused by laser beams, is the recoil energy

$$E_R = \frac{k_0^2}{2m} \qquad \left(k_0^2 = \sum_{\alpha=1}^{d} \frac{\pi^2}{a_\alpha^2} \right) . \tag{5}$$

The total number of atoms N is placed inside an optical lattice, with the number of sites N_L. The lattice vectors are $\{\mathbf{a}_j : j = 1, 2, \ldots, N_L\}$. The lattice itself is fixed in space. The filling factor

$$\nu \equiv \frac{N}{N_L} = \rho a^d \tag{6}$$

can be expressed through the average atomic density ρ and mean interatomic distance a, defined as

$$\rho \equiv \frac{N}{V} \,, \qquad a^d \equiv \frac{V}{N_L} \,, \tag{7}$$

with V being the system volume.

Atomic interactions, generally, contain two parts, represented by local and nonlocal potentials

$$\Phi(\mathbf{r}) = \Phi_{loc}(\mathbf{r}) + \Phi_{non}(\mathbf{r}) . \tag{8}$$

The local interaction potential can be written in the form

$$\Phi_{loc}(\mathbf{r}) = \Phi_d \delta(\mathbf{r}) \,, \tag{9}$$

with the interaction strength

$$\Phi_d = \frac{\Phi_{eff}}{(\sqrt{2\pi}\, l_\perp)^{3-d}} \; , \qquad \Phi_{eff} \equiv 4\pi \, \frac{a_{eff}}{m} \; , \tag{10}$$

in which $l_\perp$ is the transverse oscillator length and a_{eff} is an effective scattering length. The latter, depending on the system geometry, takes the following values: in the quasi-one-dimensional case [7], it is

$$a_{eff} = \frac{a_s}{1 - 0.46 a_s / l_\perp} \qquad (d = 1) \; .$$

In the quasi-two-dimensional case [8], one has

$$a_{eff} = \frac{a_s}{1 - (a_s/\sqrt{2\pi}\, l_\perp) \ln[(2\pi)^{3/2} \rho l_\perp a_s]} \qquad (d = 2) \; .$$

And in three dimensions, it is just the scattering length,

$$a_{eff} = a_s \qquad (d = 3) \; .$$

Nonlocal interactions can correspond to dipolar forces [9, 10]

In an insulating state, atoms are localized at the points of minima of an effective potential formed by the combination of the given optical lattice and an effective potential created by other atoms. At each given moment of time the points of minima $\{\mathbf{r}_j\}$ do not necessarily coincide with the lattice sites $\{\mathbf{a}_j\}$, since atoms fluctuate. The set $\{\mathbf{r}_j\}$ of the effective minima is to be treated as a random set. As far as atoms are localized in the vicinity of the points $\{\mathbf{r}_j\}$, it is possible to expand the field operators over localized orbitals [11, 12] centered at the corresponding spatial points,

$$\psi(\mathbf{r}) = \sum_{nj} c_{nj} \psi_n(\mathbf{r} - \mathbf{r}_j) \; , \tag{11}$$

where n is a set of quantum numbers defining energy levels. Substituting expansion (11) into Hamiltonian (1), we consider only the lowest energy level, assuming that the gap between the energy levels is sufficiently large, being much larger than an average phonon energy. Then we obtain the Hamiltonian

$$\hat{H} = -\sum_{i \neq j} J(\mathbf{r}_{ij}) c_i^\dagger c_j + \sum_j \left(\frac{\mathbf{p}_j^2}{2m} + V_L \right) c_j^\dagger c_j + \frac{U}{2} \sum_j c_j^\dagger c_j^\dagger c_j c_j + \frac{1}{2} \sum_{i \neq j} U(\mathbf{r}_{ij}) c_i^\dagger c_j^\dagger c_j c_i \; , \tag{12}$$

in which $J(\mathbf{r}_{ij})$ is a hopping term, $\mathbf{p}_j^2/2m$ is a kinetic-energy term, V_L is an average of the lattice potential, U is an on-site interaction parameter, and $U(\mathbf{r}_{ij})$ describes interactions between different sites. Here $\mathbf{r}_{ij} \equiv \mathbf{r}_i - \mathbf{r}_j$. The above quantities are the matrix elements over the localized orbitals, whose calculation can be found in Refs. [6, 13, 14].

Instead of the localized orbitals, it is possible to use the so-called maximally localized Wannier functions [15, 16]. These functions have been successfully employed not only for crystals with ideally periodic lattices, but also for crystals with defects, disordered networks, amorphous solids, and even liquids [15, 16]. The definition of the maximally localized Wannier functions, which can be used for the present consideration, is given in the Appendix.

Since the vector $\mathbf{r}_j$, by assumption, is a random vector, close to $\mathbf{a}_j$, it is convenient to introduce the notation

$$\mathbf{r}_j = \mathbf{a}_j + \mathbf{u}_j \; , \tag{13}$$

where

$$\mathbf{a}_j \equiv \langle \mathbf{r}_j \rangle \; , \qquad \langle \mathbf{u}_j \rangle = 0 \; . \tag{14}$$

Here the angle brackets mean statistical averaging with respect to the total system Hamiltonian. In a stable equilibrium state, random oscillations around the sites $\mathbf{a}_j$ imply zero average deviation $\langle \mathbf{u}_j \rangle$. A nonzero value of the latter would signify a structural phase transition [17].

For what follows, it is convenient to define the relative-distance vectors $\mathbf{a}_{ij} \equiv \mathbf{a}_i - \mathbf{a}_j$ and the relative deviations $\mathbf{u}_{ij} \equiv \mathbf{u}_i - \mathbf{u}_j$. The deviations are assumed to be small, which makes it possible to resort to the second-order expansions

$$U(\mathbf{r}_{ij}) \simeq U_{ij} + \sum_\alpha U_{ij}^\alpha u_{ij}^\alpha - \frac{1}{2} \sum_{\alpha\beta} U_{ij}^{\alpha\beta} u_{ij}^\alpha u_{ij}^\beta \; ,$$

$$J(\mathbf{r}_{ij}) \simeq J_{ij} + \sum_\alpha J_{ij}^\alpha u_{ij}^\alpha - \frac{1}{2} \sum_{\alpha\beta} J_{ij}^{\alpha\beta} u_{ij}^\alpha u_{ij}^\beta \; , \tag{15}$$

in which

$$U_{ij} \equiv U(\mathbf{a}_{ij}) \; , \qquad U_{ij}^\alpha \equiv \frac{\partial U_{ij}}{\partial a_{ij}^\alpha} = \frac{\partial U_{ij}}{\partial a_i^\alpha} \; , \qquad U_{ij}^{\alpha\beta} \equiv - \frac{\partial^2 U_{ij}}{\partial a_{ij}^\alpha \partial a_{ij}^\beta} = \frac{\partial^2 U_{ij}}{\partial a_i^\alpha \partial a_j^\beta} \; ,$$

with similar notations used for the hopping term.

Keeping in mind different physical nature of deviations and atomic operators, we employ the decoupling

$$u_{ij}^\alpha u_{ij}^\beta c_i^\dagger c_j = \langle u_{ij}^\alpha u_{ij}^\beta \rangle c_i^\dagger c_j + u_{ij}^\alpha u_{ij}^\beta \langle c_i^\dagger c_j \rangle - \langle u_{ij}^\alpha u_{ij}^\beta \rangle \langle c_i^\dagger c_j \rangle \; . \tag{16}$$

Phonon operators are introduced by the nonuniform canonical transformation [14, 18]

$$\mathbf{u}_j = \vec{\delta}_j + \frac{1}{\sqrt{2N}} \sum_{ks} \sqrt{\frac{\nu}{m\omega_{ks}}}\, \mathbf{e}_{ks} \left(b_{ks} + b_{-ks}^\dagger \right) e^{i\mathbf{k}\cdot\mathbf{a}_j} \; ,$$

$$\mathbf{p}_j = - \frac{i}{\sqrt{2N}} \sum_{ks} \sqrt{\frac{m\omega_{ks}}{\nu}}\, \mathbf{e}_{ks} \left(b_{ks} - b_{-ks}^\dagger \right) e^{i\mathbf{k}\cdot\mathbf{a}_j} \tag{17}$$

differing from the standard transformation by the presence of the vector $\vec{\delta}_j$ that is required for cancelling in the Hamiltonian the terms linear in the phonon operators [19].

The phonon spectrum is defined by the eigenproblem

$$\frac{\nu}{m} \sum_{j(\neq i)} \sum_\beta \Phi_{ij}^{\alpha\beta} e^{i\mathbf{k}\cdot\mathbf{a}_{ij}} e_{ks}^\beta = \omega_{ks}^2 e_{ks}^\alpha \; , \tag{18}$$

in which $\mathbf{e}_{ks}$ are the polarization vectors, s being a polarization index, and the renormalized dynamical matrix is

$$\Phi_{ij}^{\alpha\beta} = U_{ij}^{\alpha\beta} \langle c_i^\dagger c_j^\dagger c_j c_i \rangle - 2 J_{ij}^{\alpha\beta} \langle c_i^\dagger c_j \rangle \; . \tag{19}$$

In the case of a cubic lattice, it is convenient to define the effective dynamical matrix

$$D_{ij} \equiv - \frac{1}{d} \sum_{\alpha=1}^d \Phi_{ij}^{\alpha\alpha} \qquad (D_0 \equiv D_{\langle ij \rangle}) \; , \tag{20}$$

denoting the matrix in Eq. (20) for the nearest neighbour sites by D_0 .

In this way, we obtain the Hamiltonian

$$\hat{H} = E_N + \hat{H}_{at} + \hat{H}_{ph} + \hat{H}_{ind} \; , \tag{21}$$

where E_N is a non-operator quantity, the second term is the renormalized atomic Hamiltonian, the third term is the phonon Hamiltonian, and the last term describes effective atomic interactions induced by the phonon existence. Expressing these terms, we use the correlation functions

$$\langle u_{ij}^\alpha u_{ij}^\beta \rangle = 2(1-\delta_{ij})\langle u_j^\alpha u_j^\beta \rangle \; ,$$

$$\langle u_i^\alpha u_j^\beta \rangle = \frac{\delta_{ij}}{2N_L} \sum_{ks} \frac{e_{ks}^\alpha e_{ks}^\beta}{m\omega_{ks}} \coth\left(\frac{\omega_{ks}}{2T}\right) \; .$$

The first term reads as

$$E_N = -NK_N \; , \tag{22}$$

where K_N is the mean kinetic energy per atom

$$K_N \equiv \frac{1}{N_L} \sum_j \left\langle \frac{\mathbf{p}_j^2}{2m} \right\rangle = \frac{1}{4\nu N} \sum_{ks} \omega_{ks} \coth\left(\frac{\omega_{ks}}{2T}\right) \; .$$

The atomic Hamiltonian becomes

$$\hat{H}_{at} = -\sum_{i\neq j} \widetilde{J}_{ij} c_i^\dagger c_j + \frac{U}{2} \sum_j c_j^\dagger c_j^\dagger c_j c_j + \frac{1}{2} \sum_{i\neq j} \widetilde{U}_{ij} c_i^\dagger c_j^\dagger c_j c_i + K_N \sum_j c_j^\dagger c_j \; , \tag{23}$$

with the renormalized hopping and interaction terms

$$\widetilde{J}_{ij} = J_{ij} - (1-\delta_{ij}) \sum_{\alpha\beta} J_{ij}^{\alpha\beta} \langle u_j^\alpha u_j^\beta \rangle \; , \qquad \widetilde{U}_{ij} = U_{ij} - (1-\delta_{ij}) \sum_{\alpha\beta} U_{ij}^{\alpha\beta} \langle u_j^\alpha u_j^\beta \rangle \; .$$

The phonon Hamiltonian is

$$\hat{H}_{ph} = \sum_{ks} \omega_{ks} \left(b_{ks}^\dagger b_{ks} + \frac{1}{2} \right) \; , \tag{24}$$

in which the phonon spectrum is defined in Eq. (18) through the renormalized dynamical matrix (19). And the induced atomic interactions result in the term

$$\hat{H}_{ind} = \sum_{i\neq j} \sum_{\alpha\beta} F_i^\alpha \gamma_{ij}^{\alpha\beta} F_j^\beta \; , \tag{25}$$

where

$$F_i^\alpha = \sum_{j(\neq i)} \left(2J_{ij}^\alpha c_i^\dagger c_j - U_{ij}^\alpha c_i^\dagger c_j^\dagger c_j c_i \right) \; , \qquad \gamma_{ij}^{\alpha\beta} = \frac{1}{N_L} \sum_{ks} \frac{e_{ks}^\alpha e_{ks}^\beta}{m\omega_{ks}^2} \, e^{i\mathbf{k}\cdot\mathbf{a}_{ij}} \; .$$

3. Lindemann criterion of stability

An insulating state, by its definition, presupposes that atomic wave packets are well localized close to their lattice sites, so that the packets, corresponding to the nearest neighbours, practically do not overlap. As soon as the overlap becomes essential, the system cannot anymore be treated as localized. In that case, the insulating state is not stable, but there occurs the delocalization of atoms. This behavior can be discussed in terms of correlation functions: The correlation of the fluctuations in the localized state should decay exponentially on a finite correlation length ξ [20]. Then an instability due to an increasing overlap is indicated by a divergency of the correlation length. A similar description is based on the Lindemann criterion, which will be used subsequently.

The oscillations of an atomic wave packet are characterized by the mean-square deviation

$$r_0^2 \equiv \sum_{\alpha=1}^{d} \langle u_j^\alpha u_j^\alpha \rangle \; . \tag{26}$$

In an insulating state, this deviation is usually much smaller than the distance between the nearest neighbours a. The delocalization transition happens, when the mean square deviation increases approaching a. This is the Lindemann criterion of delocalization [21]. The weak form of the criterion implies that

$$\frac{r_0}{a} < 1 \; . \tag{27}$$

We calculate the mean-square deviation for a cubic lattice, with nearest-neighbour interactions, and employ the Debye approximation. The lattice is assumed to be sufficiently large, so that $N_L \gg 1$. Then, at finite temperature, we have in one dimension

$$r_0^2 \simeq \frac{T}{2\pi^2 \nu D_0}\, N_L \qquad (d = 1,\; T > 0) \; , \tag{28}$$

in two dimensions

$$r_0^2 \simeq \frac{T}{(2\pi)^2 \nu D_0}\, \ln N_L \qquad (d = 2,\; T > 0) \; , \tag{29}$$

and in three dimensions

$$r_0^2 \simeq \frac{9T}{mT_D^2}\, N_L \qquad (d = 3,\; T > T_D) \; , \tag{30}$$

where the Debye temperature is

$$T_D = \sqrt{4\pi\, \frac{\nu D_0}{m}}\, \left[\frac{d}{2}\, \Gamma\left(\frac{d}{2}\right) \right]^{1/d} \; . \tag{31}$$

This tells us that, at finite temperature, only three-dimensional insulating states can be stable.

At zero temperature, we find

$$r_0^2 = \frac{d^2}{2(d-1)mT_D} \qquad (T = 0) \; , \tag{32}$$

which shows that, at $T = 0$, two-dimensional and three-dimensional insulating states are admissible.

Summarizing, we see that insulating optical lattices can be stable with respect to collective phonon excitations: in two dimensions at zero temperature, if

$$\frac{E_R}{T_D} < \frac{\pi^2}{2} \qquad (d = 2,\; T = 0) \; , \tag{33}$$

in three dimensions at zero temperature when

$$\frac{E_R}{T_D} < \frac{2\pi^2}{3} \qquad (d = 3,\; T = 0) \; , \tag{34}$$

and in three dimensions at finite temperature, provided that

$$\frac{E_R T}{T_D^2} < \frac{\pi^2}{6} \qquad (d = 3,\; T > T_D) \; . \tag{35}$$

The meaning of these conditions is rather clear. They require that kinetic energy be smaller than potential energy by the amount in the right-hand side of these inequalities.

Even if the insulating state in an optical lattice is not stable, it can be metastable, living quite long time. The lifetime of a metastable state can be estimated by the formula

$$t_{met} = \frac{2\pi}{\omega_0} \exp\left(\frac{V_0}{E_R}\right) ,$$

in which ω_0 is the effective oscillator frequency corresponding to a potential well at a lattice site, V_0 is the optical lattice barrier height, and E_R is the recoil energy. For example, in the case of a cubic optical lattice filled by ^{87}Rb atoms [22], the lifetime of the insulating state is quite long, being $t_{met} > 200$ s, which is longer than the lifetime of atoms in a trap, which allows one to accomplish the necessary measurements. The lifetime of insulating states for atoms, such as ^{52}Cr, ^{162}Er, and ^{164}Dy, possessing long-range dipole interactions [9, 10], can be even longer.

In conclusion, taking into account collective phonon degrees of freedom can essentially change the region of stability of insulating states in optical lattices [6]. The phonon instability can be characterized by the Lindemann criterion [21]. It looks that the insulating optical lattices, studied at the present time in experiments with trapped atoms, are not stable with respect to phonon excitations, but correspond to only metastable states that, although, can live sufficiently long time for being experimentally observed.

We have introduced the phonons in complete analogy with their introduction in the theory of quantum crystals [23], where, first, one considers atoms, localized in some randomly distributed spatial points close to the sites of a periodic lattice. Then, defining small deviations from these lattice sites, one introduces phonon excitations. The sole difference between the optical lattices and quantum crystals is that the lattice sites in the former are prescribed by laser beams creating the lattice, while in the case of quantum crystals, the lattice vectors have to be defined by minimizing the system free energy.

Note that sound waves exist in all interacting systems of many particles, whether in ideally periodic crystals or amorphous solids, in normal or superfluid liquids, in bulk samples or finite systems [24, 25].

Appendix: Maximally localized Wannier functions

The localized Hamiltonian (12) has been derived by employing the field-operator expansion (11) over the localized orbitals describing single-particle states of atoms localized in the vicinity of the related lattice sites. We have also mentioned that, instead of the localized orbitals, we could used the maximally localized Wannier functions [15, 16]. Here we give a brief definition of these functions in the form that would be convenient for the purpose of our paper.

Suppose, first, that atoms are localized in a lattice described by a set of lattice vectors $\{\mathbf{r}_j : j = 1, 2, \ldots, N_L\}$. Because of atomic interactions, the minima of an effective potential, defining the lattice $\{\mathbf{r}_j\}$, do not coincide with the sites of the optical lattice $\{\mathbf{a}_j\}$, although each $\mathbf{r}_j$ is close to $\mathbf{a}_j$. One can define a Bloch function

$$\varphi_{nk}(\mathbf{r}) = u_{nk}(\mathbf{r})e^{i\mathbf{k}\cdot\mathbf{r}} , \qquad u_{nk}(\mathbf{r} + \mathbf{r}_j) = u_{nk}(\mathbf{r}) .$$

However, Bloch functions are strongly nonunique, since the new function

$$\overline{\varphi}_{nk}(\mathbf{r}) = \sum U_{mn}(\mathbf{k})\varphi_{mk}(\mathbf{r})$$

is also another Bloch function, provided that the matrix $[U_{mn}]$ is unitary, such that

$$\sum_j U^*_{jm}(\mathbf{k})U_{jn}(\mathbf{k}) = \sum_j U_{mj}(\mathbf{k})U^*_{nj}(\mathbf{k}) = \delta_{mn}.$$

It is straightforward to check that the new Bloch functions are orthonormal and form a complete basis,

$$\int \overline{\varphi}^*_{mk}(\mathbf{r})\overline{\varphi}_{np}(\mathbf{r})\, d\mathbf{r} = \delta_{mn}\delta_{kp}\,, \qquad \sum_{nk} \overline{\varphi}_{nk}(\mathbf{r})\overline{\varphi}^*_{nk}(\mathbf{r}') = \delta(\mathbf{r}-\mathbf{r}')\,.$$

Respectively, Wannier functions are also strongly nonunique, enabling the introduction of the form

$$\psi_n(\mathbf{r}-\mathbf{r}_j) \equiv \frac{1}{\sqrt{N_L}} \sum_k \overline{\varphi}_{nk}(\mathbf{r})e^{-i\mathbf{k}\cdot\mathbf{r}_j} = \frac{1}{\sqrt{N_L}} \sum_{mk} U_{mn}(\mathbf{k})\varphi_{mk}(\mathbf{r})e^{-i\mathbf{k}\cdot\mathbf{r}_j} =$$

$$\frac{1}{\sqrt{N_L}} \sum_{mk} U_{mn}(\mathbf{k})u_{mk}(\mathbf{r})e^{i\mathbf{k}\cdot(\mathbf{r}-\mathbf{r}_j)} = \frac{1}{\sqrt{N_L}} \sum_{mk} U_{mn}(\mathbf{k})u_{mk}(\mathbf{r}-\mathbf{r}_j)e^{i\mathbf{k}\cdot(\mathbf{r}-\mathbf{r}_j)}\,.$$

Maximally localized Wannier functions are defined as the functions, with the matrix $[U_{mn}]$ minimizing the variance functional

$$\sum_n \left(\langle r^2 \rangle_n - \langle r \rangle_n^2 \right)\,,$$

where the brackets $< \cdots >_n$ imply the notation

$$\langle A(\mathbf{r}) \rangle_n \equiv \int \psi_n^*(\mathbf{r})A(\mathbf{r})\psi_n(\mathbf{r})\, d\mathbf{r}\,.$$

The so-defined maximally localized Wannier functions are orthogonal to each other, being strongly localized, and exponentially decaying outside of their related centers $\mathbf{r}_j$.

These well localized Wannier functions can be used for deriving the localized model (11). Then, taking into account that each location $\mathbf{r}_j$ is close to the site $\mathbf{a}_j$, small deviations are introduced as in Eqs. (13) and (14). The deviations describe atomic fluctuations corresponding to collective phonon excitations.

Acknowledgement
Financial support from RFBR (grant #14-02-00723) and from the University of Augsburg is appreciated.

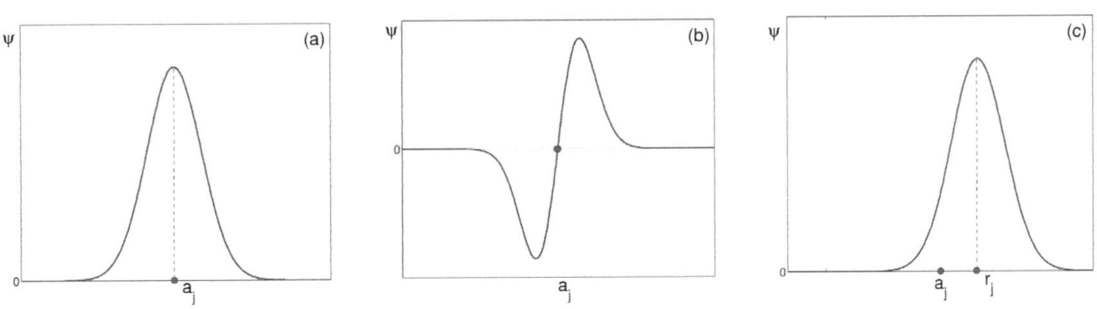

Figure 1. (a) The wave function for the ground state of an atom localized at a lattice site a_j; (b) The wave function for a localized atom in an excited single-particle state; (c) The wave function of an atom displaced from the lattice site a_j to a close position r_j, such that $|\mathbf{r}_j - \mathbf{a}_j| \ll a$, with a being the nearest neighbour lattice distance.

[1] Morsh O, Oberthaler M 2006 *Rev. Mod. Phys.* **78** 179
[2] Moseley C, Fialko O, and Ziegler K 2008 *Ann. Phys. (Berlin)* **17** 561
[3] Bloch I, Dalibard J, and Zwerger W 2008 *Rev. Mod. Phys.* **80** 885
[4] Yukalov V I 2009 *Laser Phys.* **19** 1
[5] Yukalov V I 2013 *Cond. Matter Phys.* **16** 23002
[6] Yukalov V I and Ziegler K 2015 *Phys. Rev. A* **91** 023628
[7] Olshanii M 1998 *Phys. Rev. Lett.* **81** 938
[8] Petrov D S and Shlyapnikov G V 2001 *Phys. Rev. A* **64** 012706
[9] Griesmaier A 2007 *J. Phys. B* **40** R91
[10] Baranov M A 2008 *Phys. Rep.* **464** 71
[11] Harrison W A 1970 *Solid State Theory* (New York: McGraw-Hill)
[12] Slater J C 1974 *The Self-Consistent Field for Molecules and Solids* (New York: McGraw-Hill)
[13] Yukalov V I 2012 *Laser Phys.* **22** 1070
[14] Yukalov V I 2012 *Laser Phys.* **22** 1145
[15] Silvestrelly P L, Marzari N, Vanderbilt D, and Parrinello M 1998 *Solid State Commun.* **107** 7
[16] Marzari N, Mostofi A A, Yates J R, Souza I, and Vanderbilt D 2012 *Rev. Mod. Phys.* **84** 1419
[17] Ziegler K and Kogan E 2011 *Eur. Phys. Lett.* **95** 36003
[18] Yukalov V I 1989 *Physica A* **155** 519
[19] Yukalov V I 2013 *Laser Phys.* **23** 062001
[20] Ziegler K, Berman O L, and Kezerashvili R Y 2015 arXiv:1508.03904
[21] Lindemann F A 1910 *Z. Phys.* **11** 609
[22] Greiner M, Mandel O, Eslinger T, Hänsch T V, and Bloch I 2002 *Nature* **415** 39
[23] Guyer R 1969 *Solid State Phys.* **23** 413
[24] Srivastava G P 1990 *The Physics of Phonons* New York: Taylor and Francis
[25] Birman J L, Nazmitdinov R G, and Yukalov V I 2013 *Phys. Rep.* **526** 1

Spatially-partitioned many-body vortices

S Klaiman[1] **and O E Alon**[2]

[1] Theoretische Chemie, Heidelberg University, Heidelberg, Germany
[2] Department of Physics, University of Haifa at Oranim, Tivon, Israel

E-mail: ofir@research.haifa.ac.il

Abstract. A vortex in Bose-Einstein condensates is a localized object which looks much like a tiny tornado storm. It is well described by mean-field theory. In the present work we go beyond the current paradigm and introduce many-body vortices. These are made of *spatially-partitioned* clouds, carry definite total angular momentum, and are *fragmented* rather than condensed objects which can only be described beyond mean-field theory. A phase diagram based on a mean-field model assists in predicting the parameters where many-body vortices occur. Implications are briefly discussed.

1. Introduction

A vortex is a fundamental object in many branches of physics and engineering. Vortices have been created in experiments with Bose-Einstein condensates (BECs) made of ultracold quantum gases [1-4]. There are ample theoretical works discussing vortices, variants thereof, and their properties in BECs, see, e.g., [5-22]. For a comprehensive review see [23] and references therein.

BECs are made of many identical particles. As such, they are subject to the Schrödinger equation and are described by a *many-body* wavefunction. In BECs, a vortex is commonly described by a complex order parameter (single-particle function) which is the solution of the non-linear Schrödinger equation, or, as it is explicitly known to in the field, Gross-Pitaevskii equation [23-26]. This implies that a vortex is a quantum object for which the many-body wavefunction is well approximated by the product

$$\phi(\mathbf{r}) = e^{il\varphi} f(r) \quad \longrightarrow \quad \Phi(\mathbf{r}_1, \ldots, \mathbf{r}_N) = \Pi_{j=1}^{N} \phi(\mathbf{r}_j). \tag{1}$$

In (1), l is an integer which equals to the angular momentum carried by each particle and N is the number of particles. For simplicity, the coordinate $\mathbf{r} = (r, \varphi)$ is in two spatial dimensions. Clearly, the wavefunction Φ is an eigenfunction of the many-particle angular-momentum operator $\hat{L}_z$ with the total angular momentum $L = Nl$.

In the present work we go beyond the current paradigm for vortices and introduce vortices of the many-body kind, or many-body vortices (MBVs) as they shall be referred to below. MBVs are objects made of *spatially-partitioned* clouds and carry definite total angular momentum. Two kinds of MBVs are discussed. MBVs of the first kind are at the global minimum of the energy for states with definite total angular momentum and where all the bosons carry the same angular momentum. MBVs of the second kind are excited states in which macroscopic fractions of bosons carry different angular momenta. We devise a phase diagram for MBVs based on a simple model, and use it to predict and analyze the parameters of their occurrences. MBVs

Journal of Physics: Conference Series **691** (2016) 012015 doi:10.1088/1742-6596/691/1/012015

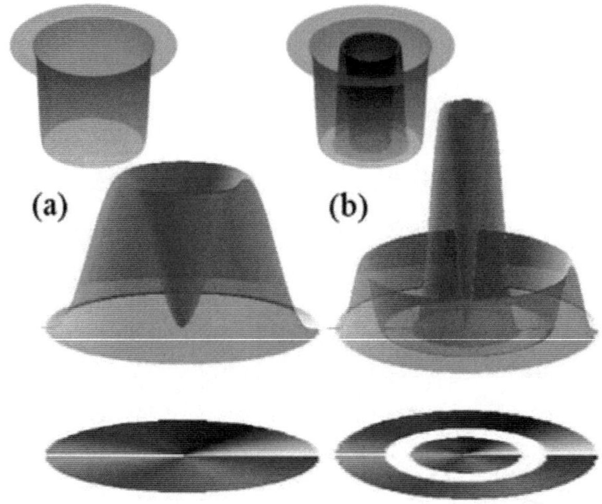

Figure 1. (a) The standard vortex in a two-dimensional circular trap potential. The bosons carry the same angular momentum, $L/N = 1$. (b) Many-body vortex of the first kind in a circular trap split by a radial barrier. Although the bosons carry the same angular momentum, $L/N = 1$, the wavefunction is fragmented, many-body in nature and composed of an inner and outer clouds. See text for more details. Shown are the densities, the bottom panels plot the phase. The insets (in gray) depict the trapping potentials.

are *fragmented* rather than condensed objects, and can only be described beyond mean-field theory. Fragmentation is a many-body property of BECs derived from the eigenvalues of the reduced one-body density matrix [27-29]. If there is only one macroscopic eigenvalue the system is referred to as condensed, whereas if there are two or more such eigenvalues the system is said to be fragmented. Fragmentation of BECs has drawn much attention, see, e.g., [30-43].

2. Setup and condensed vortices

Consider a repulsive trapped BEC made of $N = 100$ bosons in two spatial dimensions. The many-boson Hamiltonian is given by $\hat{H}(\mathbf{r}_1, \ldots, \mathbf{r}_N) = \sum_{j=1}^{N}[\hat{T}(\mathbf{r}_j) + V(\mathbf{r}_j)] + \sum_{j<k} W(\mathbf{r}_j - \mathbf{r}_k)$. $\hat{T}(\mathbf{r})$ is the kinetic-energy operator, $V(\mathbf{r})$ the trap potential, and $W(\mathbf{r} - \mathbf{r}')$ the interparticle interaction. The short-range interaction between the bosons is modeled by a Gaussian function [44, 45] $W(\mathbf{r}-\mathbf{r}') = \lambda_0 \frac{e^{-(\mathbf{r}-\mathbf{r}')/2\sigma^2}}{2\pi\sigma^2}$ with a width $\sigma = 0.25$. The interaction parameter is $\lambda_0 = 0.02$. We employ dimensionless units, which are readily obtained by dividing the Hamiltonian by $\frac{\hbar^2}{md^2}$, where m is the mass of a boson and d is a length scale. Realistic experimental parameters for the systems studied below are given in [46].

We begin with the standard vortex in the potential of Fig. 1a, computed with the Gross-Pitaevskii wavefunction (1) for angular momentum $L/N = 1$. Here $V(\mathbf{r}) = V_c(\mathbf{r})$, where $V_c(\mathbf{r}) = \{200e^{-(r-r_c)^4/2}, r \leq r_c = 9; 200, r > r_c\}$ is a flat trap which has the shape of a "crater" of radius r_c. We have chosen a flat potential $V_c(\mathbf{r})$ in order to allow the BEC to fill the full area. This can readily be seen in Fig. 1a.

The central question we address in this work is how to go beyond the standard mean-field vortex (1)? The answer, as we shall see below, has to do with partitioning the BEC in space into two parts, thereby introducing many-body effects. In as much as splitting a BEC into two, left–and–right clouds leads to many-body physics and even fragmentation, see, e.g., [32], we hereby propose to analogously spatially-split a vortex. To this end, the BEC is now placed in

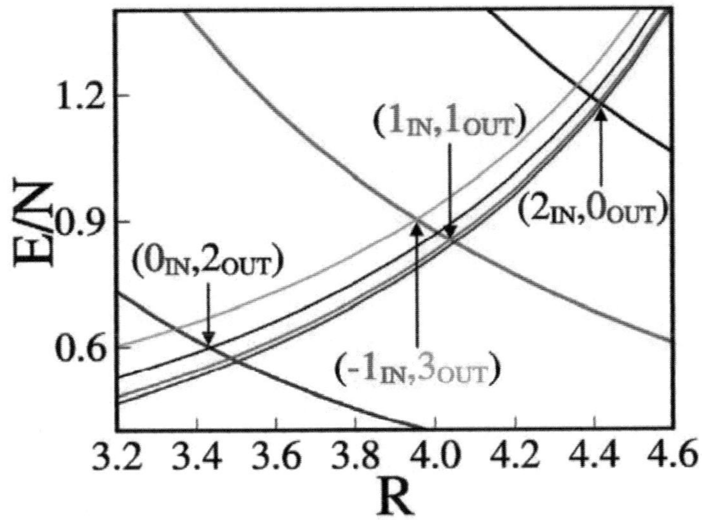

Figure 2. Ground-state phase diagram of many-body vortices in a circular trap. A model based on the Gross-Pitaevski mean-field theory is employed. The number of bosons is $N = 100$ and the interaction parameter $\lambda_0 = 0.02$. The energies of states in the inner (IN) disk and outer (OUT) annulus parts of the trap are, respectively, decreasing and increasing as a function of the radius R. The four (three) curves represent, from bottom to top, states in the OUT (IN) part with total angular momentum $L/N = 0$ (solid blue), 1 (solid red), 2 (solid black), and $L/N = 3$ (solid orange). States with angular momentum $\pm L/N$ are degenerate. The crossings of the IN and OUT curves in the mean-field diagram marks the parameters where the MBVs occur. For these radii, maximal fragmentation on the many-body level is encountered. See text for more details. All quantities are dimensionless.

the radially-split trap shown in Fig. 1b. The explicit form of the one-body potential is given by $V(\mathbf{r}) = V_c(\mathbf{r}) + V_b(\mathbf{r})$, where $V_b(\mathbf{r}) = 200e^{-2(r-R)^4}$ is a ring-shaped radial barrier of radius R. The BEC has now, in principle, two spatially-partitioned areas to occupy.

The split potential $V(\mathbf{r})$ can be considered as made of two distinct parts, an inner disk and an outer annulus, separated by a radial barrier centered at $r = R$. In the forthcoming analysis, this observation serves in forming a model. Obviously, the energy of the system of total angular momentum $L/N = 1$ changes with the barrier's radius R. Furthermore, the location of the vortex in the split potential also changes with R. To facilitate our discussion, the energies of a vortex in the inner disk and outer annulus as a function of R are plotted in Fig. 2. These energies are computed numerically using the ansatz (1), i.e., by solving the Gross-Pitaevskii mean-field equation in the disk and, separately, in the annulus [48]. For larger R, the system is located in the inner disk, and looks much like the vortex shown in Fig. 1a. For smaller R, it is energetically favorable for the system to be located in the outer annulus. Now, the system's lowest state of definite angular momentum $L/N = 1$ is a vortex state in the outer annulus rather than in the inner disk.

The inner disk and outer annulus become energetically equivalent at the barrier's radius $R = 4.04$, see the crossing point marked as $(1_{\mathrm{IN}}, 1_{\mathrm{OUT}})$ in Fig. 2. One might expect then the BEC to occupy the entire trap and, consequently, for many-body effects to set in. We are now in a need for a suitable theoretical framework to cope with such many-body effects.

Journal of Physics: Conference Series **691** (2016) 012015 doi:10.1088/1742-6596/691/1/012015

3. Fragmented vortices

3.1. Theoretical and computational framework

To go beyond the standard paradigm for vortices (1), let us suppose that the system of many identical bosons occupies not one but two, orthogonal one-particle functions, $\phi_1(\mathbf{r})$ and $\phi_2(\mathbf{r})$. These functions carry the angular momenta l_1 and l_2, respectively. The system's most general wavefunction takes on the form

$$\phi_1(\mathbf{r}) = e^{il_1\varphi} f_1(r), \quad \phi_2(\mathbf{r}) = e^{il_2\varphi} f_2(r), \quad \longrightarrow$$
$$|\Psi\rangle = \sum_{n_1=0}^{N} C_{n_1} |n_1, n_2\rangle, \quad n_1 + n_2 = N. \tag{2}$$

In (2), the C_{n_1} are expansion coefficients and $|n_1, n_2\rangle$ are Fock states (permanents) with n_1 bosons in the one-particle function $\phi_1(\mathbf{r})$ and n_2 bosons in $\phi_2(\mathbf{r})$. For the wavefunction $|\Psi\rangle$ to be an eigenfunction of the many-particle angular-momentum operator $\hat{L}_z$, let $l_1 = l_2 = l$. In this case, there is in principle no restriction on the expansion coefficients C_{n_1}. Namely, any distribution of the bosons between the two one-particle functions $\phi_1(\mathbf{r})$ and $\phi_2(\mathbf{r})$ has the total angular momentum $L/N = l$. Let us investigate now the BEC in the split trap of Fig. 1b with the many-body ansatz (2).

To calculate the optimal many-body wavefunction (2) of the BEC, namely to determine the one-particle functions $\phi_1(\mathbf{r})$ and $\phi_2(\mathbf{r})$ and expansion coefficients C_{n_1}, we employ the multiconfigurational time-dependent Hartree for bosons (MCTDHB) method [51-53]. The MCTDHB method has been shown to produce accurate many-body solutions in various applications, see, e.g., [54-66]. Explicitly, we employ imaginary-time propagation within the the recursive MCTDHB (R-MCTDHB) [67] and MCTDHB [68] software packages, augmented by angular-momentum projection operators. For our investigations, a square box of size $[-12, 12) \times [-12, 12)$ and spatial grid of size 128×128 are used and found to converge the results to the accuracy given below.

3.2. Many-body vortex of the first kind

As a first application, we have computed with the many-body ansaz (2) the standard vortex and found the mean-field description (1) to be valid in the flat potential trap of Fig. 1a. Explicitly, the BEC is condensed to more than 99.9%. We then turned to the split potential of Fig. 1b, and to the crossing point at radius $R = 4.04$ marked as $(1_{\text{IN}}, 1_{\text{OUT}})$ in Fig. 2. Indeed, as was expected, the density is spread now in the disk *and* the annulus, see Fig. 1b, while the entire BEC has one and the same phase. We remind that the total angular momentum is $L/N = 1$.

The object shown in Fig. 1b has appealing properties and deserves a deeper examination. First, its energetics. Of all states carrying definite total angular momentum $L/N = 1$, it has the lowest energy in the trap. Second, it is an 'extended' object which combines two spatially-partitioned parts. The inner part looks much like the standard vortex, the outer part as an annulus vortex state. We stress that this is one and the same quantum object with a single many-body wavefunction! Third, analysis of the many-body wavefunction reveals that the object in Fig. 1b is actually fragmented rather than condensed, with natural orbitals' occupations of $n_1 = 50.45\%$ and $n_2 = 49.55\%$. Fourth, the two fragments are coupled, see for more details the appendix. Summarizing all these properties, we shall henceforth refer to this novel quantum object as an MBV of the first kind.

3.3. Many-body vortices of the second kind

The idea of an MBV – a spatially-partitioned and fragmented quantum object, carrying macroscopic definite total angular momentum – can be further pursued. This is because now a degree-of-freedom to distribute the angular momentum between two parts is available. Let us come back to the many-body wavefunction (2). The second possibility to the requirement from $|\Psi\rangle$ to carry definite total angular momentum L can be met for $l_1 \neq l_2$. This dictates that only a single term in (2) can have a non-vanishing expansion coefficient C_{n_1}, say the term with $n_1 = n$. In that case the total angular momentum of $|\Psi\rangle$ is $L/N = (n/N)l_1 + (1-n/N)l_2$. This is a general argument. We shall now demonstrate how it manifests itself with a BEC in the split trap of Fig. 1b.

We return to the model of the split trap in terms of an inner disk and outer annulus, and compute the states of angular momentum $L/N = 0, 1, 2, 3$ using the Gross-Pitaevskii wavefunction (1). Fig. 2 plots the energies of these states as a function of the radius R. As the radius increases, the energies of states in the disk decrease, and those of the states in the annulus increase. States with higher angular momentum L/N are higher in energy, and those with $\pm L/N$ are degenerate.

There are many intersection points of the Gross-Pitaevskii energy curves for the inner and outer states. We concentrate on three such intersection points, marked as $(0_{\text{IN}}, 2_{\text{OUT}})$, $(-1_{\text{IN}}, 3_{\text{OUT}})$, and $(2_{\text{IN}}, 0_{\text{OUT}})$ in Fig. 2. These occur at completely different radii, $R = 3.43$, 3.955, and 4.42, respectively. What is common to these three intersection points? First, by definition, that the Gross-Pitaevskii energy of a BEC in the inner disk is equal to that in the outer annulus. Second, that the BEC in the inner part has *different* angular momentum than the BEC in the outer part. And third, that the total angular momentum of a fragmented state equally distributed between such states would be $L/N = 1$.

With the above preparatory analysis at hand, we have employed the many-body ansatz (2) at the respective radii $R = 3.43$, 3.955, and 4.42. We have searched for the states lowest-in-energy when allowing the bosons to be distributed between two one-particle functions, $\phi_1(\mathbf{r})$ and $\phi_2(\mathbf{r})$, with angular momenta $l_1 = 0$ and $l_2 = 2$, $l_1 = -1$ and $l_2 = 3$, and $l_1 = 2$ and $l_2 = 0$, respectively. In all three cases, as expected from the conservation of angular momentum, a single term of the many-body expansion (2) emerges. Furthermore, in all three cases, in line with the predictions of the phase diagram in Fig. 2, the term lowest-in-energy is the one with roughly $n_1 = n_2 = 50\%$ natural orbitals' occupations (the respective numerical values are: $n_2 = 48.0\%$, $n_1 = 52.0\%$ for the orbitals with $l_1 = 0$ and $l_2 = 2$; $n_1 = 50.0\%$, $n_2 = 50.0\%$ for the orbitals with $l_1 = -1$ and $l_2 = 3$; and $n_2 = 47.0\%$, $n_1 = 53.0\%$ for the orbitals with $l_1 = 2$ and $l_2 = 0$). Fig. 3 depicts the densities of these novel objects, all spatially-partitioned, macroscopically fragmented, and carrying definite total angular momentum of essentially $L/N = 1$. Note the phases of each object which are distinct in the inner and outer parts. Nonetheless, each of them is a quantum object with a single many-particle wavefunction, see the appendix for more details. We shall henceforth refer to these quantum objects as MBVs of the second kind.

3.4. Discussion

A comparative discussion on the energetics and stability of MBVs is in place. The MBVs of the first kind (Fig. 1b) are at the global minimum of energy; They correspond to a fragmented ground state of a BEC with total definite angular momentum L/N in the split trap of Fig. 1b. The MBVs of the second kind (Fig. 3) are macroscopically fragmented excited states [35]. Consider the MBV associated with the crossing point $(0_{\text{IN}}, 2_{\text{OUT}})$ in the phase diagram of Fig. 2. The global minimum of energy for the corresponding barrier's radius $R = 3.43$ and angular momentum $L/N = 1$ is the annulus vortex state in the outer part of the trap. The wavefunctions of the MBV and annulus vortex state are, of course, orthogonal. Furthermore, if trap imperfections are to introduce coupling between the two states, we expect the tunneling of $N/2$ of the particles from the inner disk to the outer annulus, along with the exchange of angular momentum between the two parts, to be a very slow process. This would render even in case of trap imperfections the MBV a long-lived quantum state. The same analysis holds for the MBV associated with

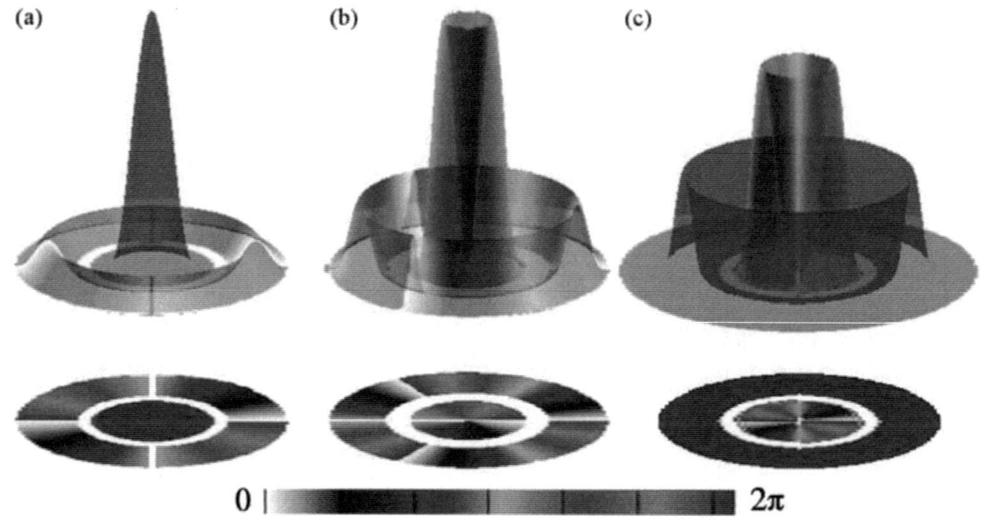

Figure 3. Many-body vortices of the second kind. The total angular momentum is essentially $L/N = 1$. The angular momenta l_1 and l_2 of the fragments in the disk and annulus parts are, respectively: (a) $l_1 = 0$ and $l_2 = 2$; (b) $l_1 = -1$ and $l_2 = 3$; and (c) $l_1 = 2$ and $l_2 = 0$. Shown are the densities, the bottom panels plot the phases. Observe the different phases in the inner and outer parts. See text for more details.

the crossing point $(-1_{\text{IN}}, 3_{\text{OUT}})$ at barrier's radius 3.955. Finally, for the MBV associated with the crossing point $(2_{\text{IN}}, 0_{\text{OUT}})$ at barrier's radius 4.42, the global minimum of energy for angular momentum $L/N = 1$ is a vortex state in the inner disk part of the trap. Here as well, in case of trap imperfections we expect the tunneling of $N/2$ particles from the outer annulus to the inner disk to be a slow process, rendering thereby the MBV a long-lived quantum state.

4. Concluding Remarks

In conclusion, in the present work we go beyond the standard paradigm of a vortex in Bose-Einstein condensates, which is a localized object well described by mean-field theory, and introduce vortices of the many-body kind. The MBVs are made of *spatially-partitioned* clouds, carry definite total angular momentum, and are *fragmented* rather than condensed quantum objects, describable only beyond mean-field theory. Two kinds of MBVs are discussed. MBVs of the first kind are at the global minimum of the energy for states of definite total angular momentum and comprised of bosons carrying the same angular momentum. MBVs of the second kind are fragmented excited states in which macroscopic fractions of bosons carry different angular momenta. A phase diagram based on the solutions of the Gross-Pitaevskii equation in the inner and outer parts of the trap is instrumental in predicting the parameters where MBVs occur. The most recent experiment [69] utilizing a circular trap split by a radial barrier like in Fig. 1 encourages us to anticipated that many-body vortices will be further studied theoretically and searched for experimentally.

Acknowledgments

We thank Alexej Streltsov and Lorenz Cederbaum for discussions. Partial financial support by the DFG is acknowledged. Computation time on the Cray XE6 system Hermit and the NEC Nehalem cluster Laki at the HLRS, and the bwGRiD cluster are gratefully acknowledged. OEA acknowledges funding by the Israel Science Foundation (grant No. 600/15).

Appendix A. Properties of spatially-partitioned many-body vortices

We collect in this appendix further properties of the spatially-partitioned many-body vortices. The information is illustrated in figures. Figures A1 and A2 discuss the natural orbitals of spatially-partitioned many-body vortices of both kinds. Figure A3 compares the second-order correlation functions [70, 71] of the standard, mean-field vortex and the spatially-partitioned many-body vortex of the first kind. Finally, Fig. A4 depicts the so-called pathway from condensation to fragmentation [37, 72] between the last two objects.

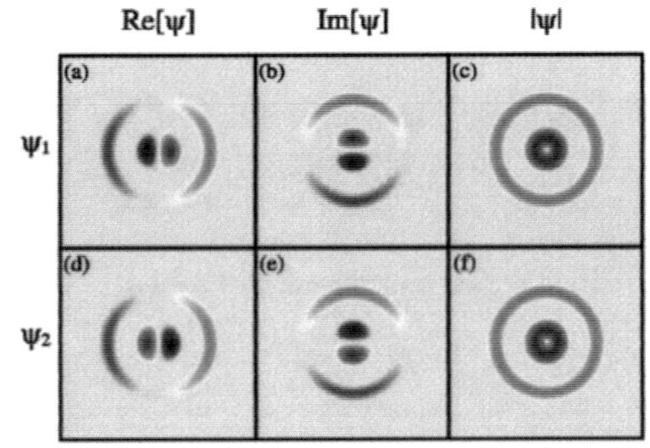

Figure A1. The natural orbitals of the many-body vortex of the first kind delocalize over the inner disk and outer annulus regions of the trap. Shown are the real, imaginary, and absolute value of the natural orbitals. Each fragment carries the same angular momentum per particle, $l_1 = l_2 = 1$. All quantities are dimensionless.

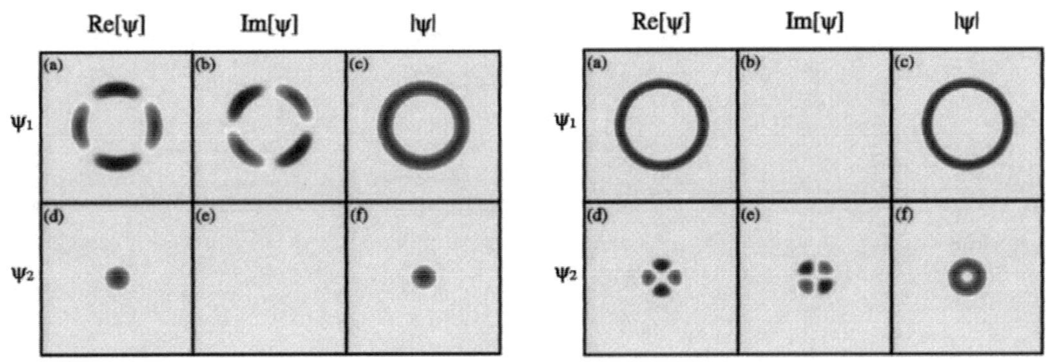

Figure A2. The natural orbitals of the many-body vortex of the second kind (left: $l_1 = 0$, $l_2 = 2$; right: $l_1 = 2$, $l_2 = 0$) localize in the inner disk and outer annulus. Shown are the real, imaginary, and absolute value of the natural orbitals. All quantities are dimensionless.

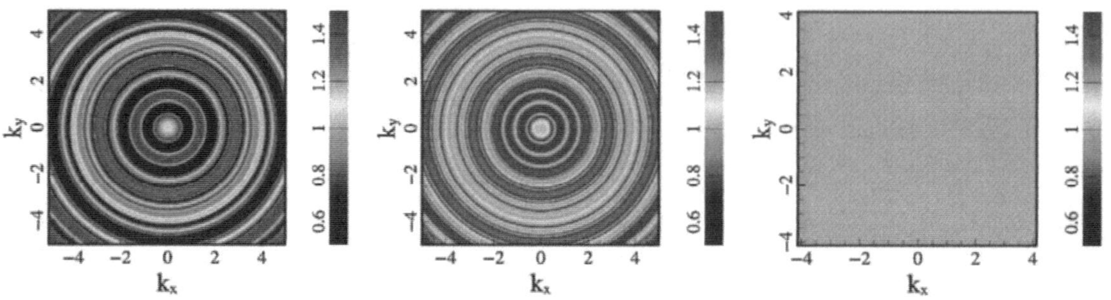

Figure A3. The second-order momentum correlation function $g^{(2)}(\mathbf{k},\mathbf{k}',\mathbf{k},\mathbf{k}')$ of the many-body vortex of the first kind. (left) Shown is the projection for $\mathbf{k}'=(2,2)$ and $\mathbf{k}=(k_x,k_y)$. (middle) Shown is the diagonal, $\mathbf{k}=\mathbf{k}'=(k_x,k_y)$. One can clearly see that there are significant two-body correlations. (right) For comparison, the second-order momentum correlation function for the standard (mean-field) vortex, which exhibits no two-body correlations, is displayed. All quantities are dimensionless.

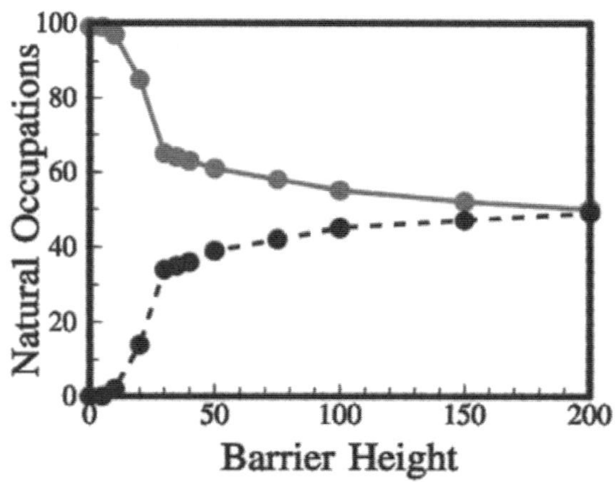

Figure A4. The ground state of the Bose-Einstein condensate in the circular trap as a function of the barrier height. All bosons carry the same angular momentum, $l_1=l_2=1$. Shown are the occupation numbers (in percents). The system gradually develops fragmentation with increasing barrier's height. All quantities are dimensionless.

References

[1] Matthews M R, Anderson B P, Haljan P C, Hall D S, Wieman C E and Cornell E A 1999 *Phys. Rev. Lett.* **83** 2498
[2] Madison K W, Chevy F, Wohlleben W and Dalibard J 2000 *Phys. Rev. Lett.* **84** 806
[3] Abo-Shaeer J R, Raman C, Vogels J M and Ketterle W 2001 *Science* **292** 476
[4] Raman C, Abo-Shaeer J R, Vogels J M, Xu K and Ketterle W 2001 *Phys. Rev. Lett.* **87** 210402
[5] Dalfovo F and Stringari S 1996 *Phys. Rev. A* **53** 2477
[6] Marzlin K P and Zhang W 1998 *Phys. Rev. A* **57** 4761
[7] Butts D A and Rokhsar D S 1999 *Nature (London)* **397** 327
[8] Cooper N R, Wilkin N K and Gunn J M F 2001 *Phys. Rev. Lett.* **87** 120405
[9] Tempere J, Devreese J T and Abraham E R I 2001 *Phys. Rev. A* **64** 023603

[10] Martikainen J P and Stoof H T C 2003 *Phys. Rev. A* **68** 013610
[11] Martikainen J P and Stoof H T C 2004 *Phys. Rev. A* **70** 013604
[12] Kapale K T and Dowling J P 2005 *Phys. Rev. Lett.* **95** 173601
[13] Kaurov V M and Kuklov A B 2005 *Phys. Rev. A* **71** 011601(R)
[14] Cozzini M, Jackson B and S. Stringari S 2006 *Phys. Rev. A* **73** 013603
[15] Wright K C, Leslie L S and Bigelow N P 2008 *Phys. Rev. A* **77** 041601(R)
[16] Romanovsky I, Yannouleas C and Landman U 2008 *Phys. Rev. A* **78** 011606(R)
[17] Dagnino D, Barberán N, Lewenstein M and Dalibard J 2009 *Nature Physics* **5** 431
[18] Brand J, Haigh T J and Zülicke U 2009 *Phys. Rev. A* **80** 011602(R)
[19] Salgueiro J R, Zacarés M, Michinel H and Ferrando A 2009 *Phys. Rev. A* **79** 033625
[20] Ramanathan A, Wright K C, Muniz S R, Zelan M, Hill W T, Lobb C J, Helmerson K, Phillips W D and Campbell G K 2011 *Phys. Rev. Lett.* **106** 130401
[21] Fialko O, Bradley A S and Brand J 2012 *Phys. Rev. Lett.* **108** 015301
[22] Cremon J C, Kavoulakis G M, Mottelson B R and Reimann S M 2013 *Phys. Rev. A* **87** 053615
[23] Fetter A L 2009 *Rev. Mod. Phys.* **81** 647
[24] Pitaevskii L and Stringari S 2003 *Bose-Einstein Condensation* (New York: Oxford University Press)
[25] Leggett A J 2008 *Quantum Liquids: Bose Condensation and Cooper Pairing in Condensed-Matter Systems* (New York: Oxford University Press)
[26] Pethick C J and Smith H 2008 *Bose-Einstein Condensation in Dilute Gases* 2nd ed (New York: Cambridge University Press)
[27] Löwdin P O 1955 *Phys. Rev.* **97** 1474
[28] Penrose O and Onsager L 1956 *Phys. Rev.* **104** 576
[29] Coleman A J and Yukalov V I 2000 *Reduced Density Matrices: Coulson's Challenge (Lectures Notes in Chemistry vol 72)* (Berlin: Springer)
[30] Nozières P and Saint James D 1982 *J. Phys. (Paris)* **43** 1133
[31] Nozières P 1996 *Bose-Einstein Condensation* ed Griffin A, Snoke D W and Stringari S (Cambridge: Cambridge University Press)
[32] Spekkens R W and Sipe J E 1999 *Phys. Rev. A* **59** 3868
[33] Pitaevskii L and Stringari S 1999 *Phys. Rev. Lett.* **83** 4237
[34] Cederbaum L S and Streltsov A I 2003 *Phys. Lett. A* **318** 564
[35] Cederbaum L S and Streltsov A I 2004 *Phys. Rev. A* **70** 023610
[36] Klaiman S, Moiseyev N and Cederbaum L S 2006 *Phys. Rev. A* **73** 013622
[37] Streltsov A I, Alon O E and Cederbaum L S 2006 *Phys. Rev. A* **73** 063626
[38] Mueller E J, Ho T L, Ueda M and Baym G 2006 *Phys. Rev. A* **74** 033612
[39] Bader P and Fischer U R 2009 *Phys. Rev. Lett.* **103** 060402
[40] Fischer U R and Bader P 2010 *Phys. Rev. A* **82** 013607
[41] Zhou Q and Cui X 2013 *Phys. Rev. Lett.* **110** 140407
[42] Kawaguchi Y 2014 *Phys. Rev. A* **89** 033627
[43] Song S W, Zhang Y C, Zhao H, Wang X and Liu W M 2014 *Phys. Rev. A* **89** 063613
[44] Christensson J, Forssén C, Åberg S and Reimann S M 2009 *Phys. Rev. A* **79** 012707
[45] Doganov R A, Klaiman S, Alon O E, Streltsov A I and Cederbaum L S 2013 *Phys. Rev. A* **87** 033631
[46] As a concrete example we choose $d = 2\mu$m and ^{87}Rb atom as a boson. We point out that other realistic choices can be made. The unit of energy $\frac{\hbar^2}{md^2}$ corresponds to 29.1Hz. The relation between the dimensionless interaction parameter λ_0, and the scattering length a_s and the transverse confinement $l_z = \sqrt{\frac{\hbar}{m\omega_z}}$ along the z axis is given by $\lambda_0 = 2\sqrt{2\pi}\frac{a_s}{l_z}$, where [47] is used. For $\omega_z = 2\pi \cdot 897$Hz (the aspect ratio is $\frac{l_z}{2r_c} = \frac{1}{100}$) and $a_s = 1.44$nm, we get $\lambda_0 = 0.02$.
[47] Petrov D S, Holzmann M and Shlyapnikov G V 2000 *Phys. Rev. Lett.* **84** 2551
[48] For noninteracting particles the energies of a single particle carrying angular momentum l in the disk and annulus are known [49, 50], and can be used to predict for which radius R the disk and annulus are energetically equivalent. Since the interaction plays a central role in the creation of MBVs, we will not exploit this analysis here.
[49] Kuttler J R and Sigillito V G 1984 *SIAM Rev.* **26** 163
[50] Alvarado C and Amore P 2011 *J. Math. Phys.* **52** 063516
[51] Streltsov A I, Alon O E and Cederbaum L S 2007 *Phys. Rev. Lett.* **99** 030402
[52] Alon O E, Streltsov A I and Cederbaum L S 2008 *Phys. Rev. A* **77** 033613
[53] Proukakis N P, Gardiner S A, Davis M J and Szymanska M H (ed) 2013 *Quantum Gases: Finite Temperature and Non-Equilibrium Dynamics (Cold Atoms Series vol 1)* (London: Imperial College Press)
[54] Sakmann K, Streltsov A I, Alon O E and Cederbaum L S 2009 *Phys. Rev. Lett.* **103** 220601

[55] Grond J, Schmiedmayer J and Hohenester U 2009 *Phys. Rev. A* **79** 021603(R)
[56] Sakmann K, Streltsov A I, Alon O E and Cederbaum L S 2010 *Phys. Rev. A* **82** 013620
[57] Lode A U J, Sakmann K, Alon O E, Cederbaum L S and Streltsov A I 2012 *Phys. Rev. A* **86** 063606
[58] Heimsoth M, Hochstuhl D, Creffield C E, Carr L D and Sols F 2013 *New J. Phys.* **15** 103006
[59] Streltsov A I 2013 *Phys. Rev. A* **88** 041602(R)
[60] Streltsova O I, Alon O E, Cederbaum L S and Streltsov A I 2014 *Phys. Rev. A* **89** 061602(R)
[61] Březinová I, Burgdörfer J, Lode A U J, Streltsov A I, Cederbaum L S, Alon O E, Collins L A and Schneider B I 2014 *J. Phys.: Conf. Ser.* **488** 012032
[62] Lode A U J, Klaiman S, Alon O E, Streltsov A I and Cederbaum L S 2014 *Phys. Rev. A* **89** 053620
[63] Klaiman S, Lode A U J, Streltsov A I, Cederbaum L S and Alon O E 2014 *Phys. Rev. A* **90** 043620
[64] Alon O E 2015 *J. Phys.: Conf. Ser.* **594** 012039
[65] Klaiman S and Alon O E 2015 *Phys. Rev. A* **91** 063613
[66] Beinke R, Klaiman S, Cederbaum L S, Streltsov A I and Alon O E *Phys. Rev. A* **92** 043627
[67] Lode A U J and Tsatsos M C 2014 The Recursive Multiconfigurational Time-Dependent Hartree for Bosons Package http://ultracold.org http://rmctdhb.org http://schroedinger.org, version 1.0
[68] Streltsov A I, Sakmann K, Lode A U J, Alon O E and Cederbaum L S 2013 The Multiconfigurational Time-Dependent Hartree for Bosons Package http://MCTDHB.org, version 2.3, Heidelberg
[69] Corman L, Chomaz L, Bienaimé T, Desbuquois R, Weitenberg C, Nascimbène S, Dalibard J and Beugnon J 2014 *Phys. Rev. Lett.* **113** 135302
[70] Naraschewski M and Glauber R J 1999 *Phys. Rev. A* **59** 4595
[71] Sakmann K, Streltsov A I, Alon O E and Cederbaum L S 2008 *Phys. Rev. A* **78** 023615
[72] Alon O E and Cederbaum L S 2005 *Phys. Rev. Lett.* **95** 140402

Superfluidity and BEC in a Model of Interacting Bosons in a Random Potential

Martin Könenberg[1], Thomas Moser[2], Robert Seiringer[3] and Jakob Yngvason[4]

[1]Department of Mathematics and Statistics, Memorial University St. John's, NL, Canada; present address: Fachbereich Mathematik, Universität Stuttgart, Germany,
[2]Faculty of Physics, University of Vienna, Boltzmanngasse 5, 1090 Vienna, Austria; present address: IST Austria, Am Campus 1, 3400 Klosterneuburg, Austria,
[3]IST Austria, Am Campus 1, 3400 Klosterneuburg, Austria,
[4]Faculty of Physics, University of Vienna, Boltzmanngasse 5, 1090 Vienna, Austria

E-mail: [1]mkonenberg@mun.ca, [2]tmoser@ist.ac.at, [3]robert.seiringer@ist.ac.at, [4]jakob.yngvason@univie.ac.at

Abstract.
We report on a mathematically rigorous analysis of the superfluid properties of a Bose-Einstein condensate in the many-body ground state of a one-dimensional model of interacting bosons in a random potential.

1. Introduction

The relation between Bose-Einstein condensation (BEC) and superfluidity has been an intriguing question since the discovery of these quantum phenomena in the 1920's and 30's. Although both often occur simultaneously, they are definitely not the same thing: Liquid helium 4 is almost a complete superfluid near absolute zero while the BEC fraction is less than 10 % [1]. Also, a one-dimensional hard-core Bose gas (Tonks-Girardeau gas) is an example of a complete superfluid where BEC is absent [2]. On the other hand there is both theoretical and numerical evidence that a random external potential may destroy superfluidity while BEC prevails [3, 4, 5, 6]. A mathematically study of this assertion is, however, hampered by the fact that rigorous proofs of BEC for systems of interacting particles are notoriously difficult and have only been achieved in a few special cases, see in particular [7, 8, 9, 10]. Also, rigorous results on the effects of randomness on interacting many-body systems are scarce, contrary to the situation for single particle random Schrödinger operators where the literature is wast. We refer to [11, 12, 13] for lists of representative references on these subjects.

The present contribution summarizes results obtained in [12] (see also [11, 14]) on the effects of a random potential on the superfluid behavior of an interacting Bose gas in a one-dimensional model. Our main result is that in a certain parameter regime the superfluid fraction can be arbitrarily small while BEC is complete. In another regime there is both complete BEC and complete superfluidity, despite strong disorder. In the course of the proof we derive a general formula for the superfluid fraction in terms of the wave function of the condensate. Our results are based on this formula combined with energy estimates.

2. Concepts

Bose Einstein condensation in a many-body quantum state means, by definition, macroscopic occupation of some single one-particle state in the limit of large particles numbers [15]. Mathematically this is expressed through the *one-particle density matrix* of the (pure or mixed) many-body state $\langle \cdot \rangle$, defined as

$$\rho^{(1)}(x, x') = \langle a^\dagger(x) a(x') \rangle = \sum_{i=0}^{\infty} N_i \, \psi_i(x)^* \psi_i(x'), \tag{1}$$

where the eigenvalues $N_0 \geq N_1 \geq \cdots$, $\sum_{N_i} = N$, are the occupation numbers of the (orthonormal) natural orbitals $\{\psi_i\}$ that are the eigenfunctions of $\rho^{(1)}$.

The *condensate fraction* is the relative occupation, N_0/N, of the orbital ψ_0 with the highest eigenvalue, N_0. BEC means that $N_0/N = O(1)$ in the sense that $N_0 \geq cN$ with some $c > 0$ as $N \to \infty$. The function ψ_0 is called the *wave function of the condensate.*

It is important to note that the above definition of BEC requires some specified dependence of the parameters of the Hamiltonian and the many-body state on N as $N \to \infty$. The standard thermodynamic limit means that the volume increases proportionally to the particle number so that the particle density is kept constant. For experiments with cold atomic gases in traps another limit is relevant, however, and easier to handle but still far from trivial. This is the Gross-Pitaveskii (GP) limit [16], which can be regarded as a combination of a thermodynamic and a weak coupling limit. In this limit, which is the one considered in the present contribution, the ground state energy per particle is comparable to the energy gap of the free Hamiltonian in the trap.

Concerning the definition of *superfluidity* we note that this concept can mean different things [17]:

- Flow at a finite velocity without friction (non-equilibrium phenomenon)
- Non-classical response to an infinitesimal boost or rotation (equilibrium phenomenon)

Here we consider superfluidity only in the sense of the second definition. The *superfluid mass density* ρ_s (at rest) is then defined through the response of the free energy (or, at $T = 0$, the ground state energy) to a small boost v:

$$F(v) = F(0) + \tfrac{1}{2} mN \, (\rho_s/\rho) v^2 + o(v^2). \tag{2}$$

The boost is mathematically implemented by the substitution

$$p_i \to p_i - mv \tag{3}$$

in the Hamiltonian, assuming *periodic boundary conditions* in the direction $\vec{e}$ of the boost with period Λ, say. Experimentally, the boost can conveniently be realized in a thin tube bent to a circular container that is brought into slow rotation. The superfluid fraction can be derived from the moment of inertia of the system.

Equivalently one can consider the original Hamiltonian in a box (interval) but with *twisted boundary conditions*:

$$\Psi(\ldots, x_{i-1}, \Lambda\vec{e}, x_i, \ldots) = e^{-i\varphi} \Psi(\ldots, x_{i-1}, 0, x_i, \ldots) \tag{4}$$

with $\varphi = mv\Lambda/\hbar$.

In the following we discuss a simple model of interacting particles where the effect of a random potential on BEC and superfluidity can be investigated mathematically.

Journal of Physics: Conference Series **691** (2016) 012016 doi:10.1088/1742-6596/691/1/012016

3. The Model

As in [11, 12] the model is a one-dimensional gas of bosons with contact interaction (Lieb-Liniger model [18]) on the unit interval but with an additional external random potential V_ω. The Hamiltonian on the Hilbert space $L^2([0,1],dz)^{\otimes^N_{\mathrm{symm}}}$ is

$$H = \sum_{i=1}^{N} \left(-\partial_{z_i}^2 + V_\omega(z_i)\right) + \frac{\gamma}{N}\sum_{i<j}\delta(z_i - z_j) \tag{5}$$

with a coupling constant $\gamma \geq 0$ and *periodic* boundary conditions. This can also be regarded as a model of a gas on a ring (or a thin tube bent to a circle) of radius $1/2\pi$. Units have been chosen so that $\hbar = 1$ and the mass is $m = \frac{1}{2}$.

The random potential will be taken to be

$$V_\omega(z) = \sigma \sum \delta(z - z_j^\omega) \tag{6}$$

with $\sigma \geq 0$ independent of the random sample ω while the point obstacles $\{z_j^\omega\}$ are Poisson distributed with density $\nu \gg 1$. i.e., their mean distance is $\nu^{-1} \ll 1$.

Besides N, the model has three parameters:

- γ: Strength of the interaction
- ν: Density of obstacles
- σ: Strength of the random potential

In the $N \to \infty$ limit the ground state energy and the wave function of the condensate is described by a Gross-Pitaevskii (GP) energy functional (8) as discussed further below.

In [11] it was proved that the ground state energy becomes deterministic, i.e. almost surely independent of ω, if the parameters satisfy

$$\nu \gg 1, \quad \gamma \gg \frac{\nu}{(\ln\nu)^2}, \quad \sigma \gg \frac{\nu}{1 + \ln(1 + \nu^2/\gamma)}. \tag{7}$$

We shall refer to these conditions as the *standard conditions* and assume them throughout.

4. The Main Results

Our main results about the model (5) are as follows:

1. *In the whole parameter range there is complete BEC in the ground state.*
2. *If $\gamma \lesssim \nu^2$ the superfluid fraction is arbitrarily small, i.e., it goes to zero under the standard conditions* (7).
3. *The same holds for $\nu^2 \ll \gamma \ll \nu^4$ provided $\sigma \gg (\gamma/\nu^2)^2\gamma^{1/2}$.*
4. *If $\gamma \gg (\sigma\nu)^2$ there is complete superfluidity, i.e., the superfluid fraction tends to 1.*

These results are illustrated in Fig. 1. It should be noted that our results describe asymptotic properties for large values of the parameters. The boundaries of the red and green areas in the figure should therefore not be interpreted as sharp phase boundaries.

The general heuristic picture is that

- Strong repulsive interaction between the particles (large γ) tends to make the density uniform and favors superfluidity.
- Strong randomness (large ν and σ) leads to fragmentation of the density that is unfavorable for superfluidity.

Journal of Physics: Conference Series **691** (2016) 012016 doi:10.1088/1742-6596/691/1/012016

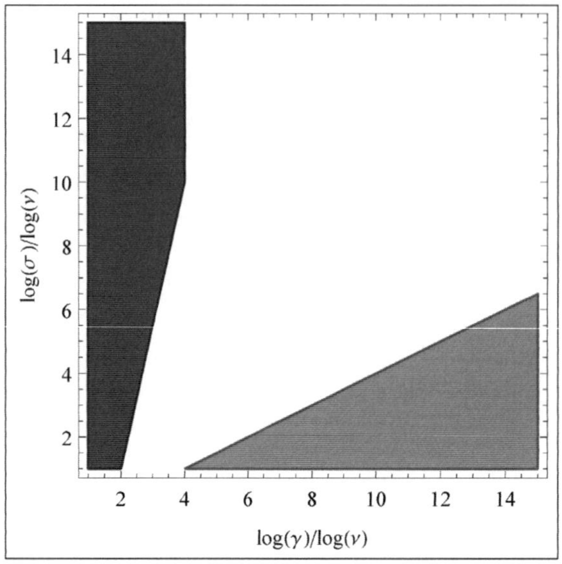

Figure 1. Red: Absence of superfluidity. Green: Complete superfluidity.

This heuristics has been confirmed in [12] by rigorous mathematical analysis. It is notable that

- BEC survives the fragmentation of the density, i.e., long range correlations prevail although superfluidity may be strongly suppressed.
- Superfludity may be suppressed even in cases when the density is close to being uniform on the average and thus not "visibly" fragmented (case 3 above).

We note also that complete absence of both BEC and superfluidity has recently been proved for the Tonks-Girardeu gas, i.e. for $\gamma/N = \infty$, in a generic random potential [13].

5. Bose-Einstein Condensation

We summarize here the basic findings of [11] concerning Bose-Einstein condensation in the ground state of the many-body Hamiltonian (5).

BEC is proved to hold in the GP limit, where $N \to \infty$ and γ is fixed, or does not grow too fast with N. This applies to an arbitrary positive external potential V.

The wave function of the condensate is the (unique) nonnegative minimizer ψ_0 of the Gross-Pitaevskii (GP) energy functional

$$\mathcal{E}^{\rm GP}[\psi] = \int_0^1 \left(|\psi'(z)|^2 + V(z)|\psi(z)|^2 + \frac{\gamma}{2}|\psi(z)|^4\right) {\rm d}z \tag{8}$$

with the normalization $\int_0^1 |\psi|^2 = 1$. The minimizer ψ_0 is also the ground state of the mean field one-particle Hamiltonan

$$h = -\partial_z^2 + V(z) + \gamma|\psi_0|^2 - \frac{\gamma}{2}\int_0^1 |\psi_0|^4 \tag{9}$$

with eigenvalue $e_0 = \mathcal{E}^{\rm GP}[\psi_0]$.

The average occupation of the one-particle state ψ_0 in the many-body ground state Ψ_0 is (by definition)

$$N_0 = \langle \Psi_0, a^\dagger(\psi_0)a(\psi_0)\Psi_0\rangle. \tag{10}$$

Bose-Einstein condensation in the GP limit follows from an estimate of the depletion of the condensate that was derived in [11]:

$$\left(1-\frac{N_0}{N}\right) \leq (\text{const.})\frac{e_0}{e_1-e_0}N^{-1/3}\min\{\gamma^{1/2},\gamma\} \tag{11}$$

where e_1 is the second lowest eigenvalue of the mean field Hamiltonian h. Moreover, the ground state energy of H per particle converges to the GP energy e_0.

6. Superfluidity

With an imposed velocity field v (moving walls) the Hamiltonian becomes:

$$H_v = \sum_{j=1}^{N}\{(\mathrm{i}\partial_{z_j}+v)^2 + V(z_j)\} + \frac{\gamma}{N}\sum_{i<j}\delta(z_i-z_j) \tag{12}$$

on $L^2([0,1],\mathrm{d}z)^{\otimes_{\rm symm}^N}$ with periodic boundary conditions. We denote by $E_0^{\rm QM}(v)$ its ground state energy and by $e_0(v)$ the corresponding energy of the modified GP functional

$$\mathcal{E}_v^{\rm GP}[\psi] = \int_0^1 \left(|\mathrm{i}\psi'(z)+v\psi(z)|^2 + V(z)|\psi(z)|^2 + \frac{\gamma}{2}|\psi(z)|^4\right)\mathrm{d}z. \tag{13}$$

For small enough v, $\mathcal{E}_v^{\rm GP}$ has a unique minimizer, denoted by ψ_v, and $e_0(v)$ is equal to the ground state energy of the boosted mean field Hamiltonian

$$h_v = (\mathrm{i}\partial_z+v)^2 + V(z) + \gamma|\psi_v(z)|^2 - \frac{\gamma}{2}\int_0^1 |\psi_v|^4. \tag{14}$$

Using the diamagnetic inequality, $|(\mathrm{i}\partial_z+v)\psi|^2 \geq |\partial_z|\psi||^2$, one shows in an analogous way to the $v=0$ case:

$$E_0^{\rm QM}(v)/N \geq e_0(v)(1-(\text{const.})N^{-1/3}\min\{\gamma^{1/2},\gamma\}). \tag{15}$$

We conclude that in the GP limit the superfluid fraction at $v=0$

$$\rho_{\rm s}/\rho = \lim_{v\to 0}\frac{1}{v^2}\lim_{N\to\infty}\frac{1}{N}(E_0^{\rm QM}(v)-E_0^{\rm QM}(0)) \tag{16}$$

is the same as the corresponding quantity derived from the GP energy, i.e.,

$$\rho_{\rm s}/\rho = \lim_{v\to 0}\frac{1}{v^2}(e_0(v)-e_0(0)). \tag{17}$$

7. A closed formula for the superfluid fraction

We claim that

$$\rho_{\rm s}/\rho = \left(\int_0^1 |\psi_0(z)|^{-2}dz\right)^{-1} \tag{18}$$

This provides and explicit connection between the wave function of the condensate and the superfluid fraction at $v=0$. The proof is short enough to be reproduced here in full:

We start with the variational equation for ψ_v:

$$(i\partial_z+v)^2\psi_v(z) + V(z)\psi_v(z) + \gamma|\psi_v(z)|^2\psi_v(z) = \mu\psi_v(z). \tag{19}$$

Multiplying by the conplex conjugate ψ_v^* and taking the imaginary part gives

$$\partial_z \left(v|\psi_v(z)|^2 - \Im[\psi_v^*(z)d\psi_v(z)/dz]\right) = 0. \tag{20}$$

Hence there exists a constant $C \in \mathbb{R}$ such that

$$\Im[\psi_v^*(z)d\psi_v(z)/dz] = v|\psi_v(z)|^2 - C. \tag{21}$$

Since

$$de_0(v)/dv = 2v - 2\int_0^1 \Im[\bar{\psi}_v(z)d\psi_v(z)/dz]dz \tag{22}$$

we actually see that $C = \frac{1}{2}de_0(v)/dv$.

Now ψ_v has no zeroes for small v so we can divide by $|\psi_v(z)|^2$ and obtain

$$S'(z) := \frac{\Im[\psi_v^*(z)d\psi_v(z)/dz]}{|\psi_v(z)|^2} = v - \frac{C}{|\psi_v(z)|^2}. \tag{23}$$

Since S' is, in fact, the derivative of the phase of ψ_v we have, due to the periodic boundary conditions,

$$\int_0^1 S'(z)dz = 2\pi n \tag{24}$$

with $n \in \mathbb{Z}$, and in fact $n = 0$ for small enough v. Therefore

$$v = C\int_0^1 |\psi_v(z)|^{-2}dz. \tag{25}$$

This gives

$$e_0'(v) = 2C = 2v\left(\int_0^1 |\psi_v(z)|^{-2}dz\right)^{-1} \tag{26}$$

and thus

$$\rho_s/\rho = \lim_{v\to 0}\frac{e_0'(v)}{2v} = \left(\int_0^1 |\psi_0(z)|^{-2}dz\right)^{-1}. \tag{27}$$

□

8. Complete superfluidity for $\gamma \gg (\sigma\nu)^2$

The formula (27) leads to a proof of complete superfluidity for $\gamma \gg (\sigma\nu)^2$ in the following way. First, taking 1 as a trial function for the GP functional (13) .we see that the GP energy is bounded from above by $\int V_\omega + \frac{\gamma}{2}$. Using this bound and the fact that ψ_0 is a GP minimizer a simple calculation leads to

$$\frac{\||\psi_0|^2 - 1\|_\infty^2}{\sqrt{1 + \|\,|\psi_0|^2 - 1\|_\infty}} \le \frac{2^{3/2}}{\sqrt{\gamma}}\int_0^1 V_\omega \sim (\sigma\nu)/\gamma^{1/2} \tag{28}$$

where $\|\cdot\|_\infty$ is the sup norm and the estimate of $\int V_\omega$ holds with high probability. Hence we see that the superfluid fraction tends to 1 in probability if

$$\gamma \gg (\sigma\nu)^2.$$

In general, however, it turns out not to be sufficient that $\gamma \gg \nu^2$, although in that case it can be shown that $\||\psi_0|^2 - 1\|_1 \to 0$.

9. Absence of superfluidity

If $\mathcal{I}$ is any (measurable) subset of $[0,1]$ with length $|\mathcal{I}|$ we have

$$|\mathcal{I}|^2 = \left(\int_{\mathcal{I}} |\psi_0||\psi_0|^{-1}\right)^2 \leq \int_{\mathcal{I}} |\psi_0(z)|^2 \cdot \int_{\mathcal{I}} |\psi_0(z)|^{-2}$$

and hence, by (27),

$$\rho_s/\rho \leq \frac{\int_{\mathcal{I}} |\psi_0(z)|^2 dz}{|\mathcal{I}|^2}. \tag{29}$$

To prove that superfluidity is small we have therefore to identify subsets $\mathcal{I}$ such that $\int_{\mathcal{I}} |\psi_0(z)|^2 dz$ is small, while $|\mathcal{I}|$ is not too small.

The random points z_j^ω split the interval $[0,1]$ into subintervals $\mathcal{I}_j = [z_j^\omega, z_{j+1}^\omega]$ of various lengths $\ell_j = z_{i+1}^\omega - z_j^\omega$ that are i.i.d. random variables with probability distribution

$$dP_\nu(\ell) = \nu e^{-\nu\ell}\, d\ell. \tag{30}$$

We take

$$\mathcal{I} = \bigcup_{j:\ell_j \leq \tilde{\ell}} \mathcal{I}_j \tag{31}$$

with a suitably chosen $\tilde{\ell}$. The average length of $\mathcal{I}$ (which is a random variable because the points z_j^ω are random) is

$$L = \nu \int_0^{\tilde{\ell}} \ell dP_\nu(\ell) = 1 - (1 + (\nu\tilde{\ell}))e^{-\nu\tilde{\ell}}). \tag{32}$$

In particular it tends to 1 if and only if $\tilde{\ell} \gg \nu^{-1}$.

With the notation

$$n_j^{\mathrm{GP}} = \int_{\mathcal{I}_j} |\psi_0(z)|^2 dz \tag{33}$$

we define

$$N_{\mathrm{small}} = \int_{\mathcal{I}} |\psi_0(z)|^2 dz = \sum_{\ell_j \leq \tilde{\ell}} n_j^{\mathrm{GP}}. \tag{34}$$

Note that ψ_0 and n_j^{GP} also depend on the random sample ω for the Poisson distribution of the obstacles but we have suppressed this in the notation for simplicity.

Our estimate on N_{small} is based on estimates on the GP energy. The superfluid fraction is then bounded from above by N_{small}/L^2.

9.1. *The case* $\gamma \lesssim \nu^2$

One chooses

$$\tilde{\ell} = s/\nu \tag{35}$$

with a suitable $s > 0$ and proceeds in the following steps:

- Split the GP energy into contributions from 'large' intervals of length $\geq \tilde{\ell}$ and 'small' intervals of length $< \tilde{\ell}$.
- Estimate each contribution using estimates on an auxiliary GP functional for the density between two obstacles.
- Using that $e_0(\gamma, \nu, \sigma)/\nu^2$ stays bounded under the standard conditions (7) one obtains a bound $o(1)$ on the mass N_{small}.
- The total length L of the small intervals and its fluctuation are estimated to be $O(1)$ and $o(1)$ respectively.

The upshot is that the superfluid fraction ($\leq N_{\mathrm{small}}/L^2$) tends to 0 in probability if $\gamma \lesssim \nu^2$.

9.2. The case $\gamma \gg \nu^2$

Here both the mass and the length of the small intervals are $o(1)$, but one can prove that $N_{\text{small}}/L^2 \ll 1$ still holds if the obstacle barriers are high enough, namely for

$$\sigma \gg (\gamma/\nu)^4 \gamma^{1/2}. \tag{36}$$

In addition one needs to know that the fluctuations of L are much smaller than the average value and this can be shown to hold for $\gamma \ll \nu^4$. Altogether one obtains: If $\nu^2 \ll \gamma \ll \nu^4$ the superfluid fraction tends to 0 in probability provided (36) holds.

10. Summary

We have studied superfluidity in the ground state of a one-dimensional model of bosons with a repulsive contact interaction and in a random potential generated by Poisson distributed point obstacles.

In the Gross-Pitaevskii (GP) limit this model always shows complete BEC, but depending on the parameters, superfluidity may or may not occur. In the course of the analysis we derived the closed formula (18) for the superfluid fraction, expressed in terms of the GP wave function.

An advantage of the model considered is its amenability to a rigorous mathematical analysis leading to unambiguous statements. It has its limitations, however: Nothing is said about positive temperatures and the proof of BEC applies to the Gross-Pitaevskii limit rather than the thermodynamic limit.

Nevertheless, to our knowledge this is the only model where a Bose glass phase in the sense of [5], i.e., complete BEC but absence of superfluidity, has been rigorously established so far.

Acknowledgments

Support from the Natural Sciences and Engineering Research Council of Canada NSERC (MK, and RS under project P 27533-N27) and from the Austrian Science Fund FWF (JY, under project P 22929-N16) is gratefully acknowledged.

References

[1] K. Huang, *Bose-Einstein condensation and superfluidity*, in: *Bose-Einstein Condensation*, A. Griffin, D.W. Stroke, S. Stringari, eds., Cambridge University Press, 31–50 (1995).

[2] A. Lenard, *Momentum distribution in the ground state of the one-dimensional system of impenetrable bosons*, J. Math. Phys. **5**, 930-943, (1964).

[3] G.E. Astrakharchik, J. Boronat, J. Casulleras, and S. Giorgini, *Superfluidity versus Bose-Einstein condensation in a Bose gas with disorder*, Phys. Rev. A **66**, 023603 (2002).

[4] M. Kobayashi and M. Tsubota, *Bose-Einstein condensation and superfluidity of a dilute Bose gas in a random potential*, Phys. Rev. B **66**, 174516 (2002).

[5] V.I. Yukalov, R. Graham, *Bose-Einstein condensed systems in random potentials*, Phys. Rev. A **75**, 023619 (2007).

[6] V.I. Yukalov, E.P. Yukalova, K.V. Krutitsky, R. Graham, *Bose-Einstein condensed gases in arbitrarily strong random potentials*, Phys. Rev. A, **76**, 053623 (2007).

[7] F.J. Dyson, E.H. Lieb, B. Simon, *Phase Transitions in Quantum Spin Systems with Isotropic and Nonisotropic Interactions*, J. Stat. Phys. **18**, 335–383 (1978).

[8] T. Kennedy, E.H. Lieb, S. Shastry, *The XY Model has Long-Range Order for all Spins and all Dimensions Greater than One*, Phys. Rev. Lett. **61**, 2582–2584 (1988).

[9] M. Aizenman, E.H. Lieb, R. Seiringer, J.P. Solovej, and J. Yngvason, *Bose-Einstein quantum phase transition in an optical lattice model*, Phys. Rev. A **70**, 023612-1-12 (2004).

[10] E.H. Lieb, R. Seiringer, *Proof of Bose-Einstein Condensation for Dilute Trapped Gases*, Phys. Rev. Lett. **88**, 170409-1–4 (2002).

[11] R. Seiringer, J. Yngvason, V.A. Zagrebnov, *Disordered Bose-Einstein condensates with interaction in one dimension*, J. Stat. Mech. P11007 (2012).

[12] M. Könenberg, T. Moser, R. Seiringer and J. Yngvason, *Superfluid behavior of a Bose-Einstein condensate in a random potential*, N. J. Phys. **17**, 013022 (2015).

[13] R. Seiringer and S. Warzel, *Decay of correlations and absence of superfluidity in the disordered Tonks-Girardeau gas*, preprint, arXiv:1512.05282.
[14] R. Seiringer, J. Yngvason, V.A. Zagrebnov, *Disordered Bose-Einstein condensates with interaction*, in: XVIIth International Congress in Mathematical Physics, pp. 610–619, A. Jensen (ed.), World Scientific 2013. arXiv:1209.4046.
[15] O. Penrose, L. Onsager, *Bose-Einstein Condensation and Liquid Helium*, Phys. Rev. **104**, 576–84 (1956).
[16] E.H. Lieb, R. Seiringer, J.P. Solovej and J. Yngvason, *The Mathematics of the Bose Gas and its Condensation*, Birkhäuser, 2005; arXiv:cond-mat/0610117
[17] A.J. Leggett, *Superfluidity*, Rev. Mod. Phys. **71**, S318–S322 (1999).
[18] E.H. Lieb, W. Liniger, *Exact Analysis of an Interacting Bose Gas. I.*, Phys. Rev. **130**, 1605-1616 (1963).

Journal of Physics: Conference Series **691** (2016) 012017 doi:10.1088/1742-6596/691/1/012017

Development of a Magneto-Optical Trap System of Francium Atoms for the Electron Electric-Dipole-Moment Search

K Harada[1], T Aoki[2], K Kato[1], H Kawamura[3], T Inoue[3], T Aoki[1], A Uchiyama[1], K Sakamoto[1], S Ito[1], M Itoh[1], T Hayamizu[1], A Hatakeyama[4], K Hatanaka[5], T Wakasa[6] and Y Sakemi[1]

[1] Cyclotron and Radioisotope Center (CYRIC), Tohoku University, Sendai, Japan
[2] Graduate School of Arts and Sciences, University of Tokyo, Tokyo, Japan
[3] Frontier Research Institute for Interdisciplinary Sciences, Tohoku University, Sendai, Japan
[4] Department of Applied Physics, Tokyo University of Agriculture and Technology, Tokyo, Japan
[5] Research Center for Nuclear Physics, Osaka University, Osaka, Japan
[6] Department of Physics, Kyushu University, Fukuoka, Japan

E-mail: `kenichi.harada.a1@tohoku.ac.jp`

Abstract. The finite value of an electron electric dipole moment (eEDM) provides the direct evidence for the violation of time reversal symmetry. Fr atoms, whose enhancement factor is 895, trapped by laser cooling and trapping techniques are one of the strongest candidates for measuring the eEDM. We are constructing a beamline for measuring the eEDM using laser-cooled Fr atoms at the Cyclotron and Radioisotope Center. We have developed laser light sources including the frequency stabilization system and a magneto-optical trap system for Fr atoms. As the Fr production requires the cyclotron operation, we also use Rb atoms whose chemical properties are similar to those of the Fr atoms. Thus, the Rb beam is utilized for optimizing the operation parameters of the entire apparatus. We have also developed the laser light sources for Rb atoms and observed the beat signal for frequency stabilization of the source using the frequency offset locking method.

1. Introduction

A permanent electric dipole moment of an electron requires both violations of time-reversal (T) and parity (P) invariance. The finite value of an electron electric dipole moment (eEDM) also suggests a charge-parity (CP) violation on the basis of the CPT theorem, and it is crucial for explaining the observed matter-antimatter asymmetry of the universe. The Standard Model (SM) predicts a value of $\sim 10^{-38}$ $e$cm for the eEDM [1]; however, this value is extremely small and far below the current detection limit. On the other hand, several theoretical models such as the supersymmetric model beyond the SM predict much larger eEDM values [2], which can be observed using modern technologies.

The eEDM measurements using thorium monoxide (ThO) molecules [3] are well known as the most precise measurement in polar molecules. The upper limit of the eEDM obtained from the experiments is $\leq 8.7 \times 10^{-29}$ $e$cm. A pulsed beam of the ThO molecules cooled in cold neon carrier gas are employed in this experiment. The interaction time between the ThO molecules and the external electric field applied using the plates having a length of about 22 cm is almost 1 ms due to the effect of a buffer-gas cooling technique. The main advantage of the experiments using polar molecules such as ThO and YbF [4] is that the internal effective electric field $\mathcal{E}_{\rm eff}$ can be much higher than that in atoms. The $\mathcal{E}_{\rm eff}$ in a ThO molecule is about 84 GV/cm, and it is approximately one thousand times larger than that in a Tl atom. However, in the atomic or molecular beam experiment, the significant systematic effects are caused by the motional magnetic field $v \times E/c^2$ that is obtained from the atoms moving with velocity v through an electric field E and by the geometric phase shifts generated by the complicated field gradients. These systematic errors mimic the true eEDM signal and limit the accuracy of measuring the eEDM.

The laser cooling and trapping techniques of atoms are useful to overcome these drawbacks. The velocity of laser cooled atoms is dramatically low; and therefore, the motion-induced magnetic field is strongly suppressed. In particular, an optical lattice suppresses the detrimental effects induced by the atomic collision interaction, which reduces the systematic errors of the eEDM signal [5]. Additionally, the coherence time increases to about a few second by using the laser cooled atoms, which is much longer than that of the ThO molecular beam experiment. Moreover, since the atomic ensembles are confined in a small region, it is unsusceptible to the magnetic and electric field inhomogeneities. Thus, there are several advantages of using optically cooled and trapped atoms for measuring the eEDM.

Fr being the heaviest alkali atom has a large enhancement factor of about 900 [6, 7]. The EDM experiment using the laser cooled Fr atoms can achieve sensitivities even with their upper bounds, which is better than that of the ThO molecular beam experiment. However, the Fr atom poses an experimental challenge as there is no stable isotope that has a half-life longer than 22 min and is radioactive. As the Fr production requires the cyclotron operation, we also use stable Rb atoms having chemical properties similar to that of the Fr atom. The Rb beam is utilized for optimizing the operation parameters of the entire apparatus. We are constructing a beamline for measuring the eEDM using laser-cooled Fr atoms at the Cyclotron and Radioisotope Center (CYRIC), as reported in [8, 9, 10, 11].

The frequency reference and stabilization of laser light sources using several methods are required for trapping of atoms. A reference cell containing stable atoms is generally used for frequency reference and stabilization when atoms are laser-cooled and trapped. However, the Fr atom has no stable isotopes, and the methods of stabilizing the laser frequency using Fr atoms are not available. Therefore, the frequency stabilization using I_2 [12], a Fabry-Pérot cavity with a frequency-stabilized He-Ne laser light [13], and a frequency offset locking (FOL) technique are employed.

In this paper, we report the current development status of the experimental setup for trapping Fr and Rb atoms at CYRIC. We have two rooms for the experiment at CYRIC. One is a laser experimental room (lab A) where the development of laser light sources and its frequency stabilization for Fr and Rb magneto-optical traps (MOTs) are performed. The other is an EDM measurement room (lab B), which is in the radiation controlled area. In this room, we have constructed a MOT system with the neutralizer for the Fr and Rb ions as they are transported through the beamline as ions from the production area. Optical fibers having a length of 150 m are placed between lab A and lab B, which transport the laser light.

2. Experimental setup

2.1. Laser light sources

The experimental setup is schematically illustrated in Fig. 1. As a trapping light source for ^{210}Fr, we use a single frequency Ti:sapphire (Ti:S) laser source, tuned to the $7S_{1/2}$ $(F = 13/2) \rightarrow$ $7P_{3/2}$ $(F' = 15/2)$ D2 transition of Fr atoms at 718.216 nm. F and F' represent the total angular momentum of the ground and excited states, respectively. The maximum output power of the

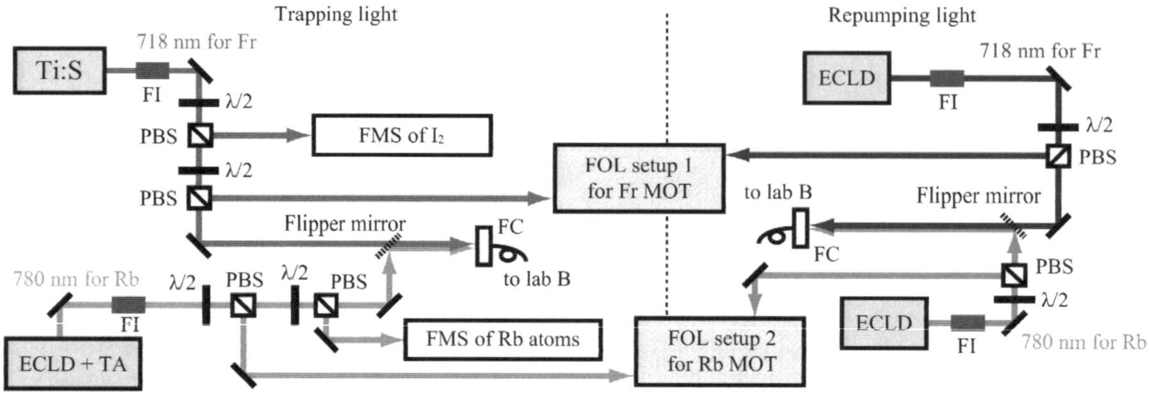

Figure 1. Experimental setup of trapping and repumping lights for the Fr and Rb atoms in lab A. The laser lights having a wavelength of 718 nm and 780 nm are exchanged using flipper mirrors and are transported to lab B using optical fibers. $\lambda/2$, half-wave plate.

Ti:S laser is 3.5 W when the output power of the pump laser light at a wavelength of 532 nm is 18 W. After the laser light passes through a Faraday isolator (FI), a part of it is used for frequency stabilization and reference using I_2 and the FOL setup divided using polarizing beam splitters (PBSs), respectively. Fr atomic vapor cells are not available as frequency references because no stable Fr isotopes exist. Therefore, I_2 is employed for the frequency stabilization and reference of trapping light. Frequency modulation spectroscopy (FMS) is a technique for precisely observing the spectrum of atoms and molecules without the Doppler effect due to their velocity. The details of the setup are reported in [11]. On the other hand, the FOL technique is used for frequency stabilization of repumping light. The error signal for stabilization is obtained by using the beat signal between the 10th-order sideband component and the repumping frequency. The details of the FOL setup 1 will be also reported in a forthcoming paper. The remaining light is transported by using a polarization-maintaining fiber (PMF) using a fiber coupler (FC) for the Fr MOT to lab B.

We use the repumping laser light with a wavelength of 718 nm generated using the custom-made external-cavity laser diode (ECLD, TOPTICA Photonics) for $7S_{1/2}$ $(F = 11/2) \rightarrow 7P_{3/2}$ $(F' = 13/2)$ D2 transition of Fr atoms. The output power of the repumping light is 36 mW. The light from the ECLD passes through an isolator, and is divided into two for the FOL setup and Fr MOT using the PBS.

On the other hand, a trapping laser light generated using the ECLD, tuned to the $5S_{1/2}$ $(F = 2) \rightarrow 5P_{3/2}$ $(F' = 3)$ transition of ^{87}Rb atoms, is employed for the Rb MOT. The output power is amplified up to 1 W by using a tapered amplifier (TA). The light then is divided for FOL setup 2, and the FMS using a Rb vapor cell by PBSs for the frequency reference and stabilization. After the light passes through the PBS, it is transported to lab B for trapping Rb atoms by using the PMF. The laser lights can be exchanged between 718 nm and 780 nm wavelengths by a flipper mirror.

Further, the repumping light for the Rb MOT is generated using the ECLD. The wavelength of the light is tuned to the $5S_{1/2}$ $(F = 1) \rightarrow 5P_{3/2}$ $(F' = 2)$ transition and its output power is 38 mW. The laser light from the ECLD is split by the PBS after it passes through the FI, and the transmitted one is input into the PMF. The other is employed for the FOL setup 2. As already mentioned, we have already developed the FOL setup 1 for Fr atoms. The purpose of developing of the FOL setup 2 is to confirm the operation of the setup and benefit the FOL technique using the Rb MOT.

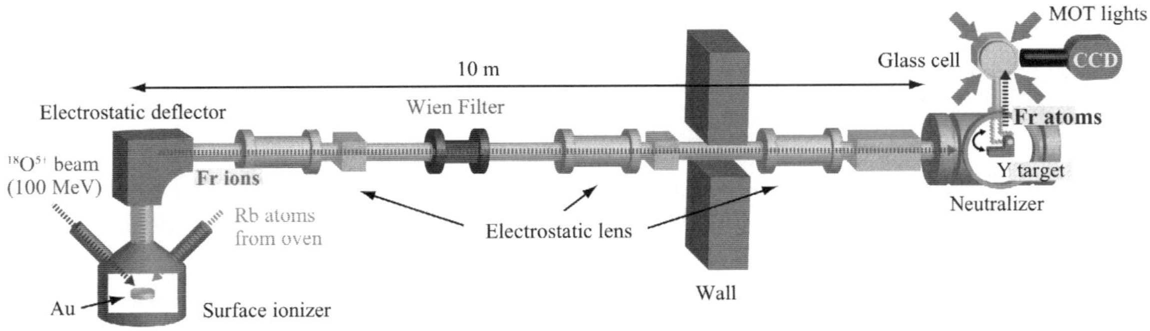

Figure 2. Schematic diagram of the beam line for the magneto-optical trapping of Fr atoms in lab B. The MOT lights are transported from lab A.

2.2. Fr and Rb MOTs

In lab B, we have developed the MOT system with a neutralizer for trapping the Fr and Rb atoms, as shown in Fig. 2. A primary beam of ^{18}O with having an energy of 100 MeV obtained from the AVF cyclotron is injected to the surface ionizer. Fr is produced by the nuclear fusion reaction, $^{18}\mathrm{O} + {}^{197}\mathrm{Au} \rightarrow {}^{210}\mathrm{Fr} + 5\mathrm{n}$. Fr is then extracted from the Au surface as an ion upward, and is transported about 10 m by the electric fields generated by an electrostatics lens and deflector to avoid the radiation from a nuclear fusion reaction. The maximum yield of the Fr beam is estimated to be about 10^6 ions/s. On the other hand, the Rb atoms from an oven are also hit on the Au surface for ionization. Rb beam is used for adjusting the parameters including the surface ionizer, beamline apparatus, and MOT system without an accelerator.

The Fr and Rb ions coming from the upstream of the beamline are hit and accumulated on the surface of the Y metal. After several tens of seconds, the Y surface was turned in the direction of the glass cell and was heated up to about 700 °C for a few seconds. The neutralized Fr atoms were released into the glass cell and trapped by the MOT. Then, the Y surface was returned to its initial position and ions were accumulated on the surface again. This process was repeated again and again. The inside wall of the glass cell was coated by octadecyltrichlorosilane (OTS) to prevent the adsorption of atoms on the surface of the glass. This effect is useful for increasing the number of trappable atoms.

The vacuum pressure of the MOT system was 1.0×10^{-7} Pa. We have already achieved the trapping of the neutralized Rb atoms in the glass cell using the Rb ion beam. The maximum number of the trapped atoms detected using a charge coupled device (CCD) camera is estimated to be approximately 10^6 when the flux of the Rb beam on the Y surface was 1 nA. In the case of Fr atoms, the probability of getting large number of Fr atoms for being trapped in MOT is less. Thus, the lens system composed by four spherical lenses was installed at a distance of 56 mm from the atom trapped at the center of the glass cell. This system is used to efficiently collect the fluorescence from atoms and to compensate for a spherical aberration [14]. As the saturation parameter of a Fr atom is 1.87, the fluorescence intensity from an atom in the MOT is calculated to be 8.3 fW, which can be detected using an avalanche photo diode. We will perform the experiment for trapping Fr atoms by using this system.

2.3. FOL for the Rb MOT

The FOL technique can lock the frequency difference between the trapping and repumping lights to a constant value [15]. The technique is used for frequency stabilization and also as a means of scanning both the lasers simultaneously, while maintaining a constant frequency difference when the servo loop for feedback is closed. We have been already achieved the FOL of the

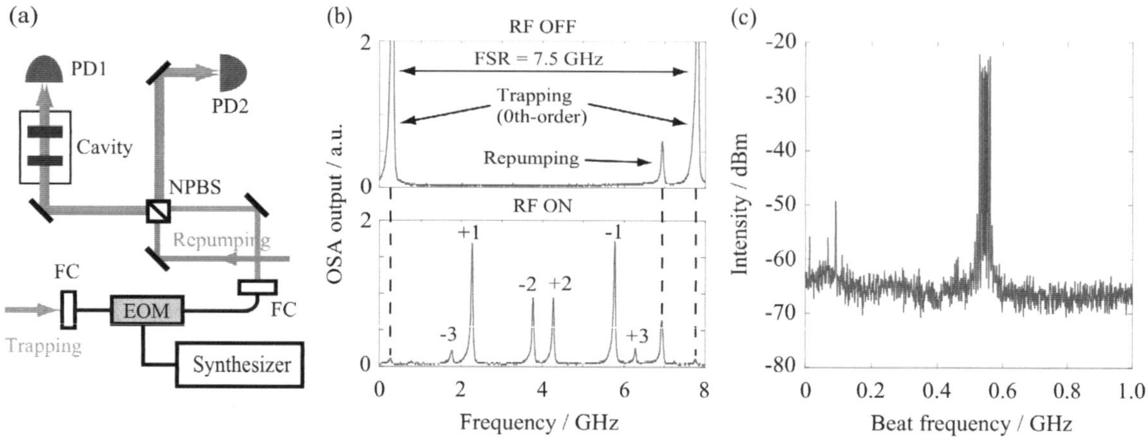

Figure 3. (a) The detailed illustration for the FOL setup 2. (b) Optical spectra measured by PD1 for (above) the trapping light without EOM modulation and repumping light, and (below) the trapping light modulated by 2.0 GHz RF with 19.0 dBm and the repumping light. (c) The beat signal between the 3rd-order sideband component and the repumping frequency.

laser source using an electro-optic modulator (EOM) for the Fr MOT. As the beat frequency between the trapping and repumping lights is about 46 GHz, it is difficult to detect a frequency of 46 GHz directly by using a photodetector and to generate the frequency components with a frequency of 46 GHz. Instead of directly generating a 46 GHz sideband as a 1st-order, we then use the EOM that can generate the 10th-order sideband components by changing the input power of a radio frequency (RF) of 4.6 GHz. The beat signal of a few hundred MHz between the 10th-order sideband component and the repumping frequency are observed for the frequency stabilization by using the delayed self-homodyne detection method. To confirm the operation of the system, we also develop the FOL system for the Rb MOT. The details of the FOL setup 2 are illustrated in Fig. 3 (a). After the trapping light passes through the EOM, the trapping and repumping lights are divided into two beams and overlapped with each other by a 50:50 non-polarizing beam splitter (NPBS). The EOM is driven at an RF of 2 GHz generated by a synthesizer. They are detected by photodetector 1 (PD1) after passing through a cavity to monitor the spectrum of the lights and by photodetector 2 (PD2) to observe the beat signal, respectively. The free spectral range (FSR) of the cavity produced by COHERENT is 7.5 GHz. The detection bandwidth of PD2 used for this detection is 1.5 GHz.

3. Result

To confirm the sideband generation for the trapping light after passing the EOM, we first measured the output powers from the cavity by PD1. The difference between the trapping and repumping frequencies for the Rb MOT is about 6.6 GHz because the frequency difference between the hyperfine splittings of $5S_{1/2}$ of a ^{87}Rb atom is about 6.83 GHz. When the RF signal of 19.0 dBm is input to the EOM, the 3rd-order sideband components appeared, as shown in Fig. 3(b). The frequency in the horizontal axis is obtained by changing the cavity length. The frequency difference between the 0th-order and 1st-order sideband component is 2.0 GHz which is generated from the synthesizer, and the frequency of the 3rd-order component is 6.0 GHz away from that of the 0th-order. As the RF power is limited by the maximum output of the synthesizer, an amplifier is needed for increasing the peak intensity of the 3rd-order sideband components. We then observed the beat signal between the 3rd-order sideband component of the trapping light and repumping light by PD2 and an RF spectrum analyzer. The beat signal of about 540 MHz was obtained, as shown in Fig. 3. However, the obtained 3dB width of the beat signal was about 30 MHz (For Fr, the obtained width of the beat signal between the 10th-order sideband component and repumping frequency was about 1.2 MHz because the Ti:S laser and commercial ECLD are used). The resolution bandwidth and Sweep time of the analyzer was 300 kHz and 10 ms, respectively. The broadened width is because of the linewidth of ECLD made by hand laser used for the Rb MOT. The improvement of the laser sources is required for narrowing the linewidth of the beat signal. Then, the error signal for frequency stabilization is obtained using the delayed self-homodyne detection technique.

4. Conclusion

We reported the current development status of the experimental setup for the Fr and Rb MOTs. We have developed the laser light sources for trapping Fr and Rb atoms. The FOL technique for the Rb MOT is developed for confirming the operation of the locking system. We have also constructed the apparatus, including the neutralizer for the Fr and Rb ions and the MOT system for trapping the neutralized Fr and Rb atoms in lab B. We have successfully observed the neutralized Rb of 10^6 atoms in the MOT by using the Rb ion beams.

In future, we will observe the fluorescence from the Fr atoms trapped in the MOT with a single atom detection technique by the experiment using the cyclotron. We are also developing the EDM measurement systems including the electrode, magnetometer, and optical lattice. After optimization of the Fr MOT, we will perform the EDM measurement using cold Fr atoms.

This research was supported by a Grant-in-Aid for Scientific Research (Nos. 21104005, 26220705) and the Murata Science Foundation. One of the authors (Y. Sakemi) is supported by JSPS and INSA under the Japan-India Research Cooperative Program.

[1] Pospelov M. E, and Khriplovich I. B, *Sov. J. Nucl. Phys.* 1991 **53**, 638
[2] Bernreuther W, Suzuki M, *Rev. Mod. Phys.* 1991 **63**, 313
[3] The ACME Collaboration, Baron J, *et al.*, *Science* 2014 **343**, 269
[4] Hudson J. J, *et al.*, *Nature* 2011 **473**, 493
[5] Chin C, *et al.*, *Phys. Rev. A* 2001 **63**, 033401
[6] Mukherjee D, *et al.*, *J. Phys. Chem. A* 2009 **113**, 12549
[7] Roberts B. M, *et al.*, *Phys. Rev. A* 2013 **88**, 042507
[8] Sakemi Y, *et al.*, *J. Phys.: Conf. Ser.* 2011 **302**, 012051
[9] Kawamura H, *et al.*, *Nucl. Instrum. Meth. B* 2013 **317**, 582
[10] Inoue T, *et al.*, *Hyperfine Interact* 2015 **231**, 157
[11] Harada K, *et al.*, *JPS Conf. Proc.* 2015 **6**, 030128
[12] Simsarian J. E, *et al.*, *Phys. Rev. Lett.* 1996 **76**, 3522
[13] Atutov S. N, *et al.*, *J. Opt. Soc. Am. B* 2003 **20**, 953
[14] Alt W, *Optik* 2002 **113**, 142
[15] Schünemann U, *et al.*, *Rev. Sci. Instrum.* 1999 **70**, 242

Journal of Physics: Conference Series **691** (2016) 012018 doi:10.1088/1742-6596/691/1/012018

Gaussian fluctuations in the two-dimensional BCS-BEC crossover: finite temperature properties

G Bighin and L Salasnich

Dipartimento di Fisica e Astronomia "Galileo Galilei", Università di Padova, Via Marzolo 8, 35131 Padova, Italy

E-mail: bighin@pd.infn.it

Abstract.
The role of fluctuations is enhanced in lower dimensionality systems: in a two dimensions off-diagonal long-range order is destroyed by the fluctuations at any finite temperature, drastically modifying the critical properties with respect to the three-dimensional counterpart. Recently two-dimensional systems of interacting fermions have been the subject of Montecarlo studies and experimental investigations, in particular an ultracold gas of attractive fermions with a widely tunable interaction due to a Feshbach resonance has been realized and the Berezinskii-Kosterlitz-Thouless transition has been observed.

The present work deals with the theoretical description of an ultracold Fermi gas: we discuss the role of Gaussian fluctuations of the order parameter in the equation of state, in particular we take into account the first sound velocity, showing that the inclusion of order parameter fluctuations is needed in order to get the correct composite-boson limit in the strong-coupling regime. The theory is also compared with experimental data. Finally we focus on the superfluid density in the weak-coupling, intermediate and strong-coupling regimes at finite temperature, through which the Berezinskii-Kosterlitz-Thouless critical temperature is obtained.

1. Introduction

The recent experimental realization of an attractive quasi-two-dimensional ultracold Fermi gas with widely tunable interactions [1–4], the experimental observation of the critical properties [5], along with many Montecarlo investigations [6, 7] renovated the interested in the two-dimensional BCS-BEC crossover and motivate further theoretical pursuits in this field.

In three dimensions a mean-field theory of the BCS-BEC crossover at $T = 0$ can provide good qualitative agreement with experiments; however in lower dimensional systems the role of fluctuations is more important [8] and one expects that a mean-field theory may not provide a reasonable agreement with experimental data. It has been indeed reported that in two dimensions at $T = 0$ an ultracold Fermi gas is correctly described by the mean-field theory only in the weak-coupling limit, the correct composite boson equation of state of the strong coupling limit being correctly recovered only when the order parameter fluctuations are taken into account [9].

In the present work, using a path integral approach [10] we derive the one-loop equation of state [11], which we use to calculate the first sound speed and the superfluid density at finite temperature. The first sound speed is compared with experimental data, showing that the Gaussian fluctuations of the order parameter are fundamental even in the zero-temperature limit.

The superfluid density as a function of the crossover allows one to determine the Berezinskii-Kosterlitz-Thouless critical temperature [12, 13] through the Kosterlitz-Nelson condition [14].

2. Mean-field results

Let us consider a two-dimensional system of fermions, interacting through an attractive s-wave contact potential, contained in a volume $V = L^2$, described within the grand canonical ensemble at fixed chemical potential μ and temperature T. Within the path integral formalism the partition function for the system can be written as [10]

$$\mathcal{Z} = \int \mathcal{D}\psi_\sigma \mathcal{D}\bar{\psi}_\sigma e^{-\frac{1}{\hbar}\int_0^{\hbar\beta} \mathrm{d}\tau \int_{L^2} \mathrm{d}^2 r \mathscr{L}} \tag{1}$$

where ψ, $\bar{\psi}$ are complex Grassmann fields, $\beta = 1/k_B T$, k_B is Boltzmann constant and the (Euclidean) Lagrangian density reads:

$$\mathscr{L} = \bar{\psi}_\sigma \left[\hbar\partial_\tau - \frac{\hbar^2}{2m}\nabla^2 - \mu\right] \psi_\sigma + g\,\bar{\psi}_\uparrow\,\bar{\psi}_\downarrow\,\psi_\downarrow\,\psi_\uparrow\,. \tag{2}$$

The quartic interaction cannot be treated exactly; the Hubbard-Stratonovich [15, 16] transformation introduces an additional auxiliary pairing field, corresponding to a Cooper pair, decoupling the quartic interaction. The transformation essentially boils down to the following identity

$$e^{-\frac{1}{\hbar}\int_0^{\hbar\beta} \mathrm{d}\tau \int_{L^2} \mathrm{d}^2 r g\,\bar{\psi}_\uparrow\,\bar{\psi}_\downarrow\,\psi_\downarrow\,\psi_\uparrow} \propto \int \mathcal{D}\Delta\mathcal{D}\bar{\Delta}\,\exp^{-\frac{1}{\hbar}S_{hs}[\psi,\bar{\psi},\Delta,\bar{\Delta}]} \tag{3}$$

where

$$S_{hs}[\psi,\bar{\psi},\Delta,\bar{\Delta}] = -\int_0^{\hbar\beta} \mathrm{d}\tau \int_{L^2} \mathrm{d}^2 r \left(\frac{|\Delta|^2}{g} + \bar{\Delta}\psi_\downarrow\psi_\uparrow + \Delta\bar{\psi}_\uparrow\bar{\psi}_\downarrow\right)\,, \tag{4}$$

which can be straightforwardly verified by completing the square and performing the Gaussian integration. The partition function now reads:

$$\mathcal{Z} = \int \mathcal{D}\psi_\sigma \mathcal{D}\bar{\psi}_\sigma \mathcal{D}\Delta\mathcal{D}\bar{\Delta} e^{-\frac{1}{\hbar}\int_0^{\hbar\beta} \mathrm{d}\tau \int_{L^2} \mathrm{d}^2 r \mathscr{L}} \tag{5}$$

with the following Lagrangian density

$$\mathscr{L} = \bar{\psi}_\sigma \left[\hbar\partial_\tau - \frac{\hbar^2}{2m}\nabla^2 - \mu\right] \psi_\sigma + \bar{\Delta}\,\psi_\downarrow\,\psi_\uparrow + \Delta\bar{\psi}_\uparrow\,\bar{\psi}_\downarrow - \frac{|\Delta|^2}{g}\,. \tag{6}$$

The integration over ψ, $\bar{\psi}$ can be carried out exactly [10], obtaining:

$$\mathcal{Z} = \int \mathcal{D}\Delta\mathcal{D}\bar{\Delta} e^{-\frac{1}{\hbar}\int_0^{\hbar\beta} \mathrm{d}\tau \int_{L^2} \mathrm{d}^2 r(-\ln(-\mathbb{G}^{-1}) - |\Delta|^2/g)} \tag{7}$$

and the physics of the system is encoded in the inverse Green's function, which in coordinate-space representation reads:

$$-\mathbb{G}^{-1} = \begin{pmatrix} \hbar\partial_\tau - \frac{\hbar^2}{2m}\nabla^2 - \mu & \Delta(\mathbf{x},\tau) \\ \bar{\Delta}(\mathbf{x},\tau) & \hbar\partial_\tau + \frac{\hbar^2}{2m}\nabla^2 + \mu \end{pmatrix}\,. \tag{8}$$

The full inverse Green function can be decomposed in a mean-field component $-\mathbb{G}_0^{-1}$, where the pairing field Δ is replaced by its uniform and constant saddle point value, plus a fluctuation part $\mathbb{F}$:

$$-\mathbb{G}^{-1} = -\mathbb{G}_0^{-1} + \mathbb{F} = \begin{pmatrix} \hbar\partial_\tau - \frac{\hbar^2}{2m}\nabla^2 - \mu & \Delta_0 \\ \Delta_0 & \hbar\partial_\tau + \frac{\hbar^2}{2m}\nabla^2 + \mu \end{pmatrix} + \begin{pmatrix} 0 & \eta(\mathbf{x},\tau) \\ \bar{\eta}(\mathbf{x},\tau) & 0 \end{pmatrix} \tag{9}$$

and clearly $\Delta(\mathbf{x},\tau) = \Delta_0 + \eta(\mathbf{x},\tau)$. We now analyze the mean-field approximation, which simply consists in neglecting the fluctuation field η and, consequently, $\mathbb{F}$. The single-particle excitation spectrum is found solving for the poles of the Nambu-Gor'kov Green's function $\mathbb{G}_0$ in momentum space [17]:

$$E_{\mathbf{k}} = \sqrt{(k^2/2m - \mu)^2 + \Delta_0^2}\,. \tag{10}$$

Using the bound state equation to relate the interaction strength g to the bound state[1] energy ϵ_B

$$-\frac{1}{g} = \frac{1}{2L^2}\sum_{\mathbf{k}} \frac{1}{\epsilon_k + \frac{1}{2}\epsilon_B} \tag{11}$$

at $T = 0$ the integrations defining $\mathcal{Z}$ can be carried out analytically in two dimensions, finally the thermodynamic grand potential reads:

$$\Omega_{mf}(\mu,\Delta_0) = -\frac{1}{\beta}\ln\mathcal{Z} = -\frac{mL^2}{4\pi\hbar^2}\left[\mu^2 + \mu\sqrt{\mu^2+\Delta_0^2} + \frac{1}{2}\Delta_0^2 - \Delta_0^2\ln\left(\frac{-\mu+\sqrt{\mu^2+\Delta_0^2}}{\epsilon_B}\right)\right]. \tag{12}$$

From the grand potential we impose the saddle-point condition for Δ_0, i.e. $(\partial\Omega_{mf}/\partial\Delta_0)_{\mu,V} = 0$, obtaining the gap equation:

$$\Delta_0 = \sqrt{2\epsilon_b(\mu + \frac{1}{2}\epsilon_B)}\,, \tag{13}$$

inserting the result from Eq. (13) into Eq. (12) one gets the mean-field equation of state [18]:

$$\Omega_{mf}(\mu) = -\frac{mL^2}{2\pi\hbar^2}(\mu + \frac{1}{2}\epsilon_B)^2\,. \tag{14}$$

The number of particles is obtained from the thermodynamic relation $N = -\partial\Omega/\partial\mu$ and reads:

$$N = \frac{mL^2}{\pi\hbar^2}(\mu + \frac{1}{2}\epsilon_B)\,. \tag{15}$$

3. Pairing fluctuations and equation of state

In the previous Section we simply discarded the fluctuation part of the inverse Green's function, here we rewrite the logarithm appearing in Eq. (7) as

$$\ln\left(-\mathbb{G}^{-1}\right) = \ln\left(-\mathbb{G}_0^{-1}\left(\mathbb{1} - \mathbb{G}_0^{-1}\mathbb{F}\right)\right) = \ln\left(-\mathbb{G}_0^{-1}\right) + \ln\left(\mathbb{1} - \mathbb{G}_0^{-1}\mathbb{F}\right) \tag{16}$$

and the Gaussian (one-loop) approximation consists in the following expansion for the second term in the r.h.s. of Eq. (16)

$$\ln\left(\mathbb{1} - \mathbb{G}_0^{-1}\mathbb{F}\right) = \sum_{m=1}^{\infty}\frac{\left(\mathbb{G}_0^{-1}\mathbb{F}\right)^m}{m} \approx -\mathbb{G}_0^{-1}\mathbb{F} - \frac{1}{2}\mathbb{G}_0^{-1}\mathbb{F}\mathbb{G}_0^{-1}\mathbb{F}\,. \tag{17}$$

[1] A bound state is present in two-dimension for every value of the attractive interaction.

The pairing field has been written as $\Delta(\mathbf{x},\tau) = \Delta_0 + \eta(\mathbf{x},\tau)$ where Δ_0 the saddle point value, so the fluctuation fields η, $\bar{\eta}$ are small, justifying the present expansion in powers of $\mathbb{F}$. Re-arranging the terms and rewriting the action in momentum space [10] one gets

$$\mathcal{Z} = \int \mathcal{D}\eta\mathcal{D}\bar{\eta} e^{-\frac{1}{\hbar}S_g[\eta,\bar{\eta}]} \tag{18}$$

with the following action

$$S_g[\eta,\bar{\eta}] = \frac{1}{2}\sum_{\mathbf{q},m}(\bar{\eta}(\mathbf{q},\mathrm{i}\Omega_m),\eta(-\mathbf{q},-\mathrm{i}\Omega_m))\ \mathbb{M}(\mathbf{q},\mathrm{i}\Omega_m)\begin{pmatrix}\eta(\mathbf{q},\mathrm{i}\Omega_m)\\ \bar{\eta}(-\mathbf{q},-\mathrm{i}\Omega_m)\end{pmatrix}, \tag{19}$$

having introduced the Fourier transform of the fluctuation field η, along with the bosonic Matsubara frequencies $\Omega_m = (2m+1)\,\pi/\beta$; the matrix elements of the inverse pair fluctuation propagator $\mathbb{M}$, along with a complete derivation can be found in Ref. [10]. The poles of the propagator determine the collective mode spectrum, which in the strong-coupling regime can be parametrized as

$$\hbar\omega_q = \sqrt{\epsilon_q\left(\lambda\epsilon_q + 2mc_s^2\right)}\,, \tag{20}$$

where λ and c_s are a function of the crossover. In order to write the one-loop contribution to the equation of state [11, 19] one needs to introduce a regularization procedure [11, 19], as follows:

$$\Omega_g(\mu) = \frac{1}{2\beta}\sum_{\mathbf{q},m}\ln\left[\frac{\mathbb{M}_{11}(\mathbf{q},\mathrm{i}\Omega_m)}{\mathbb{M}_{22}(\mathbf{q},\mathrm{i}\Omega_m)}\det(\mathbb{M}(\mathbf{q},\mathrm{i}\Omega_m))\right]e^{\mathrm{i}\Omega_m 0^+}\,. \tag{21}$$

In Fig. 1 we plot the mean-field value of the chemical potential and of the pairing gap as a function of the crossover, calculated using the mean-field equation of state in Eq. (14) and the full one-loop equation of state given by the sum of Eq. (14) and Eq. (21).

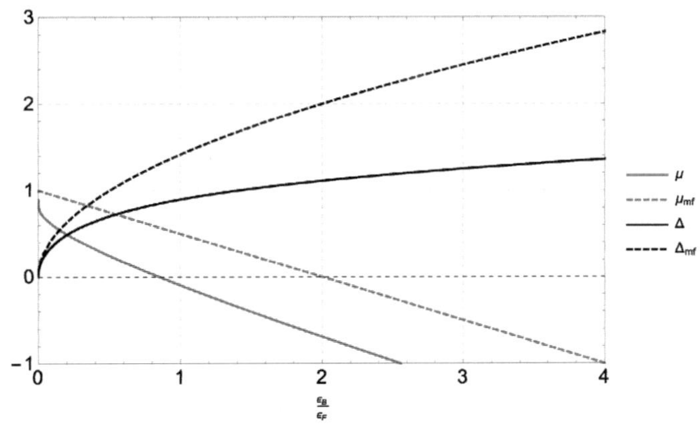

Figure 1. The chemical potential μ (red lines) and the pairing gap Δ_0 (black lines), as calculated from the mean-field equation of state (dashed lines) and from the one-loop equation of state (solid lines), as a function of the scaled binding energy ϵ_B/ϵ_F.

4. First sound

Within this framework the first sound speed at $T = 0$ can be calculated using the elegant thermodynamic relation [20] due to Landau

$$c_s^2 = \frac{n}{m}\frac{\partial\mu}{\partial n}\,. \tag{22}$$

Using the mean-field equation of state in Eq. (14) one finds the result

$$c_s^2 = \frac{v_f^2}{2} \, , \tag{23}$$

where v_F is the Fermi velocity, this result being plotted in dashed blue in Fig. 2. On the other hand, introducing the one-loop equation of state in Eq. (21) one finds that the mean-field result is accurate only in the deep-BCS regime: when the coupling is increased the sound speed greatly deviates from its mean-field value tending to the composite boson limit [9]

$$c_s^2 = \frac{4\pi\hbar^2}{m_B^2} \frac{n_B}{\ln\left(\frac{1}{n_B a_B^2}\right)} \tag{24}$$

shown in dashed red in Fig. 2, $m_B = 2m$ being the mass of each composite boson, $n_B = n/2$ being the bosonic density and $a_B = 0.551 a_s$ being the boson-boson scattering length [9]. Our theoretical findings are also compared with preliminary experimental data [21], showing good agreement, see Fig 2. The results can be readily extended to finite temperature [22] using the thermodynamic approach in [23], however the temperature dependance is very weak across the whole crossover between $T = 0$ and the Berezinskii-Kosterlitz-Thouless critical temperature T_{BKT} [22].

We stress that the present result in Fig. 2 clearly shows that the inclusion of Gaussian fluctuations in the equation of state is of fundamental importance in the two-dimensional BCS-BEC crossover, the mean-field equation of state being accurate only in the deep-BEC regime.

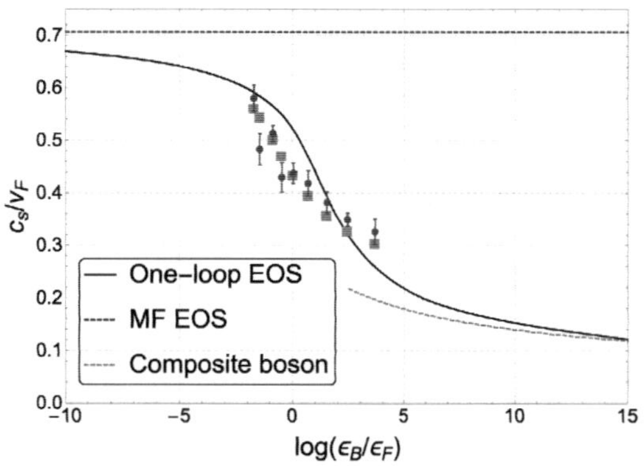

Figure 2. The first sound speed at $T = 0$ from the one-loop equation of state (solid black line), compared with the theoretical prediction using the mean-field equation of state (blue dashed line) and the composite boson limit valid for $\epsilon_B \gg \epsilon_F$ (red dashed line). The data points are experimental data reported in [21], blue dots for data measured from the speed of density waves and red squares for data measured from the equation of state.

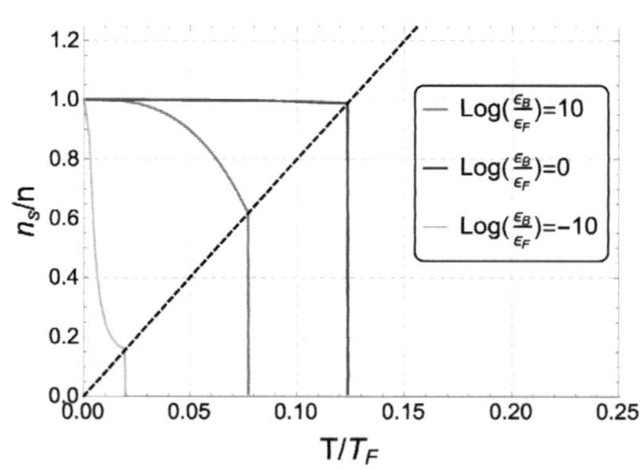

Figure 3. Superfluid density for three different values of the scaled binding energy, from the weakly-coupled regime to the strong interacting one. The black dashed line marks the Nelson-Kosterlitz condition, setting the Berezinskii-Kosterlitz-Thouless critical temperature T_{BKT}; a more detailed analysis of T_{BKT} across the whole crossover is reported in [22].

5. Superfluid density

The fermionic and bosonic contributions to the normal density are calculated using Landau's quasiparticle excitation formulas [24]

$$n_{n,f} = \beta \int \frac{\mathrm{d}^2 k}{(2\pi)^2} k^2 \frac{e^{\beta E_k}}{(e^{\beta E_k}+1)^2} \tag{25}$$

and

$$n_{n,b} = \frac{\beta}{2} \int \frac{\mathrm{d}^2 q}{(2\pi)^2} q^2 \frac{e^{\beta \omega_q}}{(e^{\beta \omega_q}-1)^2} \, . \tag{26}$$

The superfluid density is then $n_s = n - n_{n,f} - n_{n,b}$, having neglected the hybridization between the single-particle and collective excitations which gets relevant at higher temperature scales [22, 25]. As mentioned in the Introduction the role of the fluctuations is enhanced in lower dimensionalities [8], in particular in one- and two-dimensional systems there cannot the spontaneous breaking of a continuos symmetry at finite temperature [26, 27]; as a consequence off-diagonal long-range order is achieved strictly only at $T = 0$. Nonetheless in a two-dimensional systems a finite-temperature transition is to be observed, between the normal state and a state characterized by *algebraic* off-diagonal long-range order [12, 13]. Figure 3 shows three superfluid density profiles for $\log(\epsilon_B/\epsilon_F) = 10$, $\log(\epsilon_B/\epsilon_F) = 0$ and $\log(\epsilon_B/\epsilon_F) = -10$, respectively in the deep-BEC, intermediate and deep-BCS regimes, along with the line corresponding to the Nelson-Kosterlitz condition [14]

$$k_B T_{BKT} = \frac{\hbar^2 \pi}{8m} n_s(T_{BKT}) \tag{27}$$

determining the critical temperature T_{BKT}.

With respect to the mean-field result [28] a full Gaussian-level equation of state here predicts a lower critical temperature in the deep-BEC regime. This treatment can be extended to the whole crossover, a full comparison with the recent experimental observation of T_{BKT} [5] in a quasi-two-dimensional ultracold Fermi gas is reported in Ref. [22].

6. Conclusions

The BCS-BEC crossover in a two-dimensional Fermi gas is the subject of great interest, both from an experimental and theoretical point of view. In this work, using a one-loop equation of state, we have calculated the first sound speed and the superfluid density, showing that the Gaussian fluctuations are needed in order to recover the correct composite boson limit even in the zero-temperature case. We also discussed the superfluid density at finite temperature, using this result to calculate the Berezinskii-Kosterlitz-Thouless critical temperature [22].

The treatment in this work can easily extended to account for the description of many properties of two-dimensional ultracold Fermi gases, e.g. the second sound, the Berezinskii-Kosterlitz-Thouless critical temperature across the whole crossover, the condensate fraction [22, 29]. Moreover the energy density, the pressure and the contact have been calculated [11] within an approach equivalent to the present one and show good agreement with Montecarlo data.

A description of a trapped non-uniform system, as well as a full calculation of the superfluid density beyond the approximation in Eqs. (26) and (25) is the subject of ongoing work.

Acknowledgements

The author gratefully acknowledges Ministero Istruzione Università Ricerca (PRIN project 2010LLKJBX) for partial support.

References

[1] K. Martiyanov, V. Makhalov, and A. Turlapov, "Observation of a Two-Dimensional Fermi Gas of Atoms", Physical Review Letters **105**, 030404–4 (2010).

[2] B. Fröhlich, M. Feld, E. Vogt, M. Koschorreck, W. Zwerger, and M. Köhl, "Radio-Frequency Spectroscopy of a Strongly Interacting Two-Dimensional Fermi Gas", Physical Review Letters **106**, 105301–4 (2011).

[3] P. Dyke, E. D. Kuhnle, S. Whitlock, H. Hu, M. Mark, S. Hoinka, M. Lingham, P. Hannaford, and C. J. Vale, "Crossover from 2D to 3D in a Weakly Interacting Fermi Gas", Physical Review Letters **106**, 105304–4 (2011).

[4] A. T. Sommer, L. W. Cheuk, M. J. H. Ku, W. S. Bakr, and M. W. Zwierlein, "Evolution of Fermion Pairing from Three to Two Dimensions", Physical Review Letters **108**, 045302–5 (2012).

[5] P. A. Murthy, I. Boettcher, L. Bayha, M. Holzmann, D. Kedar, M. Neidig, M. G. Ries, A. N. Wenz, G. Zürn, and S. Jochim, "Observation of the Berezinskii-Kosterlitz-Thouless Phase Transition in an Ultracold Fermi Gas", Physical Review Letters **115**, 010401–6 (2015).

[6] G. Bertaina and S. Giorgini, "BCS-BEC Crossover in a Two-Dimensional Fermi Gas", Physical Review Letters **106**, 110403–4 (2011).

[7] E. R. Anderson and J. E. Drut, "Pressure, Compressibility, and Contact of the Two-Dimensional Attractive Fermi Gas", Physical Review Letters **115**, 115301–5 (2015).

[8] Z. Hadzibabic and M. Köhl, "Low-dimensional atomic bose gases", in *Ultracold bosonic and fermionic gases*, Vol. 5, edited by A. L. F. Kathryn Levin and D. M. Stamper-Kurn, Contemporary Concepts of Condensed Matter Science (Elsevier, 2012), pp. 95–120.

[9] L. Salasnich and F. Toigo, "Composite bosons in the two-dimensional BCS-BEC crossover from Gaussian fluctuations", **91**, 011604–5 (2015).

[10] J. Tempere and J. P. A. Devreese, "Path-Integral Description of Cooper Pairing", in *Superconductors - materials, properties and applications* (InTech, Oct. 2012), pp. 1–32.

[11] L. He, H. Lü, G. Cao, H. Hu, and X.-J. Liu, "Quantum fluctuations in the BCS-BEC crossover of two-dimensional Fermi gases", **92**, 023620–15 (2015).

[12] V. L. Berenzinskii, "Destruction of long-range order in one-dimensional and two-dimensional systems possessing a continuous symmetry group. II. Quantum systems", Soviet Physics JETP **34** (1972).

[13] J. M. Kosterlitz and D. J. Thouless, "Ordering, metastability and phase transitions in two-dimensional systems", Journal of Physics C: Solid State Physics **6**, 1181–1203 (1973).

[14] D. R. Nelson and J. M. Kosterlitz, "Universal Jump in the Superfluid Density of Two-Dimensional Superfluids", Physical Review Letters **39**, 1201–1205 (1977).

[15] R. L. Stratonovich, "On a method of calculating quantum distribution functions", Sov. Phys. Doklady **2**, 416–419 (1958).

[16] J. Hubbard, "Calculation of partition functions", Physical Review Letters **3**, 77–80 (1959).

[17] H. T. C. Stoof, K. B. Gubbels, and D. B. M. Dickerscheid, *Ultracold Quantum Fields* (Springer, Dordrecht, 2009).

[18] M. Marini, F. Pistolesi, and G. C. Strinati, "Evolution from BCS superconductivity to Bose condensation: analytic results for the crossover in three dimensions", Eur. Phys. J. B **1**, 151–159 (1998).

[19] R. B. Diener, R. Sensarma, and M. Randeria, "Quantum fluctuations in the superfluid state of the BCS-BEC crossover", Physical Review A **77**, 023626–21 (2008).

[20] E. Lipparini, *Modern Many-Particle Physics*, 2nd ed. (World Scientific, Singapore, 2008).

[21] N. Luick, "Local probing of the Berezinskii-Kosterlitz-Thouless transition in a two-dimensional Bose gas", MSc Thesis (University of Hamburg, 2014).

[22] G. Bighin and L. Salasnich, "Gaussian fluctuations in the two-dimensional BCS-BEC crossover: first sound, second sound and critical temperature", arXiv.org (2015).

[23] L. Salasnich, "Low-temperature thermodynamics of the unitary Fermi gas: Superfluid fraction, first sound, and second sound", Physical Review A **82**, 063619–7 (2010).

[24] A. L. Fetter and J. D. Walecka, *Quantum Theory of Many-Particle Systems (Dover Books on Physics)* (Dover Publications, June 1971).

[25] E. Taylor, A. Griffin, N. Fukushima, and Y. Ohashi, "Pairing fluctuations and the superfluid density through the BCS-BEC crossover", Physical Review A **74**, 063626–15 (2006).

[26] N. D. Mermin and H. Wagner, "Absence of ferromagnetism or antiferromagnetism in one- or two-dimensional isotropic Heisenberg models", Physical Review Letters **17**, 1133–1136 (1966).

[27] S. R. Coleman, "There are no Goldstone bosons in two-dimensions", Commun. Math. Phys. **31**, 259–264 (1973).

[28] L. Salasnich, P. A. Marchetti, and F. Toigo, "Superfluidity, sound velocity, and quasicondensation in the two-dimensional BCS-BEC crossover", Physical Review A **88**, 053612–7 (2013).

[29] L. Salasnich, "Condensate fraction of a two-dimensional attractive Fermi gas", Physical Review A **76**, 015601–4 (2007).

Vortex rings and vortex ring solitons in shaken Bose-Einstein condensate

V.I. Yukalov[1], **A.N. Novikov**[1,2], **E.P. Yukalova**[3], **and V.S. Bagnato**[2]

[1]Bogolubov Laboratory of Theoretical Physics, Joint Institute for Nuclear Research, Dubna 141980, Russia

[2]Instituto de Fisica de São Calros, Universidade de São Paulo, CP 369, 13560-970 São Carlos, São Paulo, Brazil

[3]Laboratory of Informational Technologies, Joint Institute for Nuclear Research, Dubna 141980, Russia

E-mail: `yukalov@theor.jinr.ru`

Abstract. In a shaken Bose-Einstein condensate, confined in a vibrating trap, there can appear different nonlinear coherent modes. Here we concentrate on two types of such coherent modes, vortex ring solitons and vortex rings. In a cylindrical trap, vortex ring solitons can be characterized as nonlinear Hermite-Laguerre modes, whose description can be done by means of optimized perturbation theory. The energy, required for creating vortex ring solitons, is larger than that needed for forming vortex rings. This is why, at a moderate excitation energy, vortex rings appear before vortex ring solitons. The generation of vortex rings is illustrated by numerical simulations for trapped ^{87}Rb atoms.

1. Introduction

Nonlinear Schrödinger (NLS) equation exhibits different kinds of soliton-like solutions. Various solitonic solutions are well known for the NLS equation in nonlinear optics [1, 2] and in Bose-Einstein condensates [3, 4]. The behavior of solitons in both these nonlinear media possesses many common features [5, 6].

Here we concentrate on two types of solutions to the NLS equation, vortex ring solitons and vortex rings. The theoretical study of such solutions is usually based on the assumption of their existence at the initial moment of time, after which one considers their temporal dynamics starting from the given initial condition. In a realistic situation, in order to create a soliton in a Bose-condensed system, it is necessary to apply an external perturbation rendering the condensate strongly nonequilibrium. There are several ways of perturbing the system. Trap rotation or laser rotation inside a trap is known to produce quantum vortices and vortex lattices [7]. To generate a variety of other coherent modes, it has been suggested [8–10] to shake the condensed system, which can be realized either by trap shaking [8–11] or by scattering length modulation [12, 13]. A special variant of a double-loop pattern of laser stirring [14] shows the effect similar to trap shaking.

Ring dark solitons are known in optics [15] and in Bose-Einstein condensate [8–11, 16, 17]. Ring vortex solitons can be stable for weak atomic repulsion, but become unstable for sufficiently strong effective repulsion, such as is typical for trapped atoms. However, their lifetime can be rather long, of order of seconds, which allows one to treat them as metastable objects [18, 19].

Note that bright vortex solitons, existing for attractive atomic interactions, are unstable in three dimensions [20, 21].

In the presence of external perturbations, vortex ring solitons experience snake instability [22] and can split into multiple fragments [19], and finally decay into vortex rings [23, 24]. A reconnection between vortex lines can also lead to the emission of vortex rings [25, 26].

In the present paper, we consider the situation, when Bose-condensed trapped atoms are subject to an external perturbation, which corresponds to trap shaking. The setup is explained in Sec. 2. Such trap shaking can generate different nonlinear coherent modes, including vortices and ring vortex solitons [8–11]. The definition of coherent modes, as stationary solutions to the NLS equation, is given in Sec. 3. When the effective coupling parameter is asymptotically small, it is possible to approximate the solutions to the NLS equation by Gauss-Laguerre modes. We show that, even for arbitrarily strong coupling parameter, nonlinear coherent modes can be well characterized by Hermite-Laguerre modes, provided the description is done by means of optimized perturbation theory [27, 28].

In our previous papers [29–32], we have demonstrated that by shaking a trapped Bose-Einstein condensate, we can generate the following nonequilibrium states: weak nonequilibrium, vortex state, vortex turbulence, grain turbulence, and wave turbulence. Analyzing now in more detail the initial stage of weak nonequilibrium, we numerically simulate the three-dimensional NLS equation and find out that, when no special resonance conditions are involved, in the regime of moderately strong perturbations, vortex rings arise. This happens yet before vortices start appearing in the vortex stage and far before vortex turbulence develops. The results of the numerical simulation, observing vortex rings, are presented in Sec. 4.

To summarize, in the present paper we show the following:

(i) We demonstrate that vortex ring solitons, that are a particular case of nonlinear coherent modes, can be characterized as Hermite-Laguerre modes, provided optimized perturbation theory is used. By vortex ring solitons, we understand circular dark solitons with circulation around the trap axis.

(ii) A vortex ring is defined as a circular line of zero density, with circulation around each of its elements. By numerical simulations, we show that a moderate trap shaking produces vortex rings in a trapped ^{87}Rb.

(iii) We explain why vortex rings arise before vortex ring solitons and even before vortices. This is because the generation of vortex rings requires much less energy than that needed for the creation of other nonlinear coherent modes, including vortices and vortex ring solitons.

2. Shaken trap setup

We consider Bose atoms in a trap described by the harmonic trap potential

$$U(\mathbf{r}) = \frac{m}{2}\left(\omega_\perp^2 r_\perp^2 + \omega_z^2 r_z^2\right) , \tag{1}$$

where $r_\perp^2 = r_x^2 + r_y^2$ is the radial spatial variable and r_z is the longitudinal spatial variable. The trap aspect ratio is

$$\alpha \equiv \frac{\omega_z}{\omega_\perp} = \left(\frac{l_\perp}{l_z}\right)^2 . \tag{2}$$

Atoms interact through the local interaction potential

$$\Phi(\mathbf{r}) = \Phi_0\delta(\mathbf{r}) , \qquad \Phi_0 \equiv 4\pi\hbar^2\,\frac{a_s}{m} , \tag{3}$$

where m is atomic mass and a_s, scattering length. The strength of interactions is characterized by the dimensionless coupling parameter

$$g \equiv 4\pi N\,\frac{a_s}{l_\perp} , \tag{4}$$

Journal of Physics: Conference Series **691** (2016) 012019 doi:10.1088/1742-6596/691/1/012019

where N is the number of atoms.

Temperature is assumed to be low, so that almost all atoms are in the Bose condensed state. The shape of the atomic cloud in the trap can be estimated in the Thomas-Fermi approximation [7], which yields the cloud radius

$$R = 1.036\, l_\perp (\alpha g)^{1/5} \tag{5}$$

and the cloud length

$$L = \frac{2R}{\alpha} = 2.072\, \frac{l_\perp}{\alpha}\, (\alpha g)^{1/5} \; . \tag{6}$$

Hence the cloud volume is

$$V = \pi R^2 L = 6.986\, \frac{l_\perp^3}{\alpha}\, (\alpha g)^{3/5} \; . \tag{7}$$

Trap shaking is accomplished by imposing an additional time-dependent potential, resulting in the total potential

$$U(\mathbf{r}, t) = U(\mathbf{r}) + V(\mathbf{r}, t) \tag{8}$$

consisting of the stationary confining potential $U(\mathbf{r})$ and an alternating potential $V(\mathbf{r}, t)$. The latter is chosen so that to vibrate the trap without rotation. The shaking potential has the form

$$V(\mathbf{r}, t) = \frac{m}{2}\, \Omega_x^2(t)(x' - x_0')^2 + \frac{m}{2}\, \Omega_y^2(t)(y' - y_0')^2 + \frac{m}{2}\, \Omega_z^2(t)(z' - z_0')^2 \; , \tag{9}$$

with the vibration frequencies

$$\Omega_\alpha(t) = A_\alpha \omega_\alpha [1 - \cos(\omega t)] \; . \tag{10}$$

The primed variables x', y', and z' are shifted and tilted with respect to the original spatial variables x, y, and z, as is explained in Ref. [33].

3. Vortex ring solitons

Vortex ring solitons are a particular case of nonlinear coherent modes that can be introduced as solutions to the NLS equation possessing a circulation around the trap axis, characterized by a nonzero winding number. Coherent modes can be generated by strongly shaking a trap. In a cylindrical trap, coherent modes can be represented as Hermite-Laguerre modes, which can be done not only for weak interactions, but for any strong atomic interactions, provided *Optimized Perturbation Theory* [27, 28] is employed.

Stationary states of Bose-condensed atoms are described by the stationary NLS equation

$$\hat{H}[\eta]\eta(\mathbf{r}) = E\eta(\mathbf{r}) \; , \tag{11}$$

with the nonlinear Hamiltonian

$$\hat{H}[\eta] = -\frac{\hbar^2 \nabla^2}{2m} + U(\mathbf{r}) + \Phi_0 \mid \eta(\mathbf{r}) \mid^2 \; . \tag{12}$$

The normalization condition for the condensate function is

$$\int \mid \eta(\mathbf{r}) \mid^2 d\mathbf{r} = N \; . \tag{13}$$

It is convenient to introduce the dimensionless spatial variables

$$r \equiv \frac{r_\perp}{l_\perp} \; , \qquad z \equiv \frac{r_z}{l_\perp} \tag{14}$$

and the dimensionless condensate function

$$\eta(\mathbf{r}) \equiv \sqrt{\frac{N}{l_\perp^3}}\;\psi(r,\varphi,z) \; . \tag{15}$$

Then, with the dimensionless Hamiltonian

$$\hat{H}[\psi] \equiv \frac{\hat{H}[\eta]}{\hbar\omega_\perp} \; , \tag{16}$$

the eigenvalue problem (11) reads as

$$\hat{H}[\psi_{nmj}]\psi_{nmj}(r,\varphi,z) = E_{nmj}\psi_{nmj}(r,\varphi,z) \; . \tag{17}$$

Here

$$\hat{H}[\psi] \equiv -\,\frac{1}{2}\,\nabla^2 + \frac{1}{2}\,(r^2+\alpha^2z^2) + g|\,\psi\,|^2 \; , \tag{18}$$

with

$$\nabla^2 = \frac{\partial^2}{\partial r^2} + \frac{1}{r}\,\frac{\partial}{\partial r} + \frac{1}{r^2}\,\frac{\partial^2}{\partial\varphi^2} + \frac{\partial^2}{\partial z^2} \; .$$

The normalization condition (13) becomes

$$\int_0^\infty r\,dr \int_0^{2\pi} d\varphi \int_{-\infty}^\infty |\,\psi_{nmj}(r,\varphi,z)\,|^2\,dz = 1 \; . \tag{19}$$

In optimized perturbation theory [27, 28], we start with a trial Hamiltonian

$$\hat{H}_0 = -\,\frac{1}{2}\,\nabla^2 + \frac{1}{2}\,(u^2r^2+v^2z^2) \tag{20}$$

including control functions u and v to be defined from an optimization condition. The trial wave function, corresponding to Hamiltonian (20), is given by the expression

$$\psi_{nmj}(r,\varphi,z) = \left[\frac{2n!u^{|m|+1}}{(n+|m|)!}\right]^{1/2} r^{|m|}\,\exp\left(-\,\frac{u}{2}\,r^2\right)\,L_n^{|m|}(ur^2)\,\frac{e^{im\varphi}}{\sqrt{2\pi}}\,\times$$
$$\times\left(\frac{v}{\pi}\right)^{1/4}\frac{1}{\sqrt{2^j\,j!}}\,\exp\left(-\,\frac{v}{2}\,z^2\right)H_j(\sqrt{v}\,z) \; , \tag{21}$$

where L_n^m is a generalized Laguerre polynomial and H_j is a Hermite polynomial. Here $n = 0, 1, 2, \ldots$ is a radial quantum number, $m = 0, \pm1, \pm2, \ldots$ is an azimuthal quantum number or the winding number of circulation, and $j = 0, 1, 2, \ldots$ is an axial quantum number.

The optimization condition, defining the control functions, can be chosen in several forms, the simplest of which is

$$\left(\delta u\,\frac{\partial}{\partial u} + \delta v\,\frac{\partial}{\partial v}\right)\left(\psi_{nmj},\;\hat{H}[\psi_{nmj}]\psi_{nmj}\right) = \left(\delta u\,\frac{\partial}{\partial u} + \delta v\,\frac{\partial}{\partial v}\right)E_{nmj} = 0 \; . \tag{22}$$

The spectrum of coherent modes is

$$E_{nmj} = \frac{p}{2}\left(u + \frac{1}{u}\right) + \frac{q}{4}\left(v + \frac{\alpha^2}{v}\right) + \frac{u\sqrt{v}}{(2\pi)^{3/2}}\,J_{nmj}g \; , \tag{23}$$

in which

$$p \equiv p_{nm} = 2n + |m| + 1 \; , \qquad q \equiv q_j = 2j + 1 \tag{24}$$

and

$$J_{nmj} \equiv \frac{(2\pi)^{3/2}}{u\sqrt{v}} \int |\, \psi_{nmj}(r, \varphi, z) \,|^4 r dr d\varphi dz \; . \tag{25}$$

The optimization condition (22) gives the equations for the control-functions:

$$p\left(1 - \frac{1}{u^2}\right) + \frac{G}{\alpha p}\sqrt{\frac{v}{q}} = 0 \; , \qquad q\left(1 - \frac{\alpha^2}{v^2}\right) + \frac{uG}{\alpha p\sqrt{vq}} = 0 \; , \tag{26}$$

where we define the effective coupling parameter

$$G \equiv G_{nmj} = \frac{2p\sqrt{q}}{(2\pi)^{3/2}} \, J_{nmj} \alpha g \; . \tag{27}$$

In a typical situation with trapped atoms, the effective coupling is strong, such that

$$g \gg 1 \; , \qquad G \gg 1 \; . \tag{28}$$

Then the control functions are

$$u \simeq \frac{p}{G^{2/5}} \; , \qquad v \simeq \frac{\alpha^2 q}{G^{2/5}} \; . \tag{29}$$

We are interested in the radial shape of the condensate density. To this end, we set $j = 0$ and define the radial density

$$\rho_{nm}(r) \equiv \int |\, \psi_{nm0}(r, \varphi, z) \,|^2 d\varphi dz \; ,$$

for which we obtain

$$\rho_{nm}(r) = \frac{2n! u^{m+1}}{(n+m)!} \, r^{2m} \exp\left(-ur^2\right) \left[L_n^m\left(ur^2\right)\right]^2 \; , \tag{30}$$

where $m \geq 0$.

The effective coupling parameter is

$$G = \frac{2p}{(2\pi)^{3/2}} \, J_{nm} \alpha g \qquad (q = 1) \; ,$$

with

$$J_{nm} \equiv \frac{1}{u} \int_0^\infty \rho_{nm}^2(r) \, r dr \; .$$

In particular,

$$J_{00} = 1 \; , \qquad J_{01} = \frac{1}{2} \; , \qquad J_{02} = \frac{3}{8} \; ,$$

$$J_{10} = \frac{1}{2} \; , \qquad J_{11} = \frac{5}{16} \; , \qquad J_{12} = \frac{1}{4} \; ,$$

$$J_{20} = \frac{11}{32} \; , \qquad J_{21} = \frac{15}{64} \; , \qquad J_{22} = \frac{199}{1024} \; .$$

In Figs. 1, 2, and 3, we present the radial density of coherent modes for different quantum numbers. The radial number n defines the number of nodes, while the winding number m characterizes circulation. For numerical calculations, we accept the parameters as in our previous experiments with ^{87}Rb, described in Refs. [29–33], where

$$\alpha = 0.11 \; , \qquad g = 1.96 \times 10^4 \; , \qquad \alpha g = 2.156 \times 10^3 \; .$$

For comparison, we present vortex solutions, with nonzero m, and non-circulating solutions, with $m = 0$.

Journal of Physics: Conference Series **691** (2016) 012019 doi:10.1088/1742-6596/691/1/012019

4. Generation of vortex rings

The typical radii of rings appearing in a trap are much larger than the coherence length l_c and, clearly, smaller than the trap radius R,

$$r_R \gg l_c \; , \qquad r_R < R \; . \tag{31}$$

The coherence length

$$l_c \equiv \frac{\hbar}{mc} = \frac{1}{\sqrt{4\pi\rho a_s}} \qquad \left(c = \frac{\hbar}{m} \sqrt{4\pi\rho a_s} \right) \tag{32}$$

defines the vortex ring core.

Ring characteristics can be estimated resorting to the Thomas-Fermi approximation [7]. Normalizing the Thomas-Fermi condensate function yields the condensate energy

$$E_{TF} = \frac{1}{2} \left(\frac{15}{4\pi} \, \alpha g \right)^{2/5} = 0.5367(\alpha g)^{2/5}$$

expressed in units of $\hbar\omega_\perp$. The condensate density at the trap center reads as

$$\rho(0) = 0.537 \, \frac{\alpha N}{l_\perp^3 (\alpha g)^{3/5}} \; .$$

Taking the latter for estimating the coherence length gives

$$l_c = R \sqrt{ \frac{4\pi L}{15 g l_\perp} } = 1.365 \, \frac{l_\perp}{(\alpha g)^{1/5}} \; . \tag{33}$$

The vortex ring characteristics have been considered by Iordanskii [34], Amit and Gross [35], Roberts and Grant [36], and Jones and Roberts [37]. Following Refs. [35, 36], for the energy of a vortex ring, in units of $\hbar\omega_\perp$, we have

$$E_R = 2\pi^2 \rho r_R l_\perp^2 \, \ln \left(2.25 \, \frac{r_R}{l_c} \right) \; , \tag{34}$$

where the density ρ is to be taken at the location of the ring.

Since the density is smaller close the boundary of a trapped atomic cloud, we expect that it is easier to create rings close to the boundary, since this requires smaller energy. The following numerical simulations confirm this.

Generally, a vortex ring moves with velocity that, far from the cloud edges, can be estimated as

$$v_R = \frac{\hbar}{2 m r_R} \, \ln \left(6.12 \, \frac{r_R}{l_c} \right) \; , \tag{35}$$

the velocity being directed perpendicular to the ring plane. The ring possesses a stationary position (although nonequilibrium) when its radius is $r_R \sim 0.6 \; R$. But in the standard situation the ring has an inherent tendency to propagate along the z axis. In a trap, the ring oscillates [38, 39]. Estimating the period of the ring oscillation [40] one has

$$T_R \sim \frac{4\pi R / l_c}{\omega_z \sqrt{\ln(R/l_c)}} \; . \tag{36}$$

Journal of Physics: Conference Series **691** (2016) 012019 doi:10.1088/1742-6596/691/1/012019

Substituting here the expressions for R and l_c, we find

$$T_R \sim \frac{4\pi(\alpha g)^{2/5}}{\omega_z\sqrt{\ln(0.5\alpha g)}} \; . \tag{37}$$

For the accepted setup, where $\omega_z = 2\pi \times 23$ Hz, we get $T_R \sim 0.7$s.

We accomplish numerical solution of the NLS equation, with the shaking potential (9), using the method of Ref. [41]. It turns out that the first nonequilibrium stage of perturbation, classified in Refs. [30–32] as the regime of weak nonequilibrium, is not as dull. At this stage, there are yet no vortices. However, there appear vortex rings. A vortex ring is defined as a circular line of zero density, with a winding number ± 1 around each of the line elements. Numerically, zero density implies $10^{-8}\rho(0)$. Some typical vortex rings are shown in Fig. 4. Rings usually appear in pairs, so that their total number is even.

In order to understand why vortex rings appear before vortices, we need to compare their energies. The energy of a vortex of length L can be written [7,31,32] as

$$E_L = \frac{2\pi}{3}\,\rho(0) l_\perp^2 L \ln\left(0.949\,\frac{R}{l_c}\right) \; , \tag{38}$$

where $\rho(0)$ is the density at the trap center. Since the total number of atoms, in the Thomas-Fermi approximation, is

$$N \equiv \int \rho(\mathbf{r})\, d\mathbf{r} = \frac{4\pi}{15}\,\rho(0) R^2 L \; , \tag{39}$$

it is also possible to write

$$E_L = \frac{5 l_\perp^2}{2R^2}\, N \ln\left(0.949\,\frac{R}{l_c}\right) \; .$$

Comparing the vortex energy (38) with the ring energy (34), we take $r_R \sim 0.6R$, which gives the relation $\rho(r_R) \sim 0.7\rho(0)$ between the atomic density close to the ring location and the density at the center. Then we obtain

$$\frac{E_R}{E_L} \sim \alpha \; . \tag{40}$$

In the considered setup, $\alpha \sim 0.1$. Hence the ring energy is an order smaller than the vortex energy. Therefore it is much easier to generate vortex rings, requiring to pump into the system much less energy than that needed for producing vortices. It is possible to expect that up to around ten vortex rings could be generated before vortices will start arising. The energy of other coherent modes, including vortex ring solitons is larger than that of basis vortices with the winding number ± 1. Hence their appearance is even less probable than that of the simple vortices, unless special resonant conditions are imposed.

5. Conclusion

We have analyzed the possibility of creating vortex ring solitons and vortex rings in a shaken Bose-Einstein condensate of trapped atoms. The consideration is based on the NLS equation, assuming that practically all atoms are condensed. Vortex ring solitons are dark solitonic solutions, with circulation around the symmetry axis of the trap. Vortex rings are circular vortices with circulation around each of their elements. Vortex ring solitons can be described analytically by solving the NLS equation using optimized perturbation theory. This solution is valid for an arbitrary strength of atomic interactions, contrary to the simple perturbation theory admissible for only very weak interactions.

Numerical simulation of the three-dimensional NLS equation demonstrates the appearance of vortex rings. This happens in the regime of a rather moderate perturbation, yet before vortices could arise. That regime corresponds to the weak nonequilibrium stage, according to the classification of Refs. [30–32].

The energy of a vortex ring is essentially smaller than that of a basic vortex with circulation ± 1. In turn, the energy of a simple vortex is smaller than that of other coherent modes, including vortex ring solitons [42,43]. This explains why, under moderate trap shaking, vortex rings appear long before vortices, since their generation requires much less energy to be injected into the trap. Even if higher-energy coherent modes would be produced at the initial moment of time, they would disintegrate into the lower-energy modes, such as vortices and vortex rings. Vortex rings can also occur in the stage of vortex turbulence [44,45]. However, to reach this regime requires much stronger trap shaking. The investigation of the possible vortex-ring creation under vortex turbulence, realized by trap shaking, is in the progress.

Acknowledgement
Financial support from the Russian Foundation for Basic Research (grant # 14-02-00723) and from the University of São Paulo NAP Program is appreciated.

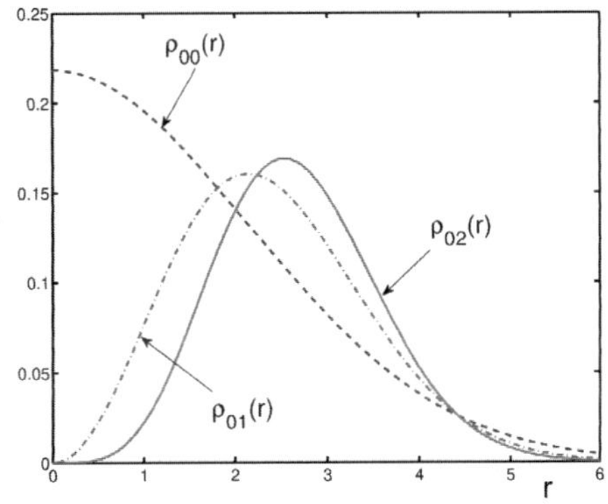

Figure 1. Radial density $\rho_{0m}(r)$ as a function of the dimensionless radius (in units of the transverse oscillator length $l_\perp$). The density of the ground state $\rho_{00}(r)$ (dashed line) is compared with the densities of the vortex states with the winding number $m = 1$ (dashed-dotted line) and $m = 2$ (solid line).

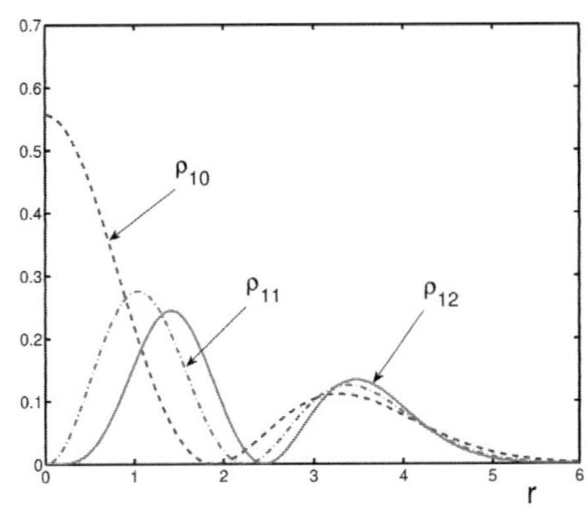

Figure 2. Radial density $\rho_{1m}(r)$ of the ring state $\rho_{10}(r)$ (dashed line), composite vortex + ring states $\rho_{11}(r)$ (dashed-dotted line) and $\rho_{12}(r)$ (solid line).

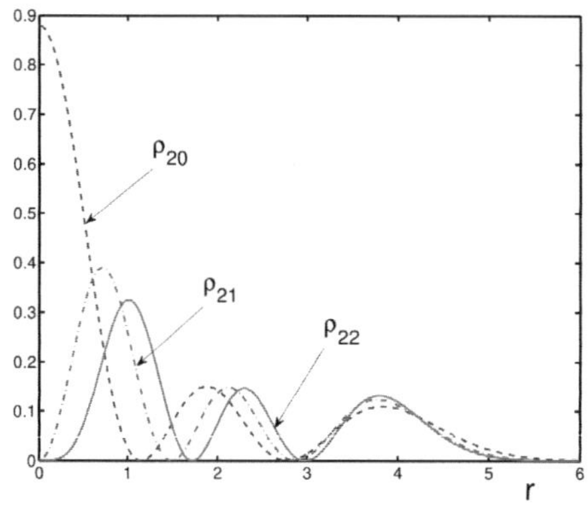

Figure 3. Radial density $\rho_{2m}(r)$ of the double ring state $\rho_{20}(r)$ (dashed line), composite vortex + double-ring states $\rho_{21}(r)$ (dashed-dotted line) and $\rho_{22}(r)$ (solid line).

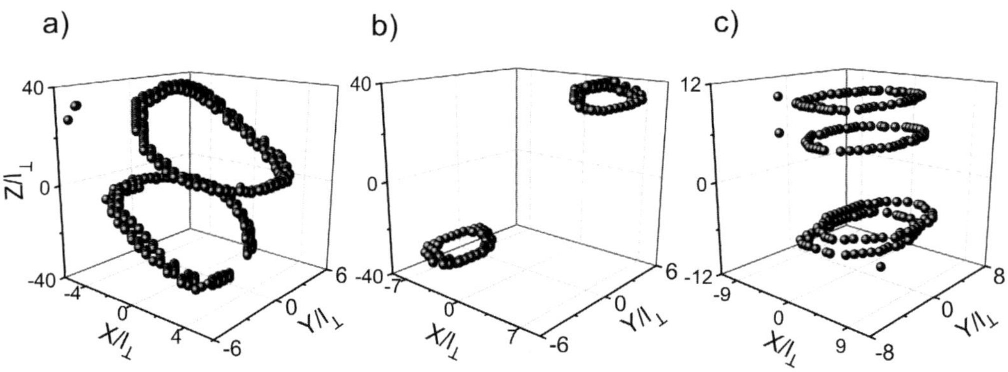

Figure 4. Spatial location of vortex rings after different shaking time: (a) $t = 10.5$ ms; (b) $t = 11$ ms; (c) $t = 14.4$ ms.

[1] Kivshar Y S and Luther-Davis B 1998 *Phys. Rep.* **298** 81
[2] Briedis D, Petersen D E, Edmundson D, Krolikowski W, and Bang O 2005 *Opt. Express* **13** 435
[3] Komineas S 2007 *Eur. Phys. J. Spec. Top.* **147** 133
[4] Frantzeskakis D J 2010 *J. Phys. A* **43** 213001
[5] Malomed B A 2006 *Soliton Management in Periodic Systems* (New York: Springer)
[6] Kartashov Y V, Malomed B A, and Torner L 2011 *Rev. Mod. Phys.* **83** 247
[7] Pethick C J and Smith H 2008 *Bose-Einstein Condensation in Dilute Gases* (Cambridge: Cambridge University)
[8] Yukalov V I, Yukalova E P, and Bagnato V S 1997 *Phys. Rev. A* **56** 4845
[9] Yukalov V I, Yukalova E P, and Bagnato V S 2000 *Laser Phys.* **10** 26
[10] Yukalov V I, Yukalova E P, and Bagnato V S 2001 *Laser Phys.* **11** 455
[11] Yukalov V I, Yukalova E P, and Bagnato V S 2002 *Laser Phys.* **12** 231
[12] Yukalov V I and Bagnato V S 2009 *Laser Phys. Lett.* **6** 399
[13] Yukalov V I 2011 *Phys. Part. Nucl.* **42** 460
[14] Allen A J, Parker N G, Proukakis N P, and Barenghi C F 2014 *Phys. Rev. A* **89** 025602
[15] Kivshar Y S and Yang X 1994 *Phys. Rev. E* **50** 40
[16] Brand J and Reinhardt W P 2001 *J. Phys. B* **34** L113
[17] Theocharis G, Frantzeskakis D J, Kevrekidis P G, Malomed B A, and Kivshar Y S 2003 *Phys. Rev. Lett.* **2003** 120403
[18] Car L D and Clark C W 2006 *Phys. Rev. A* **74** 043613
[19] Li J, Wang D S, Wu Z Y, Yu Y M, and Liu W M 2012 *Phys. Rev. A* **86** 023628

[20] Adhikari S K 2003 *New J. Phys.* **5** 137
[21] Adhikari S K 2004 *Phys. Rev. A* **69** 063613
[22] Kadomtsev B B and Petviashvily V I 1970 *Phys. Dokl.* **15** 539
[23] Anderson B P, Haljan P C, Regal C A, Feder D L, Collins L A, Clark C W, and Cornell E A 2001 *Phys. Rev. Lett.* **86** 2926
[24] Ginsberg N S, Brand J, and Hau L V 2005 *Phys. Rev. Lett.* **94** 040403
[25] Svistunov B V 1995 *Phys. Rev. B* **52** 3647
[26] Laurie J and Baggaley A W 2015 *J. Low Temp. Phys.* **180** 95
[27] Yukalov V I 1976 *Moscow Univ. Phys. Bull.* **31** 10
[28] Yukalov V I 1976 *Theor. Math. Phys.* **28** 652
[29] Shiozaki R F, Telles G D, Yukalov V I, and Bagnato V S 2011 *Laser Phys. Lett.* **8** 393
[30] Yukalov V I, Novikov A N, and Bagnato V S 2014 *Laser Phys. Lett.* **11** 095501
[31] Yukalov V I, Novikov A N, and Bagnato V S 2015 *J. Low Temp. Phys.* **180** 53
[32] Yukalov V I, Novikov A N, and Bagnato V S 2015 *Phys. Lett. A* **379** 1366
[33] Seman J A, Henn E A, Shiozaki R F, Roati G, Poveda-Cuevas F J, Magalhães K M F, Yukalov V I, Tsubota M, Kobayashi M, Kasamatsu K, and Bagnato V S 2011 *Laser Phys. Lett.* **8** 691
[34] Iordanskii S V 1965 *J. Exp. Theor. Phys.* **21** 467
[35] Amit D and Gross E P 1966 *Phys. Rev.* **145** 130
[36] Roberts P H and Grant J 1971 *J. Phys. A* **4** 55
[37] Jones C A and Roberts P H 1982 *J. Phys. A* **15** 2599
[38] Jackson B, McCann J F, and Adams C S 1999 *Phys. Rev. A* **61** 013604
[39] Reichl M D and Mueller E J 2013 *Phys. Rev. A* **88** 053626
[40] Bulgac A, Forbes M M, Kelley M M, Roche K J, and Wlazlowski G 2014 *Phys. Rev. Lett.* **112** 025301
[41] Novikov A N, Yukalov V I, and Bagnato V S 2015 *J. Phys. Conf. Ser.* **594** 012040
[42] Courteille P W, Bagnato V S, and Yukalov V I 2001 *Laser Phys.* **11** 659
[43] Bagnato V S and Yukalov V I 2013 *Prog. Opt. Sci. Photon.* **1** 377
[44] Tsubota M, Kobayashi M, and Takeushi H 2013 *Phys. Rep.* **522** 191
[45] Nemirovskii S R 2013 *Phys. Rep.* **524** 85

Radiation Trapping in a Cold and Dense Atomic Ensemble in a Magnetic Field

Igor Sokolov
Peter the Great St.Petersburg Polytechnic University, Polytechnicheskaya st. 29, Saint-Petersburg, 194251, Russia
E-mail: `ims@is12093.spb.edu`

Abstract. We report a theoretical investigation of cooperative spontaneous decay of atomic excitation in cold ensemble under condition that average interatomic separation is comparable with resonant light wavelength. We show that magnetostatic field causes some acceleration of decay at shot time interval upon turn off the excitation pulse and essential decreasing of decay rate of long-lived sub-radiant states. The dependence of this effect on duration of excitation pulse as well as on its carrier frequency is considered.

1. Introduction

Analysis of time-dependent fluorescence of cold atomic ensembles excited by pulse radiation is one of the most effective methods of studying of properties of such ensembles [1]-[7]. This analysis attracts particular interest recently in connection with cooperative effects in dense atomic clouds where both mean free path of resonant photons and averaged interatomic separation are comparable with wavelength [5, 6]. In dense and cold ensembles resonant interatomic dipole-dipole interaction can cause a number of fascinating quantum optical processes, including Anderson localization of light and atom-based random lasing. Atomic fluorescence can be an indicator of these processes.

The main goal of the present work is to study theoretically the influence of static magnetic field on cooperative spontaneous decay of excited atomic states in dense and disorder cold atomic ensembles. Magnetostatic field changes essentially properties of separate atomic scatterers in the ensemble. It leads to significant modification of multiatomic effects which take place under multiple light scattering. Such modification was studied for highly excited Rydberg states in [8]. In paper [9] it was shown that magnetic field can cause transition from extended to localized quasimodes for light in a gas of immobile two-level atoms in a magnetic field. This transition takes place either upon increasing the number density of atoms in a strong field or upon increasing the field at a high enough density. In the absence of the field the transition is impossible [10].

The conclusions about the role of static magnetic field made in the paper [9] were based on numerical analysis of collective multiatomic states. It is important to verify the predictions of the work [9] experimentally. One of the possible ways to do this is to measure dynamics of afterglow of atomic clouds located in magnetic field. In the present work we analyze such possibility. We calculate dynamics of total population of all excited atomic states of all atoms in the ensemble which determines the rate of fluorescence summarized over all angles. Comparison of results obtained for the cases with and without magnetic field gives us opportunity to reveal the role of the latter. In the present paper we restrict ourselves by the case when collective effect and particularly effects of recurrent light scattering inside considered media are important but the density is not high enough to expect localization transition. Such intermediate densities can be rather easily achieved in dipole trap.

2. Basic assumptions and approaches

At present there are several approaches to description of light interaction with atomic clouds which allow taking into consideration effects of recurrent scattering. These approaches give us opportunity to describe influence of resonant dipole-dipole interaction on optical properties of dense and cold gases [11]-[19]. In this work we use the consistent quantum-posed theoretical approach developed previously in [18]. In the framework of this approach we solve the nonstationary Schrodinger equation for the wave function of the joint system consisting of N atoms and a weak electromagnetic field. We restrict the total number of states taken into account to those with no more than one photon in the field. It allows us to obtain a finite set of equations for amplitudes of states with one excited atom in the ensemble. This set of equations is solved numerically and amplitudes of the other states are calculated through found ones. The procedure allows us to find approximately the wave function of the system and consequently to analyze both the properties of atomic system and the light. In papers [18, 20] general approach is justified and laid out in detail and we will not reproduce it here. The generalization of this approach on the case of pulse excitation was performed in the paper [5]. Influence of external magnetic field was considered in [9]. The reader is referred to those papers for the theoretical development and justifications.

For the purpose of this paper we will consider a spherically-shaped atomic cloud with a radius R. The random distribution of atoms inside the cloud is assumed to be uniform, on average. Total number of atoms is N. All atoms are identical with a ground state J = 0 and excited state J = 1. The latter is split into three Zeeman sublevels (m_e = -1,0,1). The natural linewidths of the three Zeeman sublevels of this state are the same and are equal to γ. The Zeeman splitting Δ of the three sublevels is determined by external magnetic field.

Atoms are assumed to be motionless. The influence of the residual motion typical for atomic traps is taken partially into account by means of examining the statistical ensemble of clouds with the random distribution of atoms. The main results presented below are obtained by averaging over random spatial distribution using the Monte Carlo method. The intensity of the radiation is assumed to be sufficiently small that all nonlinear effects are considered negligible. This approach, as well as the other approximations depicted above, allows us to use explicit expressions for time dependent population of total atomic excitation obtained in [5, 18, 20]. In the next section, we will use these expressions (which we do not reproduce here) to calculate the time dependent population of all excited states in dense atomic ensemble and analyze, on this foundation, the process of spontaneous decay in such ensemble.

3. Results

All calculation in this work are performed for spherical cloud with radius $R = 10$. The average atomic density is $n = 0.1$. Hereafter in this paper we use the inverse wave number of the resonance probe radiation in vacuum $k_0^{-1} = c/\omega$ as a unit of length. As it was shown in [18, 20, 21] considered density is high enough to guarantee manifestation of cooperative effect caused by recurrent scattering. At the same time it is not sufficient for localization transition even for strong magnetic field (see [9]).

We choose the value of magnetic field strength such that Zeeman splitting is much more than collective level shifts caused by dipole-dipole interaction. Specific calculations are done for $\Delta = 1000\gamma$.

In the figure 1 we show the result of our calculation of dime dependent afterglow of considered ensemble excited by short pulse. The pulse duration is $0.025\gamma^{-1}$. This pulse has a relatively big spectral width and excites a big number of collective atomic states. The currier frequency of the light coincides with transition frequency of the free atoms. The detuning of the currier frequency is zero $\delta = 0$. The solid line in the figure corresponds to zero magnetic field. It demonstrates typical behavior. Initially time dependence is determined by superradiant collective states and the decay rate exceed that for free atom (for comparison see a figure 2). The rate gradually decreases and as time passes we see practically one-exponential decay.

The other two curves in the figure 1 describe afterglow in strong magnetic field. These two curves differ from each other in conditions of excitation. Long-dashed curve corresponds to excitation of level $m_e = 1$ (exactly the same dependence take place for $m_e = -1$), whereas the short-dashed one describes the case of $m_e = 0$. It is clearly seen that in due course both these curves decrease noticeably slower that solid line. Thus we see that magnetic field increases the lifetime of subradiant, long-lived states. And the influence of the field is more essential in the case of $m_e = \pm 1$. This difference is manifestation of optical anisotropy of the ensembles induced by magnetic field and this result confirms prediction of the paper [9].

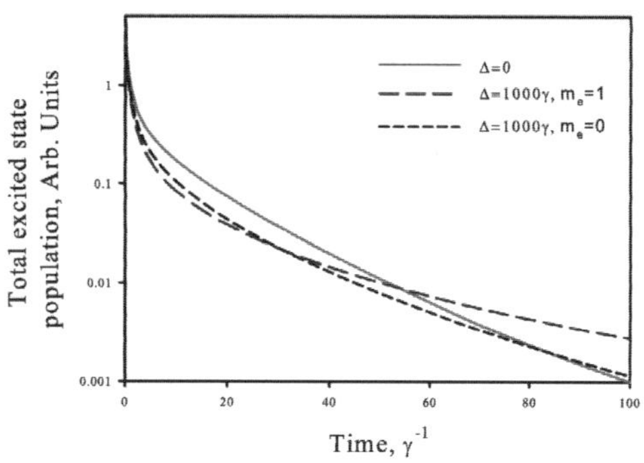

Figure 1. Time dependence of the total excited state population of a uniform spherical sample of atoms. Excitation is with a temporally short, $0.025\gamma^{-1}$ and resonant, $\delta = 0$ pulse. Curves with $\Delta = 1000\gamma$ describe the case of strong magnetic field, in the case when $\Delta = 0$ magnetic field is absent.

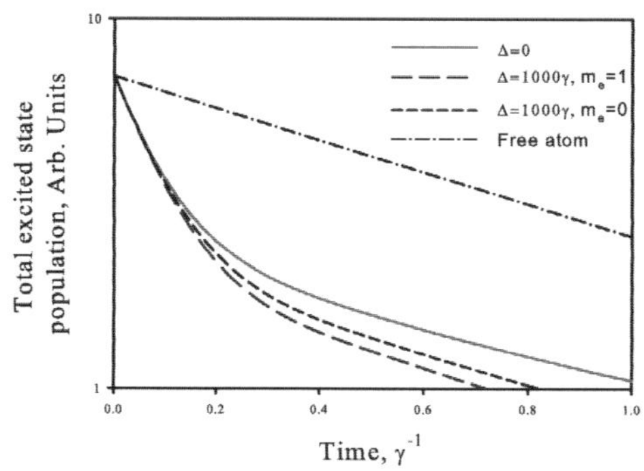

Figure 2. The same as in the figure 1, but for small time interval. Dash-dotted curve describes time dependence of free atom. Curves $m_e = 0$ and $m_e = 1$ correspond to different Zeeman sublevel excitation.

Note that decreasing of decay rate means that trapping time increases. This result seems unexpected if we analyze only far field effects. Really, magnetic field causes Zeeman splitting and thus increases resonant photon mean free path. So observed result tells us that magnetic field changes near field effects as well, i.e. dipole-dipole interatomic interaction. Moreover, this change is more important that changes in far field.

Magnetic field influences also on short-lived, superradiant states. From the figure 2 we see that magnetic field accelerates excitation decay at initial stage. And again this influence differs depending on the magnetic quantum number of the sublevels which take part in formation of corresponding collective state.

In the figures 1 and 2 we show the case of resonant excitation. For short pulse the picture varies only slightly with currier frequency of the radiation because considered pulse covers very wide spectral region with $\Delta\omega \simeq 40\gamma$. For the density $n = 0.1$ this region contains practically all excited collective states (see for example [5, 10]). The situation changes for long pulse with narrow spectrum. Corresponding dependencies are shown in figure 3-6. These figure are drawn for long, $1000\gamma^{-1}$ pulses.

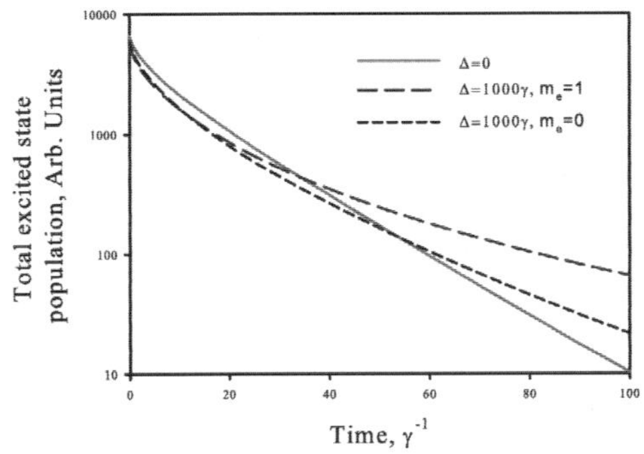

Figure 3. The same as in the figure 1, but for excitation with a temporally long, $1000\gamma^{-1}$ resonant pulse, $\delta = 0$.

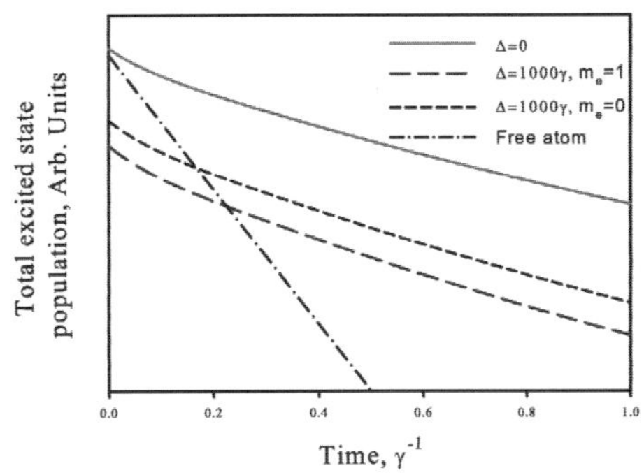

Figure 4. The same as in the figure 3, but for small time interval.

Figures 3 and 4 describe excitation with resonant light. The detuning of the currier frequency of the radiation from free atom transition frequency is zero, $\delta = 0$.

Narrowing of spectral width of the pulse leads to variation of excitation probability for both sub- and superradiant multiatomic states. Nearby resonance frequency subradiant states are excited more efficiently. It is well seen if one compares figures 3 and 1. For big time interval increasing of pulse duration causes decreasing of decay rate. For short-lived states the situation is opposite. The excitation efficiency decreases with pulse duration (see figure 4). Moreover, for considered conditions the rate of decay is less than for free atom at the beginning of decay process. It means that subradiant states do not become apparent here.

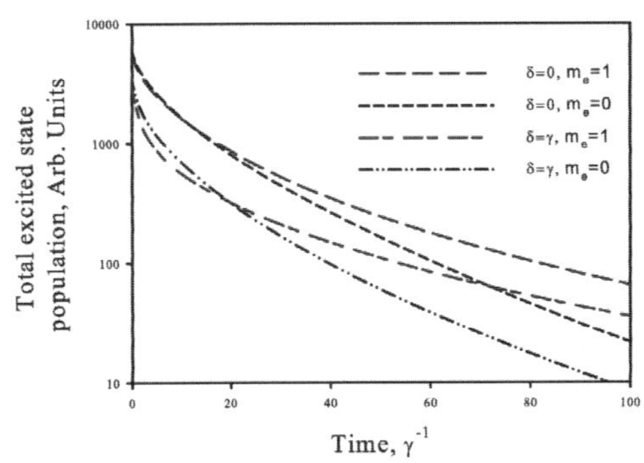

Figure 5. Comparison of resonant, $\delta = 0$, and nonresonant, $\delta = \gamma$, excitation of dense atomic clouds.

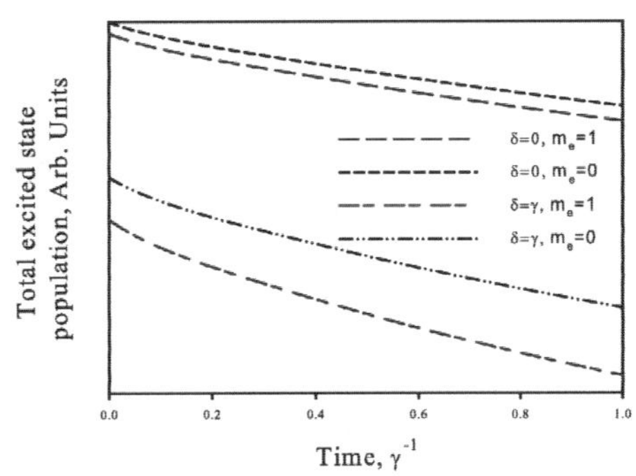

Figure 6. The same as in the figure 5, but for small time interval.

As our analysis shows (see [9, 10]) the density of long-lived states differs for different part of the spectrum. The results shown in figures 5 and 6 confirm that. In these figures we compare dynamics of spontaneous decay in ensembles excited by long pulses with different currier frequencies.

For radiation detuned from resonance at $\delta = \gamma$ the efficiency of excitation decreases but the influence of magnetic field becomes far more essenyial. Decay rate decreases for level $m_e = 0$ and even more for levels $m_e = \pm 1$. It means that in spectral region nearby $\delta = \gamma$ the density of superradiant states is higher than nearby $\delta = 0$.

4. Conclusion

In conclusion, we have found that the magnetic field has an influence on a collective spontaneous decay of cold and dense atomic ensembles. This influence manifests itself mainly by means of modification of dipole-dipole interatomic interaction. In our opinion experimental verification of predicted effects is of great interest. Measurement of frequency dependence of resonant fluorescence as well as its dependence on exciting pulse duration will allow one to get detailed information about peculiarities of collective effects in dense atomic clouds placed in magnetic field.

Acknowledgments

Acknowledgments The author acknowledges fruitful discussions with M.Havey. This work is supported by the Russian Foundation for Basic Research (Grant No. RFBR-15-02-01013).

References

[1] Labeyrie G, E. Vaujour E, Muller C A, Delande D, Miniatura C, Wilkowski D and Kaiser R 2003 *Phys. Rev. Lett.* **91** 223904
[2] Fioretti A, Molisch A F, Muller J H, Verkerk P and Allegrini M 1998 *Opt. Commun.* **149** 415
[3] Balik S, Olave R G, Sukenik C I, M. D. Havey M D, Datsyuk V M, Sokolov I M and Kupriyanov D V 2005 *Phys. Rev.* A **72** 051402(R)
[4] Sokolov I M, Kupriyanov D V, Olave R G and Havey M D 2010 *Journal of Modern Optics* **57** 1833
[5] Balik S, Win A L, Sukenik C I, M. D. Havey M D, Sokolov I M and Kupriyanov D V 2013 *Phys. Rev.* A **87** 053817
[6] Pellegrino J, Bourgain R, Jennewein S, Sortais Y R P, Browaeys A, Jenkins S D and Ruostekoski J 2014 *Phys. Rev. Lett.* **113** 133602
[7] Guerin W, Araujo M O and Kaiser R 2015 Dicke subradiance in a large cloud of cold atoms *Preprint* arXiv:1509.00227
[8] Afrousheh K, Bohlouli-Zanjani P., Carter J D, Mugford A and Martin J D D 2006 *Phys. Rev.* A **73** 063403
[9] Skipetrov S E and Sokolov I M 2015 *Phys. Rev. Lett.* **114** 053902
[10] Skipetrov S E and Sokolov I M 2014 *Phys. Rev. Lett.* **112** 023905
[11] Lehmberg R H 1970 *Phys. Rev.* A **2**, 883
[12] Milonni P W and Knight P L 1974 *Phys. Rev.* A **10**, 1096
[13] Morice O, Castin Y and J. Dalibard J 1995 *Phys. Rev.* A **51**, 3896
[14] Ruostekoski J and Javanainen J 1997 *Phys. Rev.* A **55**, 513
[15] Fu H and Berman P R 2005 *Phys. Rev.* A **72**, 022104
[16] Rusek M, Mostowski J and Orlowski A 2000 *Phys. Rev.* A **61**, 022704
[17] Pinheiro F A, Rusek M, Orlowski A and van Tiggelen B 2004 *Phys. Rev.* E **69**, 026605
[18] Sokolov I M, Kupriyanov D V and Havey M D 2011 *J. Exp. Theor. Phys.* **112** 246
[19] Kuznetsov D V, Roerich Vl K and Gladush M G 2011 *J. Exp. Theor. Phys.* **113** 647
[20] Fofanov Ya A, Kuraptsev A S, Sokolov I M and Havey M D 2011 *Phys. Rev.* A **84** 053811
[21] Fofanov Ya A, Kuraptsev A S, Sokolov I M and Havey M D 2013 *Phys. Rev.* A **87** 063839

expelled quickly and the target becomes a plasma. The high-energy electrons move around the target and the target ions stay at rest during a short period of ~fs. Between the ions and the high-energy electrons, a strong electric field is generated and accelerates protons gradually. In the ion-source stage the ion beam tends to have a lower particle energy, a broad energy spectrum and a transverse divergence. For ion cancer therapy as an example application, the ion energy spectrum should be controlled and the ion energy should be about 250-400MeV to deposit the ion energy in cancer cells inside human body.

Therefore, the laser ion accelerator would need post-acceleration devices to enhance the ion energy by a solid target as shown in Fig. 2(a) or by a near-critical gas target [6], and collimators to suppress the ion divergence (see Fig. 3), as well as beam bunchers (see Fig. 2(b)) to compress the ion beam longitudinally. Depending on the requirements for the proton energy and the beam radius, the laser ion accelerator would be designed appropriately as shown in Fig. 1.

3. Robust collimation of laser produced ion beam

Figure 3 shows the collimation device and the simulation setup employed in this paper. The sub-wavelength fine structure at the thin foil itself increases the laser absorption ratio [9-12], and the larger holes at the target rear creates the transverse TNSA (target normal sheath acceleration) field to collimate the ion beam. The ion beam diameter of 10λ is smaller than the rear hole diameter of 19.7λ (the wall width in Figs. 3 and 4) to avoid the ion beam split [7, 8]. The pre-accelerated proton beam is shown in Figs. 5(a) and (d), and is introduced to the simulation box initially. The pre-accelerated proton beam has the divergence angle shown in Fig. 6(a) and the energy and spatial distributions in Figs. 5(a) and (d). The Gaussian laser intensity is 4.0×10^{19}W/cm^2, the laser spot size is 50λ and the laser pulse length is 100fs. Figure 4 shows the transverse collimation electric field at the target during the laser illumination. Figures 5(a)-(c) show the ion beam distributions at t=0fs, 300fs and 580fs without the collimation device, respectively. Figures 5(d)-(f) present the typical results with the collimation device at t=0fs, 300fs and 580fs, respectively. Figures 5(d)-(f) demonstrate that the collimation device is effective for the ion beam collimation. Figure 6 shows the ion beam divergence angle, and presents the clear collimation of the ion beam by the structured target.

We have investigated the robustness of the collimation device against the changes in the laser intensity, the laser illumination timing, the structured target wall length and the target wall width. The definitions for the target wall length and width are presented in Fig. 3. Figures 7(a)-(d) summarize the results for the parameter studies to find the robustness of the collimation device.

Figure 7(a) shows the relation between the ion beam divergence angle and the laser intensity. When the laser intensity is weak compared with the optimal laser intensity, the ion beam collimation field is

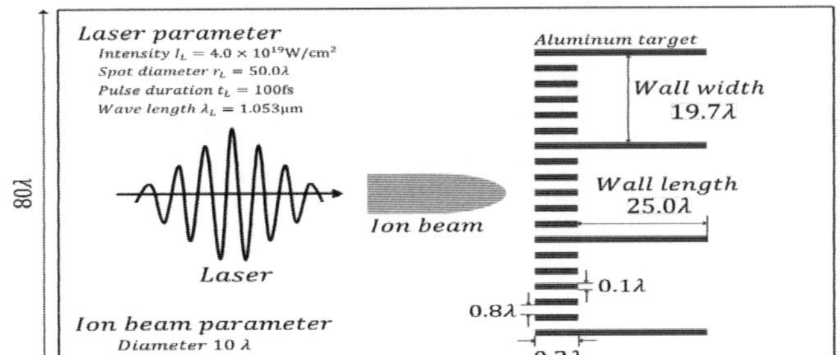

Figure 3. Structured thin-foil collimation device for ion beam. A pre-accelerated proton beam is introduced to the collimation device. The holes' walls create the transverse electric TNSA field to collimate the ion beam.

weak. On the other hand when the laser intensity is too high, the ion beam is over-focused. Figures 8 present the beam proton distributions and the proton divergence angle distributions for the laser intensity shown in the figure. The allowance of the laser intensity is rather wide as presented in Fig. 7(a). Figure 7(b) shows the relation between the ion beam divergence angle and the target wall width (see Fig. 1). For a specific laser intensity, when the wall width becomes narrow, the transverse collimation electric field becomes strong and so the collimation effect becomes too strong. When the wall width is too large, the electric field felt by the beam ions becomes small. Figures 8 shows typical example simulation results for the different target wall widths. In this paper the ion beam diameter is 10λ. Figures 7(b) and 9 show that the wall width should be the order of the beam diameter. Figure 7(c) shows the relation between the ion beam divergence angle and the target wall length (see Fig. 4). When the target wall length is too short, the time period, during which the beam protons feel the collimation transverse electric field, becomes short. So the collimation effect becomes weak for the short target wall length. When the target wall length becomes too long, a part of the long wall near the laser target interaction area in which the target has a fine structure has the collimation electric field. So the rest of the long target wall has no effect for the collimation, and the too-long wall does not degrade the collimation effect. Figures 10 present the typical simulation results in this case. Figure 7(d) presents the relation between the ion beam divergence angle and the laser timing. When the laser timing is too fast, just the head of the ion beam is collimated. When the laser timing is late, the tail of the ion beam is collimated. Figures 11 show the typical simulation results for the corresponding laser timing. The laser pulse length is 100fs in this paper, and the tolerable laser-timing shift is rather wide as shown in Fig. 7(d).

All the parameter study results are summarized in Figs. 7 and demonstrate that the collimation device of the structured target is rather robust against the changes in the laser parameter and the target parameter. The results presented in this paper show the viability of the laser-based collimation device for the collimation of the pre-accelerated ion beam, which may be produced by an intense-laser target interaction.

4. Conclusions

We have presented the robustness of the structured collimation target for the pre-accelerated ion beam against the laser and target parameter changes. In the collimation device the transverse electric field is employed to collimate the diverging beam ions and is generated by the intense-laser target interaction. The collimation device would be required for a long distance transportation of the ion beam pre-accelerated [6]. For actual applications of the ion beam, the ion beam should be transported in a long distance compared with the size of the laser target interaction area. The collimation device

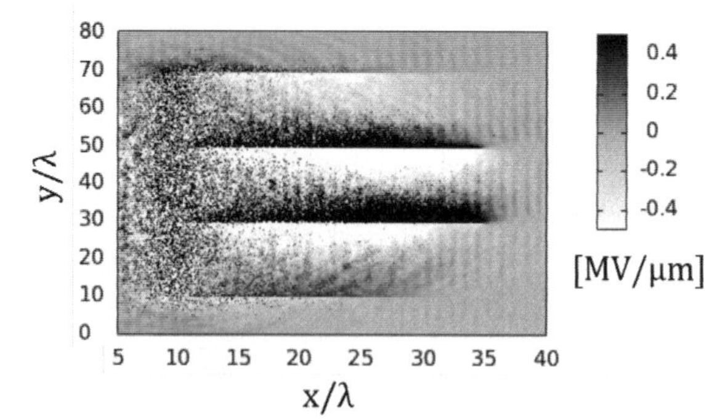

Figure 4. The transverse electric field. The Al structured target is illuminated by an intense laser. The fine structure absorbs the laser energy efficiently, and generates high-energy electrons. The TNSA transverse field is generated by the electrons collimates the proton beam.

proposed would be viable for a realistic long-distance ion beam transportation. For a practical laser-based ion beam accelerator [6], the multiple acceleration, collimation and bunching stages would be required to meet the application requirements for the ion beam quality, energy and intensity. The collimation device shown in this paper is one of the promising candidates for the pre-accelerated ion beam collimation.

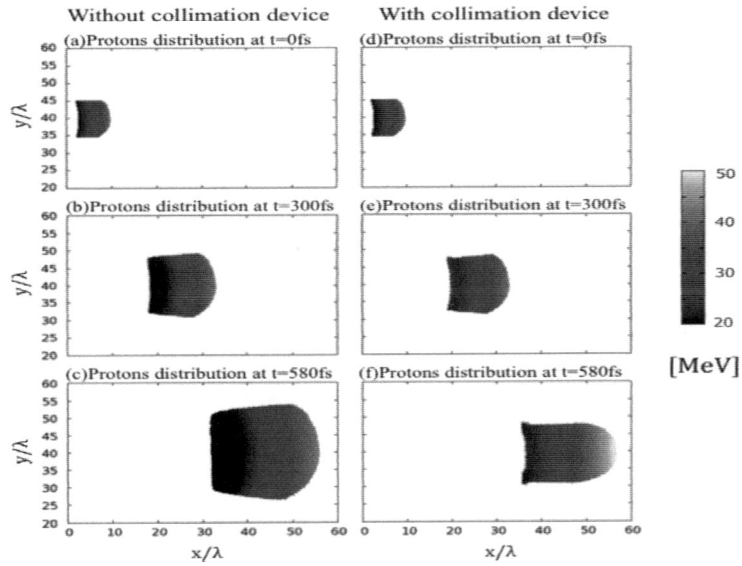

Figure 5. Spatial distributions of protons for the proton beam without the collimation device at (a)t =0fs, (b)t =300fs and (c)t=580fs, and for the proton beam with the collimation device at (d)t=0fs, (e)t=300fs and (f)t=580fs. The collimation device reduces the proton beam divergence successfully.

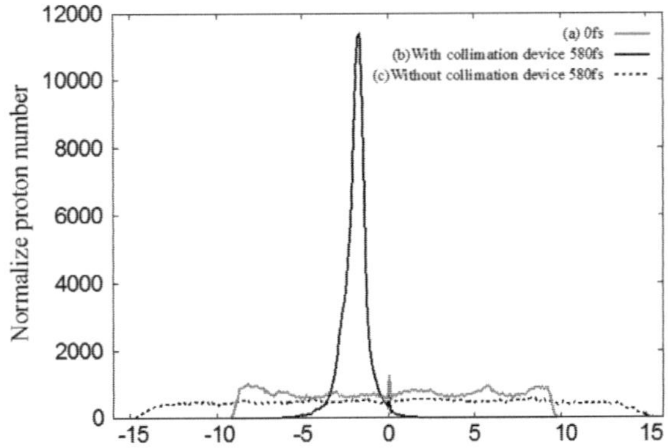

Figure 6. Divergence angle distributions for (a) the original proton beam (solid line), for (b) the proton beam without the collimation device (short-dotted line) and for (c) the proton beam with the collimation device (long-dotted line).

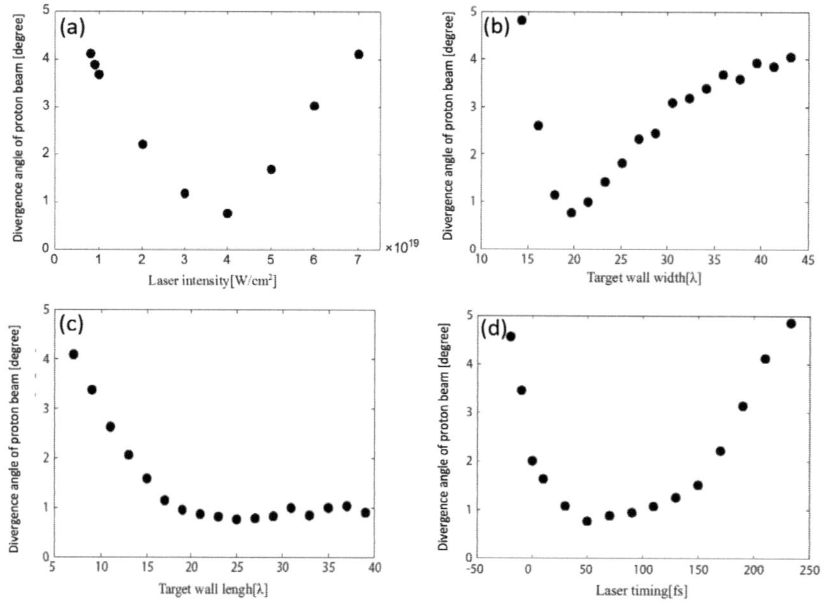

Figure 7. Divergence angle dependences of beam ions on (a) the laser intensity, (b) the target wall width, (c) the target wall length and (d) the laser timing. The collimation device shown in Fig. 3 is rather robust against the parameter changes in the laser parameters and the target structure.

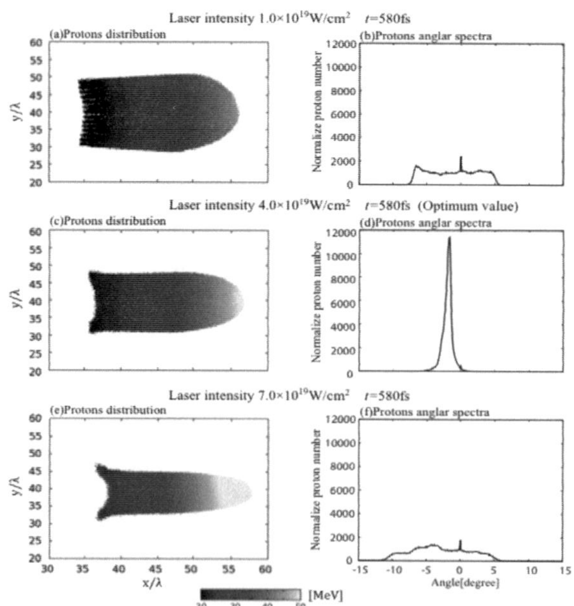

Figure 8. Distributions of the beam protons at the laser intensity of (a) $1.0\times10^{19}W/cm^2$, (c) $4.0\times10^{19}W/cm^2$ and (e) $7.0\times10^{19}W/cm^2$ at t=580fs. The corresponding distributions of the proton divergence angle are shown in (b), (d) and (f) at t=580fs.

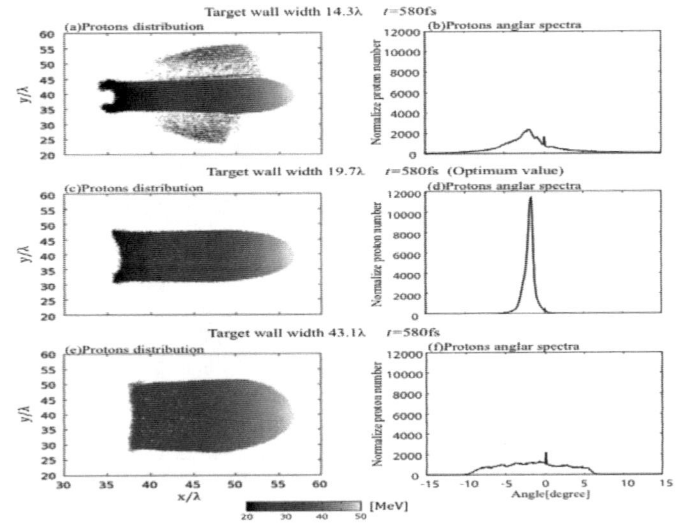

Figure 9. Distributions of the beam protons at the target wall width of (a) 14.3λ, (c) 19.7λ and (e) 43.1λ at t=580fs. The corresponding distributions of the proton divergence angle are shown in (b), (d) and (f) at t=580fs.

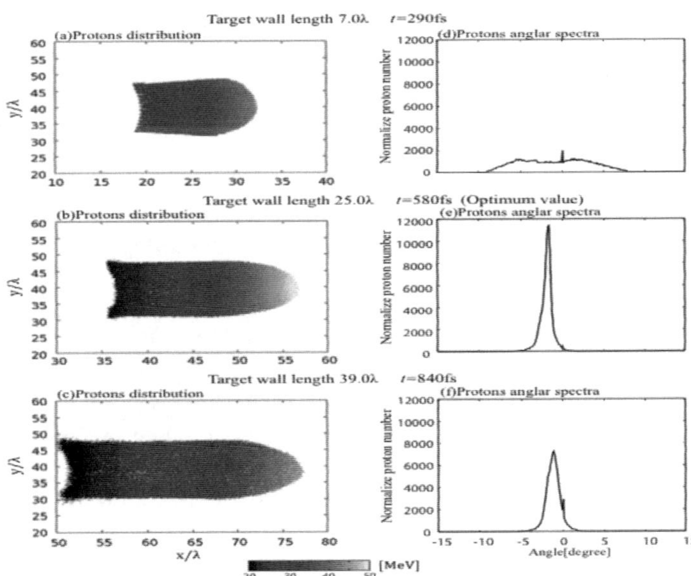

Figure 10. Distributions of the beam protons at the target wall length of (a) 7.0λ at t=290fs, (b)25.0λ at t=580fs and (c) 39.0λ at t=840fs. The corresponding distributions of the proton divergence angle are shown in (d), (e) and (f).

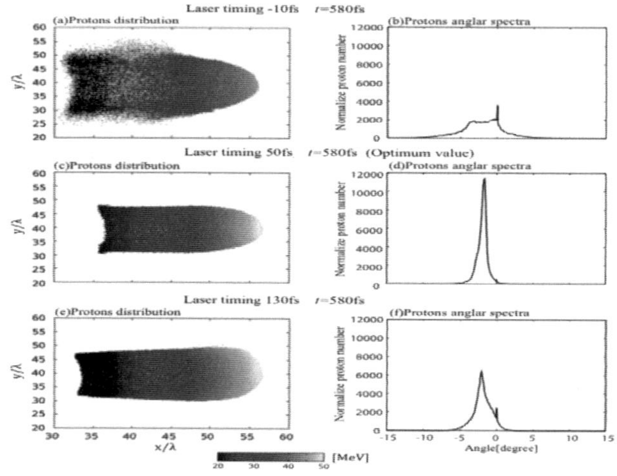

Figure 11. Distributions of the beam protons at the laser timing of (a) 10fs, (c) 50fs and (e) 130fs at t=580fs. The corresponding distributions of the proton divergence angle are shown in (b), (d) and (f) at t=580fs.

Acknowledgements
This work was partly supported by JSPS KAKENHI Grant Number 15K05359, MEXT, JSPS, the ASHULA project, ILE, Osaka University, CORE (Center for Optical Research and Education, Utsunomiya University, Japan), Fudan University and CDI (Creative Dept. of Innovation, CCRD, Utsunomiya University).

References

[1] Bulanov S V and Khoroshkov V S 2002 *Plasma Phys. Rep.* **28** 453
[2] Kaluza M, Schreiber J, Santala M I K, Tsakiris G D, Eidmann K, Meyer-ter-Vehn J and Witte K J 2004 *Phys. Rev. Lett.* **93** 045003
[3] Bulanov S V, Esirkepov T Zh, Califano F, Kato Y, Lisekina T V, Mima K, Naumova N M, Nishihara K, Pegorano F, Ruhl H, Sentoku Y and Ueshima Y 2000 *JETP Lett.* **71** 407
[4] Nakamura T and Kawata S 2003 *Phys. Rev. E* **67** 026403
[5] Nakamura T, Bulanov S V, Esirkepov T Z and Kando M 2010 *Phys. Rev. Lett.* **105** 135002
[6] Kawata S, Izumiyama T, Nagashima T, Takanao M, Barada D, Kong Q, Gu Y J, Wang P X, Ma Y Y and Wang W M 2013 *Laser Therapy* **22** 103
[7] Sonobe R, Kawata S, Miyazaki S, Nakamura M and Kikuchi T 2005 *Phys. of Plasmas* **12** 073104
[8] Nakamura M, Kawata S, Sonobe R, Kong Q, Miyazaki S, Onuma N and Kikuchi T 2007 *J. Appl. Phys.* **101** 113305
[9] Nodera Y, Kawata S, Onuma N, Limpouch J, Klimo O and Kikuchi T 2008 *Phys. Rev. E* **78** 046401
[10] Klimo O, Psikal J, Limpouch J, Proska J, Novotny F, Ceccotti T, Floquet V and Kawata S 2011 *New J. Phys.* **13** 053028
[11] Margarone D, Klimo O, et al. 2012 *Phys. Rev. Lett.* **109** 234801
[12] Dalui M, Wang W-M, et al. 2015 *Scientific Rep.* **5** 11930

Vacuum high-harmonic generation in the shock regime and photon-photon scattering dynamics

B. King[1], **P. Böhl**[2], **H. Ruhl**[2]

[1] Centre for Mathematical Sciences, Plymouth University, Plymouth PL4 8AA, United Kingdom

[2] Ludwig-Maximilians-Universität München, Theresienstraße 37, 80333 München, Germany

E-mail: b.king@plymouth.ac.uk

Abstract. The presence of charged quantum virtual states permits a nonlinear self-interaction of the electromagnetic field in vacuum. This interaction can be described as real photon-photon scattering. This effect has been calculated for the case of colliding plane waves, when the centre-of-mass energy is much less than the electron rest energy. The quantum effect is included in the classical electromagnetic field equations of motion by a standard effective approach based upon a weak-field expansion of the Heisenberg-Euler Lagrangian. Solving for the resultant electromagnetic field indicates a signal for real photon-photon scattering when the plane waves overlap, which can be significantly larger than the usually-considered asymptotic values that reach detectors. By considering arbitrary numbers of four- and six-photon scattering, the process of vacuum higher harmonic generation has been studied both analytically and numerically. A route to prolific harmonic generation is identified that does not depend on field strengths being necessarily close to the Schwinger limit. The resulting vacuum electromagnetic shock wave has been studied and a nonlinear shock parameter identified.

1. Introduction

Real photons are predicted to be able to scatter off one another in vacuum due to the presence of charged virtual states. The study of this process has a long history, with Weisskopf [1] and Heisenberg and Euler [2] producing the first comprehensive calculations, which were later reformulated by Schwinger [3]. For laser-based experiments, a useful approximation is to assume the centre-of-mass energy is much less than the rest electron energy. This permits using an approach based on the Heisenberg-Euler Lagrangian to include the quantum process of photon-photon scattering as an effective correction to the classical field equations of motion. A second useful approximation is that the peak electromagnetic field strength E is much lower than the critical, so-called "Schwinger" field strength $E_{\mathrm{cr}} = m^2c^3/e\hbar = 1.3 \times 10^{16}\,\mathrm{Vcm}^{-1}$. This permits performing a weak-field approximation of the Heisenberg-Euler Lagrangian, the leading-order of which describes four-photon scattering [4], which can be interpreted in the classical paradigm as "vacuum four-wave mixing". Higher orders in this expansion describe vacuum $2n$-wave mixing.

In vacuum four-wave mixing, the final photons can have the same energies as the initial photons, but with an altered wavevector [5, 6, 7, 8] and altered polarisation [9, 10, 11, 12]. Alternatively, final photons can be produced with an energy different to the probe, in either case equal to the sum or difference of the incoming photons [13, 14, 15]. Ongoing experiments measure real photon-photon scattering through the corresponding phase shift in a laser beam as it passes through a magnetised cavity [16, 17]. Reviews of calculations of real photon-photon scattering using intense lasers can be be found in [18, 19, 20].

One aim of the project was to include real photon-photon scattering in numerical simulation. Of particular interest was the dynamics of the electromagnetic field at finite time. The focus of this proceedings will be on the *overlap* scattered field, so-called as it is a source of scattered photons that is only present when the electromagnetic invariants are non-vanishing, such as in the overlap of counter-propagating plane waves. The generation of the zeroth, fundamental and second harmonics were simulated and showed excellent agreement with analytical calculations.

2. Low-frequency, weak-field photon-photon scattering

We consider an oscillating probe plane-wave pulse of electric field $\mathbf{E}_p$ counterpropagating with a slowly-varying background of electric field $\mathbf{E}_s$. The probe is right-moving $\mathbf{E}_p(x^-)$ whereas the background is left-moving $\mathbf{E}_s(x^+)$ where $x^\mp = t \mp z$ and we concentrate on the scattering of the probe wave. The electric (and magnetic fields $\mathbf{B}$) will be expressed in units of E_{cr}. The wave equation for the total electric field $\mathbf{E} = \mathbf{E}_p + \mathbf{E}_s$, which includes the quantum vacuum interaction can be written:

$$\left[\partial_t^2 - \partial_z^2\right] \mathbf{E} = \mathbf{T}\left[\mathbf{E}, \mathbf{B}\right] \tag{1}$$

where $\mathbf{B}$ is the total magnetic field and[1]:

$$\mathbf{T}\left[\mathbf{E}, \mathbf{B}\right] = -4\pi \left[\nabla \wedge \partial_t \mathbf{M} + \partial_t^2 \mathbf{P} - \nabla\left(\nabla \cdot \mathbf{P}\right)\right], \tag{2}$$

for dimensionless vacuum magnetisation $\mathbf{M} = (\alpha/m^4)\partial\mathcal{L}_{\text{HE}}/\partial\mathbf{B}$ and polarisation $\mathbf{P} = (\alpha/m^4)\partial\mathcal{L}_{\text{HE}}/\partial\mathbf{E}$, where the Heisenberg-Euler Lagrangian $\mathcal{L}_{\text{HE}}$ is given by [3]:

$$\mathcal{L}_{\text{HE}} = -\frac{m^4}{8\pi^2}\int_0^\infty ds\, \frac{\mathrm{e}^{-s}}{s^3}\left[s^2 ab \cot as \coth bs - 1 + \frac{s^2}{3}(a^2 - b^2)\right], \tag{3}$$

and the two electromagnetic and two secular invariants are defined thus:

$$\mathcal{F} = -F^2/4E_{\text{cr}}^2 = \frac{1}{2}(E^2 - B^2), \qquad \mathcal{G} = -FF^*/4E_{\text{cr}}^2 = \mathbf{E}\cdot\mathbf{B}, \tag{4}$$

$$a = \left[\sqrt{\mathcal{F}^2 + \mathcal{G}^2} + \mathcal{F}\right]^{1/2}, \qquad b = \left[\sqrt{\mathcal{F}^2 + \mathcal{G}^2} - \mathcal{F}\right]^{1/2}. \tag{5}$$

Here F and F^* are the Faraday tensor and its dual and we note that all fields are normalised to E_{cr}. When $E \ll 1$, one can perform a weak-field expansion of Eq. (3) to give:

$$\mathcal{L}_{\text{HE}} = \frac{m^4}{\alpha}\sum_{i=1}^{\infty}\mathcal{L}_i, \tag{6}$$

$$\mathcal{L}_1 = \frac{\mu_1}{4\pi}\left[\left(E^2 - B^2\right)^2 + 7(\mathbf{E}\cdot\mathbf{B})^2\right], \tag{7}$$

$$\mathcal{L}_2 = \frac{\mu_2}{4\pi}\left(E^2 - B^2\right)\left[2\left(E^2 - B^2\right)^2 + 13\left(\mathbf{E}\cdot\mathbf{B}\right)^2\right], \tag{8}$$

$$\mathcal{L}_3 = \frac{\mu_3}{4\pi}\left[3\left(E^2 - B^2\right)^4 + 22\left(E^2 - B^2\right)^2\left(\mathbf{E}\cdot\mathbf{B}\right)^2 + 19\left(\mathbf{E}\cdot\mathbf{B}\right)^4\right], \tag{9}$$

[1] This equation corrects the missing minus signs in [21, 22].

where $\mu_1 = \alpha/90\pi$, $\mu_2 = \alpha/315\pi$, $\mu_3 = 4\alpha/945\pi$. Then $\mathcal{L}_j$ describes photon-photon scattering involving $2+2j$ photons. For counterpropagating plane waves, $\nabla(\nabla\cdot\mathbf{P}) = 0$, and the solution to Eq. (1) can be written:

$$\mathbf{E}^{(1)} = \mathbf{E}^{(0)} + \Delta\mathbf{E}^{(0)}, \tag{10}$$

where $\mathbf{E}^{(0)} = \mathbf{E}_p^{(0)} + \mathbf{E}_s^{(0)}$ is the trivial vacuum solution, for which $\left(\partial_t^2 - \partial_z^2\right)\mathbf{E}^{(0)} = \mathbf{0}$ and the scattered field for a single photon-photon scattering event is:

$$\Delta\mathbf{E}^{(0)} = \Delta\overrightarrow{\mathbf{E}}^{(0)} + \Delta\overleftarrow{\mathbf{E}}^{(0)}, \tag{11}$$

which correspond to scattering in the forward and backwards directions respectively[2] and:

$$\Delta\overrightarrow{\mathbf{E}}^{(0)}(t,z) = -\int_{-\infty}^{z}\frac{dz'}{2}\,\mathbf{J}^{(0)}(t' = x^- + z', z') \tag{12}$$

$$\Delta\overleftarrow{\mathbf{E}}^{(0)}(t,z) = -\int_{z}^{\infty}\frac{dz'}{2}\,\mathbf{J}^{(0)}(t' = x^+ - z', z'), \tag{13}$$

where the vacuum current $\mathbf{J}^{(0)}$, fulfilling $\partial_t\mathbf{J}^{(0)} = -\mathbf{T}[\mathbf{E}^{(0)},\mathbf{B}^{(0)}]$, is the standard current term occurring in the modified Maxwell equations. Since all fields are plane waves, and the solution is also assumed to be of the form of a plane wave, the electromagnetic field can be expressed in terms of just the probe and background electric fields. Since each term of the weak-field expansion of the Heisenberg-Euler Lagrangian is expressed in terms of the electromagnetic invariants, and since these vanish identically for plane waves, only the probe-background cross terms survive so $\mathcal{L}_n \sim (E_sE_p)^{n+1}$. Since the vacuum current that occurs in the field equation of motions is acquired by differentiating $\mathcal{L}_n$ by the field, the current at each order is given by $J_n \sim (E_sE_p)^nE$. This simple power-counting argument shows how the vacuum current can be further written:

$$\mathbf{J}_n = \mathbf{J}_{n,s} + \mathbf{J}_{n,p}, \tag{14}$$

where:

$$\mathbf{J}_{n,s} = \boldsymbol{\varepsilon}_{n,s}\mu_n E_s^{n+1}(x^+)\partial_- E_p^n(x^-) \tag{15}$$

$$\mathbf{J}_{n,p} = \boldsymbol{\varepsilon}_{n,p}\mu_n E_p^{n+1}(x^-)\partial_+ E_s^n(x^+), \tag{16}$$

and the polarisation vectors $\boldsymbol{\varepsilon}_{n,s}$, $\boldsymbol{\varepsilon}_{n,p}$ are complicated functions of the wavevectors $\mathbf{k}_p$, $\mathbf{k}_s$ and polarisation vectors $\boldsymbol{\varepsilon}_p$, $\boldsymbol{\varepsilon}_s$ of the probe and background plane-wave solutions to the classical vacuum wave equation [21]. It is clearer to write quantities in terms of lightcone co-ordinates $x^\mp$ rather than t and z. Of most interest to us was the forward-scattered field, so let us specify the discussion to $\Delta\overrightarrow{\mathbf{E}}^{(0)}$. Then since we can write:

$$\Delta\overrightarrow{\mathbf{E}}(x^+,x^-) = -\int_{-\infty}^{x^+}\frac{dy}{4}\,\mathbf{J}(x'^{+} = y, x'^{-} = x^-), \tag{17}$$

the different parts of the current give *two* sources of real photon-photon scattering:

$$\Delta\overrightarrow{\mathbf{E}}_n(x^+,x^-) = -\boldsymbol{\varepsilon}_{n,p}\frac{\mu_n}{4}\partial_- E_p^n(x^-)\int_{-\infty}^{x^+}dy\;E_s^{1+n}(y) - \boldsymbol{\varepsilon}_{n,s}\frac{\mu_n}{4}E_p^{1+n}(x^-)\int_{-\infty}^{x^+}dy\;\partial_y E_s^n(y). \tag{18}$$

[2] The current has been redefined with a minus sign, correcting [21, 22]

We identify the first term with the *asymptotic* forward-scattered field as it is the term that remains when the two plane waves have passed through each other and are again well-separated:

$$\lim_{x^+\to\infty} \Delta\overrightarrow{\mathbf{E}}_n(x^+,x^-) = -\varepsilon_{n,p}\frac{\mu_n}{4}\partial_- E_p^n(x^-)\int_{-\infty}^{\infty} dy\; E_s^{1+n}(y). \tag{19}$$

The remaining term is referred to as the *overlap* forward-scattered field and can be written as a surface term:

$$\begin{aligned}\Delta\overrightarrow{\mathbf{E}}_n^{\circ}(x^+,x^-) &= -\varepsilon_{n,s}\frac{\mu_n}{4}E_p^{1+n}(x^-)\int_{-\infty}^{x^+} dy\;\partial_y E_s^n(y)\\ &= -\varepsilon_{n,s}\frac{\mu_n}{4}E_p^{1+n}(x^-)\left[E_s^n(x^+) - \lim_{y\to-\infty} E_s(y)\right],\end{aligned} \tag{20}$$

and is most significant when the probe and background most *overlap*, viz. when $E_p(x^-)E_s(x^+)$ is a maximum.

2.1. Overlap forward-scattered field

The overlap field $\Delta\overrightarrow{\mathbf{E}}_n^{\circ}(x^+,x^-)$ is only present if there is some inhomogeneity in the background field. For this reason, it does not occur in traditional treatments of photon-photon scattering that are used to define a vacuum refractive index. However, if a vacuum modified refractive index is employed at finite time to describe the outcome of photon-photon scattering experiments using high-intensity lasers, it would seem to be consistent to also include this overlap field. The modification of the vacuum refractive index due to an inhomogeneity in the background was recently derived in [23] and was found to be equivalent to the overlap field term studied here. Moreover, a similar surface term appeared also in the same year in calculations of the polarisation flip in four photon-photon scattering [11]. However, the Heisenberg-Euler Lagrangian is derived for an eternally constant electromagnetic field. Therefore, it is perhaps questionable to replace this constant background with a field of a finite duration and to make conclusions about effects at finite times. Much more satisfactory would be to derive an effective theory for a slowly-varying background, which would have some notion of what "finite time" meant. Extensions of the Heisenberg-Euler Lagrangian that contain field derivatives already exist [24, 25, 26], although a naïve application of these extended Lagrangians appear to be at a higher derivative order.

It is interesting to continue with a naïve interpretation of the overlap field. First, it would imply a difference in the number of photons scattered in the two cases: i) an ever-present constant background; ii) an *adiabatically* evolved quasi-constant background. Second, the generation of the second probe harmonic as it counterpropagates in a more slowly-varying background is possible in four-photon scattering although it vanishes identically in the standard asymptotic calculation [27]. Third, second-harmonic generation occurs at a lower order of the weak-field expansion in the overlap signal (in four-photon scattering) than in the asymptotic signal (in six-photon scattering). If the duration over which the strong background varies is of the order τ_s and the probe of frequency ω_p, then when the fields overlap, the second harmonic is dominated by the overlap signal if $E_s^2\omega_p\tau_s \ll 1$.

3. Numerical Method

The numerical treatment is based on the solution of the corresponding system of Maxwell equations following from (6) using the "pseudocharacteristic method of lines" (PCMOL) [28] matrix inversion. This allows one to convert the equations of motion into a system of ordinary

Journal of Physics: Conference Series **691** (2016) 012022 doi:10.1088/1742-6596/691/1/012022

differential equations (ODEs) and use the ODE-Solver CVODE [29]. In the linear case, Maxwell's equations in the plane wave setup considered here, can be written in matrix form:

$$\mathbb{1}_4\,\partial_t\mathbf{f} + \mathbf{Q}\,\partial_z\mathbf{f} = 0 \tag{21}$$

where $\mathbf{f} = (E_x, E_y, B_x, B_y)^T$, $\mathbb{1}_4$ is the identity matrix in four dimensions and $\mathbf{Q} = \mathrm{adiag}(1, -1, -1, 1)$ is an anti-diagonal matrix. Since the system is hyperbolic (i.e. it can be diagonalised with real eigenvalues), we can apply the PCMOL to discretise the system in space. As a first step, the system is transformed to the new diagonal basis $\mathbf{u} := \mathbf{S}\mathbf{f}$, in which one obtains four decoupled advection equations:

$$\partial_t\mathbf{u} + \mathbf{\Lambda}\partial_z\mathbf{u} = 0 \tag{22}$$

with $\mathbf{\Lambda} = \mathbf{SQS}^{-1} = \mathrm{diag}(-1, -1, 1, 1)$ being a diagonal matrix and

$$\mathbf{S} = \frac{1}{\sqrt{2}}\begin{pmatrix} -1 & 0 & 0 & 1 \\ 0 & 1 & 1 & 0 \\ 1 & 0 & 0 & 1 \\ 0 & -1 & 1 & 0 \end{pmatrix} \qquad \mathbf{u} := \mathbf{S}\,\mathbf{f} = \frac{1}{\sqrt{2}}\begin{pmatrix} B_y - E_x \\ E_y + B_x \\ E_x + B_y \\ B_x - E_y \end{pmatrix}. \tag{23}$$

The eigenvalues λ_i of $\mathbf{\Lambda}$ are called the "characteristic speeds" of the system. A positive (negative) sign of λ_i thereby corresponds to the component u_i travelling in the positive (negative) z-direction. Then the components u_i are mapped onto a $4N$-dimensional vector representing $\mathbf{u}$ on a co-located grid with N points such that the grid point with index $0 < n < N$ is assigned the field values $\left(u_1^{(n)}, u_2^{(n)}, u_3^{(n)}, u_4^{(n)}\right)$. Now the spatial derivatives are approximated using biased differences of fourth order accuracy. For $\lambda_i > 0$, $\partial_z u_i$ is calculated using backward differences and for $\lambda_i < 0$ using forward differences. At the boundary, the derivatives are calculated using values inside the box. The initial conditions are chosen such that the field invariants $\mathcal{F}$ and $\mathcal{G}$ and the field values at the boundary are essentially zero at the beginning. The flow-character of the advection equations implements open boundary conditions. The resulting set of ordinary differential equations is then integrated using the ODE-Solver CVODE. As a final step, the fields are transformed back to the original basis $\mathbf{f}$ for output.

For the nonlinear case, the resulting Maxwell equations can be written as

$$\big(\mathbb{1}_4 + \mathbf{X}\big)\partial_t\mathbf{f} + \big(\mathbf{Q} + \mathbf{Y}\big)\partial_z\mathbf{f} = 0 \tag{24}$$

where $\mathbf{X}$ and $\mathbf{Y}$ are the nonlinear corrections resulting from (6). Since the nonlinear corrections $\mathbf{X}, \mathbf{Y}$ depend only on local field values at each point, their discretised version is of block diagonal form: $\tilde{\mathbf{X}} = \oplus_{n=1}^{N}\tilde{\mathbf{X}}^{(n)}$. The upper index indicates the matrix $\mathbf{X}$ to be evaluated at point n and one has a similar expression for $\mathbf{Y}$. The main difference to the linear case is the appearance of the matrix $\mathbf{X}$ in front of the temporal derivatives. To convert the system to an ODE-form, we invert the matrices $\mathbb{1}_4 + \tilde{\mathbf{X}}^{(n)}$ at each grid point separately. The biased differencing is the same as in the linear case. The resulting system is then again integrated using CVODE with the same parameters as in the linear case. The signals are analysed using a discrete Fourier Transform in *Wolfram Mathematica* [30] under the assumption $\omega = |\mathbf{k}|$. For further details, the reader is referred to [22].

4. Single four and six photon scattering

The probe and background fields considered were of the form:

$$\mathbf{E}_p(x^-) = \boldsymbol{\varepsilon}_p\,\mathcal{E}_p\,\mathrm{e}^{-\left(\frac{x^-}{\tau_p}\right)^2}\cos\,\omega_p x^-; \qquad \mathbf{E}_s(x^+) = \boldsymbol{\varepsilon}_s\,\mathcal{E}_s\,\mathrm{e}^{-\left(\frac{x^+}{\tau_s}\right)^2}, \tag{25}$$

where $\boldsymbol{\varepsilon}_p \cdot \boldsymbol{\varepsilon}_p = \boldsymbol{\varepsilon}_s \cdot \boldsymbol{\varepsilon}_s = 1$ and we consider $\omega_p\tau_s \gg 1$ and the momentum three-vectors are $\mathbf{k}_p = \omega_p(0, 0, 1)$, $\mathbf{k}_s = -\omega_s(0, 0, 1)$. Both four- and six-photon scattering were included in the vacuum interaction. In what follows, we concentrate on the scattered probe field, $\Delta\mathbf{E}_p$.

4.1. Fundamental Harmonic

The signal of the fundamental harmonic is dominated by the usual, asymptotic signal, but acquires a small term due to the overlap field[3]:

$$\Delta \mathbf{E}_p = \mu_1 \boldsymbol{\varepsilon}_{1,s} \mathcal{E}_p \mathcal{E}_s^2 \mathrm{e}^{-\left(\frac{x^-}{\tau_p}\right)^2} \left[\sqrt{\frac{\pi}{2}}\, \omega_p \tau_s \frac{1 + \mathrm{erf}\, \frac{\sqrt{2}x^+}{\tau_s}}{2} \sin \omega_p x^- - \mathrm{e}^{-2\left(\frac{x^+}{\tau_s}\right)^2} \cos \omega_p x^- \right], \tag{26}$$

where $\boldsymbol{\varepsilon}_{1,s} = c_{1,1}\, \boldsymbol{\varepsilon}_s + c_{1,2}\, \widehat{\mathbf{k}}_s \wedge \boldsymbol{\varepsilon}_s$ with coefficients $c_{1,1} = 4\boldsymbol{\varepsilon}_s \cdot \boldsymbol{\varepsilon}_p (1 - \widehat{\mathbf{k}}_s \cdot \widehat{\mathbf{k}}_p)$ and $c_{1,2} = 7\,(\boldsymbol{\varepsilon}_s \cdot \widehat{\mathbf{k}}_p \wedge \boldsymbol{\varepsilon}_p + \boldsymbol{\varepsilon}_p \cdot \widehat{\mathbf{k}}_s \wedge \boldsymbol{\varepsilon}_s)$. When $\Delta E_p \ll E_p$, one can express the scattered field as an effective refractive index that leads to ΔE_p. Suppose we take the limit $\tau_p \to \infty$ and consider a monochromatic probe and study the case of parallel probe and background polarisations. Then for a phase change $\delta\varphi_p$,

$$\mathbf{E}_p\,(\omega_p x^- + \delta\varphi_p) = \mathbf{E}_p(\omega_p x^-) + \delta\varphi_p\, \mathbf{E}_p'(\omega_p x^-) + O(\delta\varphi_p^2), \tag{27}$$

then we see if $\delta\varphi_p^{\|} = -\omega_p \delta z^{\|}$, then since:

$$\delta z^{\|} = \frac{1}{2} \int_{-\infty}^{x^+} \delta n^{\|}(x'^+) dx'^+, \tag{28}$$

we can infer:

$$\delta n^{\|} = 16\mu_1 \left[E_s^2 + \frac{E_p(\omega_p x^-)}{E_p'(\omega_p x^-)} \frac{1}{\omega_p} \partial_z E_s^2 \right] = \delta n_\infty^{\|} + \delta n_{\mathrm{o}}^{\|}, \tag{29}$$

where we label the first term the asymptotic modification to the vacuum refractive index $\delta n_\infty^{\|}$ compared to the overlap modification of $\delta n_{\mathrm{o}}^{\|}$ that includes derivative terms. This agrees with Eq. (29) in [23] for a monochromatic probe ($\tau_p \to \infty$) when one takes $\omega_\phi = \omega_s$, $\omega = \omega_p$, $\xi(\varphi) = E_p(\varphi_p)$, $\omega_s \Gamma_2 = [\partial_+ E_s(x^+)]/E_s(x^+)$. We note the similarity with the ponderomotive force on a charge, which is proportional to the gradient of the field squared [31]. Whether the change of vacuum refractive index due to the overlap field is a true refractive index effect would have to be investigated by including higher orders of four-photon scattering.

4.2. Second Harmonic

We find two sources of the second harmonic in the overlap field, originating from single four and six photon-photon scattering. The first of these can dominate the asymptotic signal from single six photon scattering, as demonstrated in Fig. 1[4]:

$$\begin{aligned} \Delta \mathbf{E}_p \;=\; & \mu_2\, \boldsymbol{\varepsilon}_{2,s}\, \mathcal{E}_p^2 \mathcal{E}_s^3\, \mathrm{e}^{-2\left(\frac{x^-}{\tau_p}\right)^2} \left[\sqrt{\frac{\pi}{3}}\, \omega_p \tau_s \frac{1 + \mathrm{erf}\, \frac{\sqrt{3}x^+}{\tau_s}}{2} \sin 2\omega_p x^- - \frac{1}{2}\, \mathrm{e}^{-3\left(\frac{x^+}{\tau_s}\right)^2} \cos 2\omega_p x^- \right] \\ & -\frac{1}{2}\, \mu_1\, \boldsymbol{\varepsilon}_{1,p}\, \mathcal{E}_p^2 \mathcal{E}_s\, \mathrm{e}^{-\left(\frac{x^-}{\tau_p}\right)^2} \cos 2\omega_p x^-, \end{aligned} \tag{30}$$

where the polarisation vectors are now $\boldsymbol{\varepsilon}_{1,p} = c_{1,1}\, \boldsymbol{\varepsilon}_p + c_{1,2}\, \widehat{\mathbf{k}}_p \wedge \boldsymbol{\varepsilon}_p$ and $\boldsymbol{\varepsilon}_{2,s} = c_{2,1}\, \boldsymbol{\varepsilon}_s + c_{2,2}\, \widehat{\mathbf{k}}_s \wedge \boldsymbol{\varepsilon}_s$. The overlap field can dominate the second-harmonic signal when $\mathcal{E}_s^2 \omega_p \tau_s \ll 1$. Such a situation is shown in Fig. 1 for parameters $\mathcal{E}_s = 0.02$, $\mathcal{E}_p = 0.005$, $\omega_p = 0.6\,\mathrm{eV}$, $\tau_s = 6.4\lambda_p$ and $\tau_p = 5\lambda_p$, where the field is in units of the probe amplitude, $\mathcal{E}_p$. The agreement between theory and numerical simulation is displayed in Fig. 2.

[3] This corrects a minus sign in the asymptotic signal in [21]

[4] This corrects a minus sign in the asymptotic signal in [21]

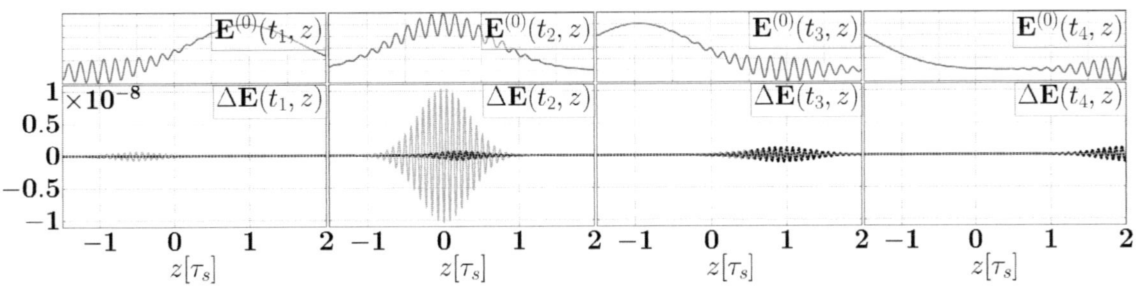

Figure 1. The thinner panels plot snapshots of the total electric field at various stages of the corresponding scattered overlap (greenish-blue solid) and asymptotic (black dashed) second harmonic field at times $t_4 > t_3 > t_2 > t_1$. Electric fields are normalised by the probe field amplitude, $\mathcal{E}_p$.

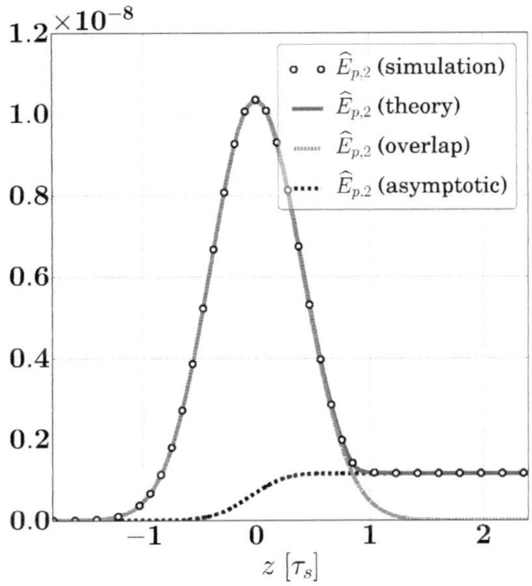

Figure 2. Agreement between numerical simulation (points) and analytical calculation (solid line) for generation of the second harmonic with $\mathcal{E}_s = 0.02$, $\mathcal{E}_p = 0.005$, $\omega_p = 0.6\,\text{eV}$, $\tau_s = 6.4\lambda_p$ and $\tau_p = 5\lambda_p$ (the field is in units of $\mathcal{E}_p$). The leading-order overlap signal from four-photon scattering (thin dashes) can dominate the leading-order asymptotic field from six-photon scattering (thick dashes).

5. Conclusion

In this conference proceeding, we have expounded upon the results in [21]. The process of four and six photon-photon scattering in counterpropagating plane waves has been studied using numerical simulation, which was shown to agree with analytical predictions. The main result was the prediction of an increased photon-photon scattering signal that occurs at finite time when probe and background pulses overlap. This signal occurs when one naïvely solves the classical wave equation in the presence of an effective vacuum current, derived from the Heisenberg-Euler Lagrangian. The standard vacuum refractive index [32] is derived for a constant electromagnetic background. If it is to be used to describe photon-photon scattering at finite time in a slowly-varying background, then it would seem to be inconsistent to neglect the interaction studied here, namely the part of the vacuum current proportional to the derivative of a power of the background. For the fundamental harmonic, the change in the field is proportional to the gradient of the background squared, and we note the similarity to the ponderomotive force on a charge in the field. If the duration over which the strong background varies is of the order τ_s and the probe of frequency ω_p, then the second harmonic is dominated by the overlap signal if $E_s^2\omega_p\tau_s \ll 1$.

6. Acknowledgments

B.K. acknowledges the hospitality of H.R. and the Arnold Sommerfeld Center for Theoretical Physics at the Ludwig Maximilian University of Munich. This work was partially funded by Deutsche Forschungsgemeinschaft DFG under contracts FOR968, RU633/1-1, SFB-TR18 project B12 and EXC-158 (cluster of excellence MAP). Plots were generated with Matplotlib [33].

References

[1] Weisskopf V 1936 *Kgl. Danske Videnskab. Selskab, Mat.-fys. Medd.* **14** 6
[2] Heisenberg W and Euler H 1936 *Z. Phys.* **98** 714
[3] Schwinger J 1951 *Phys. Rev.* **82** 664–679
[4] Karplus R and Neuman M 1950 *Phys. Rev.* **80**(3) 380–385
[5] Di Piazza A, Hatsagortsyan K Z and Keitel C H 2006 *Phys. Rev. Lett.* **97** 083603
[6] King B, Piazza A D and Keitel C H 2010 *Nature Photon.* **4** 92–94
[7] Kryuchkyan G Y and Hatsagortsyan K Z 2011 *Phys. Rev. Lett.* **107** 053604
[8] Monden Y and Kodama R 2011 *Phys. Rev. Lett.* **107** 073602
[9] Heinzl T *et al.* 2006 *Opt. Commun.* **267** 318–321
[10] King B, Di Piazza A and Keitel C H 2010 *Phys. Rev. A* **82** 032114
[11] Dinu V, Heinzl T, Ilderton A, Marklund M and Torgrimsson G 2014 *Phys. Rev. D* **89** 125003 URL `http://link.aps.org/doi/10.1103/PhysRevD.89.125003`
[12] Dinu V, Heinzl T, Ilderton A, Marklund M and Torgrimsson G 2014 *Phys. Rev. D* **90** 045025 URL `http://link.aps.org/doi/10.1103/PhysRevD.90.045025`
[13] Brodin G, Marklund M and Stenflo L 2001 *Phys. Rev. Lett.* **87** 171801
[14] Lundström E *et al.* 2006 *Phys. Rev. Lett.* **96** 083602
[15] King B and Keitel C H 2012 *New J. Phys.* **14** 103002
[16] Zavattini G *et al.* 2012 *Int. J. Mod. Phys. A* **27** 1260017
[17] Cadène A, Berceau P, Fouché M, Battesti R and Rizzo C 2014 *Eur. Phys. J.* **D68** 16 (*Preprint* `1302.5389`)
[18] Marklund M and Shukla P K 2006 *Rev. Mod. Phys.* **78** 591
[19] Di Piazza A *et al.* 2012 *Rev. Mod. Phys.* **84** 1177–1228
[20] King B and Heinzl T 2015 (*Preprint* `1510.08456`) URL `http://arxiv.org/abs/1510.08456`
[21] King B, Böhl P and Ruhl H 2014 *Phys. Rev. D* **90**(6) 065018
[22] Böhl P, King B and Ruhl H 2015 *Phys. Rev. A* **92** 032115 URL `http://link.aps.org/doi/10.1103/PhysRevA.92.032115`
[23] Hu H and Huang J 2014 *Phys. Rev. A* **90** 062111 URL `http://link.aps.org/doi/10.1103/PhysRevA.90.062111`
[24] Gusynin V P and Shovkovy I A 1996 *Can. J. Phys.* **74** 282
[25] Gusynin V P and Shovkovy I A 1999 *J. Math. Phys.* **40** 5406
[26] Dunne G V and Hall T M 1999 *Phys. Rev. D* **60** 065002
[27] Adler S L 1971 *Ann. Phys.* **67** 599–647
[28] Carver M B 1980 *Journal of Computational Physics* **35** 57–76
[29] Hindmarsh A C, Brown P N, Grant K E, Lee S L, Serban R, Shumaker D E and Woodward C S 2005 *ACM Trans. Math. Softw.* **31** 363–396 ISSN 0098-3500 URL `http://doi.acm.org/10.1145/1089014.1089020`
[30] Wolfram Research I 2012 *Mathematica* version 9.0 ed (Champaign, Illinois: Wolfram Research, Inc.)
[31] Gibbon P 2005 *Short Pulse Laser Interactions with Matter: An Introduction* (Imperial College Press)
[32] Baier R and Breitenlohner P 1967 *Acta Phys. Austriaca* **25** 212
[33] Hunter J D 2007 *Computing In Science and Engineering* **9** 90–95

Threshold Laser Intensity Refinement and Scenarios for Observation of QED Cascade Production

Alexander Fedotov, Nikolay Narozhny and Arseny Mironov
National Research Nuclear University MEPhI (Moscow Engineering Physics Institute), Kashirskoe sh. 31, 115409 Moscow, Russian Federation

E-mail: am_fedotov@mail.ru

Abstract. Previously we have proposed simple qualitative estimates for threshold laser intensity required to observe the laser-induced self-sustained QED (or A-type) cascades. They were later brilliantly confirmed by simulation of cascades arising in a rotating purely electric field. However, in view of numerous simulations performed since then for a more realistic setup our previous criterium of arising of self-sustained cascades may now seem to be too conservative thus essentially overestimating the intensity actually required. After refining our estimates and explaining the origin of assumptions and discrepancies with some new simulation results, we present and discuss in details two particular experimental schemes for their observation. These schemes in different ways overcome a non-trivial problem of injection of seed particles into the focal region and are optimized to lower the threshold intensity value as much as possible.

1. Introduction

Recently, significant attention was payed [1–13] to predictions [14, 15] of emergence of self-sustained electron-positron-photon (QED) cascades in interaction of high-intensity laser fields with matter. The self-sustained (or A-type) cascades in laser field should be distinguished from the ordinary (or S-type) cascades, see Table 1, as they typically grow up exponentially, thus probably dominating in laser plasma dynamics at high intensities. As of now, both the threshold and saturation mechanisms of A-type cascades in the context of laser-matter interactions are still under discussion.

2. Threshold laser intensity refinement

The original estimates [15] for a threshold intensity of A-type cascade generation were based on few assumptions. First, for $E \ll E_S = m^2c^3/e\hbar = 1.3 \cdot 10^{16}\mathrm{V/cm}$ ($I \ll 10^{29}\mathrm{W/cm^2}$) electron motion between the events of hard photon emission can be treated classically. Second, assuming $a_0 \gg 1$, on the one hand, the formation time of the processes of hard photon emission and pair photoproduction $t_{form} \simeq mc/eE \ll 1/\omega$, hence one can use the locally constant field approximation. On the other hand, particles are ultrarelativistic ($\gamma \sim a_0 \gg 1$) and the field in their reference frame looks as crossed field ($E \approx B$, $\vec{E} \perp \vec{B}$). Under such conditions, the probability rates of hard photon emission and pair photoproduction processes are controlled by a single parameter,

$$\chi = \frac{e\hbar}{m^3c^4}\sqrt{-(F_{\mu\nu}p^\nu)^2} \simeq \frac{\gamma\left|\left(\vec{E} + \frac{\vec{v}}{c} \times \vec{B}\right)_\perp\right|}{E_S}, \tag{1}$$

Table 1. Characterization of S(shower)- and A(avalanche)-type cascades

S(shower)-type	A(avalanche)-type
produced by energy ε_0 of seed particle	field donates energy by reaccelerating secondary particles
proceeds until particles loose their energy ($N_{e^+e^-}^{(max)} \propto \varepsilon_0$)	multiplicity is growing exponentially ($N_{e^+e^-}(t) \propto e^{\Gamma t}$) until: • field depletion [3], or • particles escape, or • thermolization of e^+e^- plasma • ...(?)
similar to Extensive Air Showers	similar to gas or dielectric breakdown
almost observed in E144 SLAC experiment	never observed in laboratory

which we call the dynamical quantum parameter. For $\chi \gtrsim 1$ these rates are given by

$$W_\gamma(\chi_e \gtrsim 1) \simeq 1.64 \frac{\alpha m^2 c^4}{\hbar \varepsilon_e} \chi_e^{2/3}, \quad W_{e^+e^-}(\chi_\gamma \gtrsim 1) \simeq 0.23 \frac{\alpha m^2 c^4}{\hbar \varepsilon_\gamma} \chi_\gamma^{2/3}, \tag{2}$$

but for $\chi_\gamma \ll 1$ pair photoproduction process is exponentially supressed $[W_{e^+e^-} \simeq \mathcal{O}(e^{-8/3\chi_\gamma})]$.

Analysis of classical equation of motion $\dot{\vec{p}} = \vec{F}_L$ shows that if the particle is initially at rest and the field direction is time-varied, then the angle ϑ between the particle momentum $\vec{p}$ and the Lorentz force $\vec{F}_L$ is typically growing initially, $\vartheta(t) \simeq \omega t$. This conjecture was proved for a model of uniformly rotating electric field [1] and the general proof is sketched in [16]. This means that the particle energy and the dynamical quantum parameter are initially growing as

$$\varepsilon(t) \simeq eEct, \quad \chi(t) \simeq \frac{E_\perp}{E_S}\gamma \sim \frac{E\vartheta}{E_S}\gamma \simeq \frac{E\omega t}{E_S} \cdot \frac{eEt}{mc} \sim \left(\frac{E}{E_S}\right)^2 \frac{\omega}{\hbar/mc^2} t^2. \tag{3}$$

The latter approaches unity $[\chi(t_{acc}) \sim 1]$ during the time interval

$$t_{acc} \simeq \frac{1}{\omega} \cdot \frac{\alpha E_S}{E} \cdot \sqrt{\frac{\hbar\omega}{\alpha^2 mc^2}} \simeq \frac{1}{\omega} \cdot \frac{\alpha E_S}{E}, \tag{4}$$

where, for simplicity, we have used the fact that for optical lasers ($\hbar\omega \sim 1\text{eV}$) the square root is accidentally of the order of unity[1]. In what follows, we call (4) the acceleration time.

Since the rates (2) differ only by numerical factor, the free path time t_{free} of electrons, positrons and hard photons with respect to the processes of hard photon emission and pair photoproduction, respectively, can be derived by equating $\frac{1}{t_{free}} \sim W(t_{free}) \sim \frac{\alpha m^2 c^4}{\hbar \varepsilon(t_{free})} \chi^{2/3}(t_{free})$, where $\varepsilon(t)$ and $\chi(t)$ are given by (3). In this way we obtain

$$t_{free} \simeq \frac{1}{\omega} \left(\frac{\alpha E_S}{E}\right)^{1/4} \sqrt{\frac{\hbar\omega}{\alpha^2 mc^2}} \simeq \frac{1}{\omega} \left(\frac{\alpha E_S}{E}\right)^{1/4}. \tag{5}$$

[1] In fact, $Ry = \alpha^2 mc^2$ is nothing but the typical atomic energy scale for a hydrogen (Rydberg).

Since all the particles are ulrarelativistic on average, the time of their escape from the focus in case of extremely tight focusing ($w_0 \sim 1/\omega$) can be estimated as

$$t_{esc} \simeq \frac{1}{\omega}. \tag{6}$$

Hence, by comparing Eqs. (4), (5) and (6) we conclude that the condition

$$E \gtrsim \alpha E_S \quad (\text{or} \quad a_0 \gtrsim 3700, \quad I \gtrsim 5 \cdot 10^{25} \text{W/cm}^2), \tag{7}$$

should define the threshold intensity for the onset of cascade production. Indeed, under the condition (7) two inequalities

$$t_{acc} \lesssim t_{free} \lesssim t_{esc}$$

are respected simultaneously. The first one ensures that namely hard photons, which are capable for further pair production, are mainly emitted. The second one guarantees that the number of generations of secondary particles arising from each of the seed particles, is larger than unity ($N_{e^+e^-} \simeq e^{t_{esc}/t_{free}} \gtrsim 1$), i.e. that the process is a cascade in nature. Under these condition the average quantities characterizing the population of secondary particles, are given by

$$\frac{\Gamma}{\omega} \simeq \left(\frac{E}{\alpha E_S}\right)^{1/4} \gg 1, \quad \chi \simeq \left(\frac{E}{\alpha E_S}\right)^{3/2} \gg 1, \quad \varepsilon \simeq \frac{mc^2}{\alpha}\left(\frac{E}{\alpha E_S}\right)^{3/4}, \quad \vartheta \simeq \left(\frac{\alpha E_S}{E}\right)^{1/4} \ll 1. \tag{8}$$

The scalings (8) were brilliantly approved by simulations in a uniformly rotating electric field model [1]. However, most of later simulations adopting more realistic setup reported observation of cascade production at somewhat lower intensity level ($10^{23 \div 24}$W/cm^2, i.e. $1 \div 2$ orders of magnitude lower than our prediction (7)). We emphasize that determination and understanding of an actual intensity threshold value for A-type cascade production is of ultimate importance for such projects as ELI or XCELS.

Here, let us point out several possible reasons that could explain why the criterium (7) may in fact overestimate the actual threshold (as observed in realistic simulations):

(i) First of all, according to practice, different independently designed codes typically show (at least, slightly) different results. This may originate in either different account for soft part of electron radiation, or implementation of different event generators, or other issues (including possible bugs). Hence, we call the community to work out together some minimal set of tests to benchmark the codes;

(ii) An obvious origin of possible discrepancy comes from using the naive assumption (6) about the time of residence of particles in the strong field region. First, in most realistic simulations presented by now focusing was not as tight (typically, $w_0 \simeq 3 \div 10\lambda$). Second, the escape time may be essentially larger because of complicated motion of electrons in a standing-wave field, as well as due to possible radiation trapping effect [17]. Obviously, increasing of t_{esc} due to any reason would be always favorable for cascade production.

(iii) Finally, let us note that any estimate of average Γ would unavoidably overestimate the threshold due to purely statistical reasons. First, we always have $\langle N \rangle \simeq \langle e^{\Gamma t} \rangle \geq e^{\langle \Gamma \rangle t}$ and moreover, this inequality becomes stronger the larger is computation time t. Secondly, the threshold value of photon dynamical parameter triggering pair photoproduction is not sharp but fuzzy. We essentially assumed in estimation that pairs can be only created when $\chi_\gamma \gtrsim 1$, however in simulations it was observed that photons with even $\chi_\gamma \simeq 0.1$ were capable for pair creation.

We think that these observations and refinements could explain an apparent contradiction between our early estimates [15] and some recent simulations.

3. Possible experimental scenarios

We assumed above that the initial seed particle is already located somewhere near the center of the focal region. However, it turns out that in real life injection of seed particles into the strong field region is by no means a trivial task. In a typical setup the field is turned on gradually. Under such conditions the electrons of the target usually have well enough time to be expelled by ponderomotive force from the region where the field is increasing long before it achieves its peak values [18].

On the other hand, injection of seed charged particles (electrons) into the focus from outside may be prevented by radiation reaction (Pomeranchuk theorem, [19]). Indeed, assuming a high-energy particle is approaching the strong field region so that radiation reaction force dominates over the Lorentz force in the Landau-Lifshitz equation, we have

$$m\frac{d\gamma}{dt} \approx -\frac{2}{3}\frac{e^4}{m^2}\underbrace{\left[(\vec{E}+\vec{v}\times\vec{H})^2-(\vec{v}\cdot\vec{E})^2\right]\Big|_{\vec{r}=\vec{\rho}+\vec{v}t}}_{F_\perp^2(t)}\gamma^2,$$

so that

$$-\int\limits_{\gamma_i}^{\gamma_f}\frac{d\gamma}{\gamma^2}=\frac{1}{\gamma_f}-\frac{1}{\gamma_i}=\frac{2}{3}\frac{e^4}{m^3}\int\limits_{t_i}^{t_f}F_\perp^2(t)\,dt,$$

hence

$$\gamma_f^{(max)}=\frac{3}{2}\frac{m^3}{\alpha^2\int\limits_{-\infty}^{0}F_\perp^2(t)\,dt}\sim\frac{3}{2}\frac{m^3}{e^4\frac{1}{2}\left(\frac{m\omega a_0}{e}\right)^2R}=\frac{3m}{e^2a_0^2\omega^2R},$$

meaning that its γ-factor in the strong field region is bounded exclusively by the parameters of the field (namely, the field strength and typical size R of the strong field region) independently on its initial energy. Such a particle will never be able to overcome ponderomotive repulsion of the strong field region if

$$\gamma_f^{(max)}\lesssim a_0,\quad\text{or}\quad a_0\gtrsim\left(\frac{3m}{e^2\omega^2R}\right)^{1/3}. \tag{9}$$

This conclusion was confirmed by numerical simulations [17, 20] and remains qualitatively true when radiation reaction is treated as quantum (even though Pomeranchuk theorem is no more valid literally in such a case).

3.1. *Seed particles injection scenario*

Still, hard γ-quanta emitted by such high-energy particles approaching the focus can penetrate inside and initiate cascades. In our initial estimations, we assumed that A-type cascade is seeded by a charged particle, which is initially slow or at rest in the strong field region. In the scenario now under consideration, secondary hard γ-quanta approaching the strong field region typically initiate an S-type cascade. However, if the field is strong enough to support development of A-type cascades (whatever the actual threshold value is) and, in addition, the deplition time of the S-cascade

$$t_S\simeq t_{free}\cdot n\simeq\frac{\hbar\varepsilon_i}{\alpha m^2c^4}\chi_i^{-2/3}\cdot\log_2\left(\frac{\chi_i}{\chi_{min}}\right)$$

(where n is the number of generations and $\chi_{min}\simeq 0.1$ is the effective threshold for pair photoproduction, see above) is tuned so that the S-type cascade decays due to energy losses exactly when the secondary particles are about to approach the focal region,

$$t_S\sim\tau_L/2, \tag{10}$$

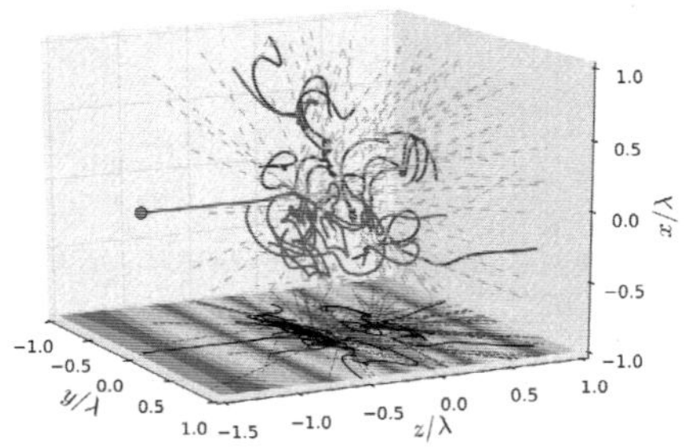

Figure 1. Visual tracks of electrons (green), positrons (blue) and hard photons (dashed dark yellow) in the cascade (seed electron is marked by a red circle) initiated by 3Gev electron in the field of two counterpropagating circularly polarized laser pulses. of strength $E_0 = 3.2{\cdot}10^{-2}E_S$ ($a_0 = 1600$, $I \simeq 5 \cdot 10^{24}\mathrm{W/cm^2}$). On bottom, distribution of electric field is shown.

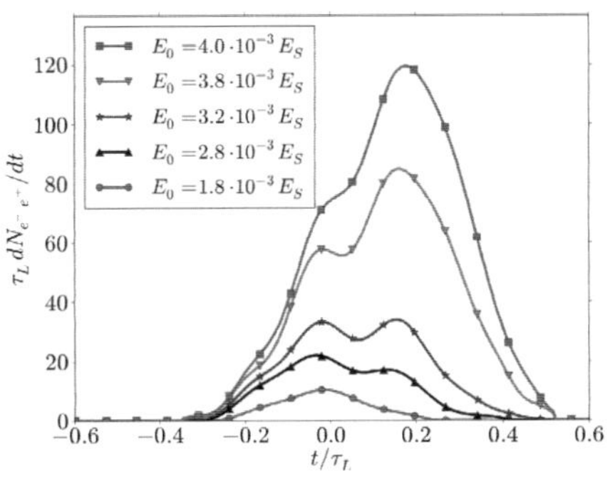

Figure 2. Positron production rate for a setup with two counterpropagating circularly polarized 10fs laser pulses. Cascades are seeded by a 3GeV copropagating electron beam (as in Fig. 1) [21].

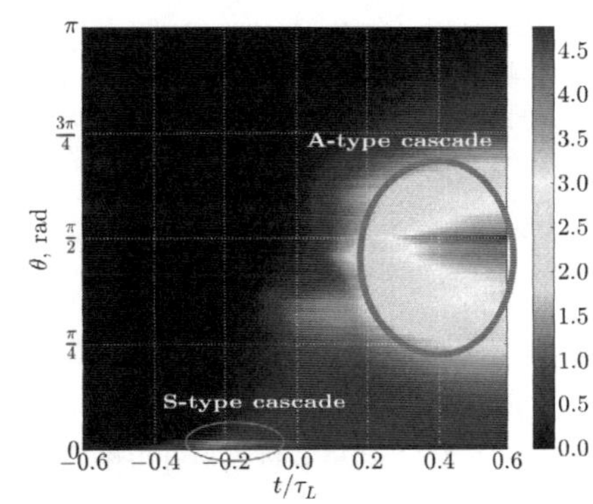

Figure 3. Typical time scanned angular distribution of hard γ-quanta in simulation (parameters are the same as in Fig. 2).

then these secondary particles penetrating the focal region can later trigger an A-type cascade as well [21].

Simulation results for such a setup are demonstrated in Figs. 1–3. Emergence of both (S- and later A-) types of cascades can be observed from either the distinctive two-hump profile of time dependence of the pair creation rate (Fig. 2), or from time scanned angular distribution of emitted hard photons (Fig. 3), the latter as they are all emitted only in forward direction, and as A-type cascade is formed by relatively slow particles, accelerating and radiating mostly in perpendicular plane. As the field is increased, the second (right) peak in Fig. 2, corresponding to an A-type cascde, is growing. Interestingly, the outgoing secondary particles of the arising A-type cascade are propagating almost along the nodes of the electric field (Fig. 1).

3.2. Multibeam scenario

An alternative scenario is to compress the seed by multiple coherent colliding beams, preventing it from fast escape outwards. The idea is similar to the one already used to enhance spontaneous pair production in [22]. As in [22], the additional benefit of such setup arises because the field strength at focal center is essentially increased due to constructive interference. However, unlike [22], here we start from differently elliptically polarized individual beams and search for their optimal polarizations. Intuitively, the beams should be focused as tightly as possible (this

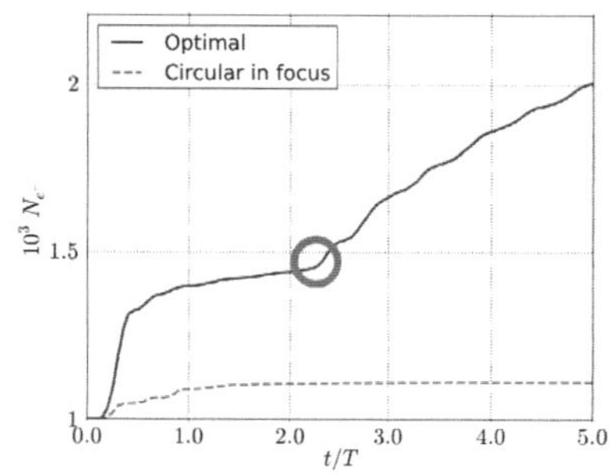

Figure 4. Number of electrons in a cascade vs time (in laser periods) for optimal (blue solid) and circular (red dashed) polarization of the total field (8 beam plane setup, $P = 7.9$PW, adopted from [16]).

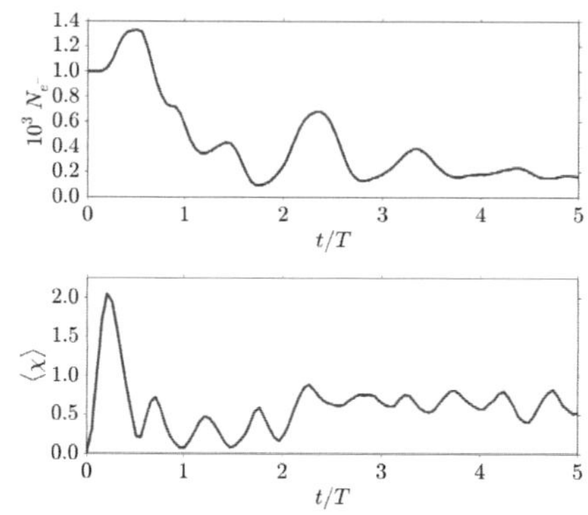

Figure 5. Number of electrons located at a distance smaller than wavelength from the origin (top) and their mean value of parameter χ (bottom). [16].

should also minimize their required number, which is obviously also preferable experimentally). In case of focusing aperture 0.2π (which is today's state-of-the-art) the number of colliding beams is limited by avoidance of mutual peripheral intersections to 8 in plane geometry and to 16 in a spatial one.

Optimization with respect to polarizations of individual beams, targeting at minimizing t_{acc} as much as possible, reveals the non-trivial results [16]. In the optimal case all the beams should be polarized individually in such a way that the total field is polarized elliptically, with $\sqrt{2}:1$ axes ratio. This result was tested by Monte Carlo simulations based on a particular 3D model of Gaussian beams. Growth of the number of pairs in a 8-beam plane setup is compared for the optimal and circular polarizations of the beams of the same net power in Fig. 4. It is evident that optimal polarization is indeed superior. An interesting feature is a bend (marked by a red circle) of the optimal curve at $t \approx 2.2T$. At exactly the same moment the mean value of the parameter χ of those particles which are near the focus turns up, see Fig. 5. This shows that particle dynamics is complicated and is yet far from clear. The threshold power and intensity found in this way are $P_{th} \simeq 7.9$PW ($I_{th} \simeq 5.6\cdot10^{23}$W/cm^2) for the (optimal) 8-beam plane setup and $P_{th} \simeq 6$PW ($I_{th} \simeq 4\cdot10^{23}$W/cm^2) for 16-beam spatial setup. Because several assumptions were made prior to optimization, it would be probably possible to decrease these threshold values even further.

4. Conclusion

Two types (S- vs. A-) of QED cascades in laser filed should be distinguished. Qualitative theory [15] predicted threshold laser intensity level $\simeq 5\cdot10^{25}$W/cm^2 for A-type cascades generation. This theory was reviewed and several reasons for simulations-based refinement of the threshold have been discussed. In spite of 'radiative impenetrability' of the strong field region (laser focus), A-type cascades can be still seeded by secondary hard photons, demonstrating cascade 'collapse and revival' effect at $I \simeq 5\cdot10^{24}$W/cm^2. With a more sophisticated (coherent) multibeam setup it is even possible to initiate cascades at the intensity level $5\cdot10^{23}$W/cm^2.

Acknowledgments

The work was supported by the Russian Foundation for Basic Research and RF President Program for Support of Leading Research Schools. A.A.M. acknowledges support from the Dynasty Foundation.

References

[1] Elkina N, Fedotov A, Kostyukov I Y, Legkov M, Narozhny N, Nerush E and Ruhl H 2011 *Phys. Rev. ST Accel. Beams* **14** 054401
[2] Kirk J G, Bell A and Arka I 2009 *Plasma Phys. Contr. Fusion* **51** 085008
[3] Nerush E, Kostyukov I Y, Fedotov A, Narozhny N, Elkina N and Ruhl H 2011 *Phys. Rev. Lett.* **106** 035001
[4] Duclous R, Kirk J G and Bell A 2011 *Plasma Phys. Contr. Fusion* **53** 015009
[5] Ridgers C, Brady C S, Duclous R, Kirk J, Bennett K, Arber T, Robinson A and Bell A 2012 *Phys. Rev. Lett.* **108** 165006
[6] Kirk J, Bell A and Ridgers C 2013 *Plasma Phys. Contr. Fusion* **55** 095016
[7] Lobet M, d'Humières E, Grech M, Ruyer C, Davoine X and Gremillet L 2013 *arXiv:1311.1107*
[8] Bashmakov V, Nerush E, Kostyukov I Y, Fedotov A and Narozhny N 2014 *Phys. Plasmas* **21** 013105
[9] Brady C, Ridgers C, Arber T and Bell A 2014 *Phys. Plasmas* **21** 033108
[10] Gonoskov A, Bastrakov S, Efimenko E, Ilderton A, Marklund M, Meyerov I, Muraviev A, Sergeev A, Surmin I and Wallin E 2015 *Phys. Rev.* E **92** 023305
[11] Jirka M, Klimo O, Bulanov S, Esirkepov T Z, Gelfer E, Bulanov S, Weber S and Korn G 2015 *arXiv:1511.04982*
[12] Grismayer T, Vranic M, Martins J L, Fonseca R and Silva L O 2015 *arXiv:1511.07503*
[13] Chang H, Qiao B, Xu Z, Xu X, Zhou C, Yan X, Wu S, Borghesi M, Zepf M and He X 2015 *Phys. Rev.* E **92** 053107
[14] Bell A and Kirk J G 2008 *Phys. Rev. Lett.* **101** 200403
[15] Fedotov A, Narozhny N, Mourou G and Korn G 2010 *Phys. Rev. Lett.* **105** 080402
[16] Gelfer E G, Mironov A A, Fedotov A M, Bashmakov V F, Nerush E N, Kostyukov I Y and Narozhny N B 2015 *Phys. Rev.* A **92**(2) 022113
[17] Fedotov A, Elkina N, Gelfer E, Narozhny N and Ruhl H 2014 *arXiv:1408.0362*
[18] Tamburini M, Di Piazza A and Keitel C H 2015 *arXiv:1511.03987*
[19] Pomeranchuk I Y 1940 *J. Phys. USSR* **2** 65
[20] Elkina N, Fedotov A, Herzing C and Ruhl H 2014 *Phys. Rev.* E **89** 053315
[21] Mironov A, Narozhny N and Fedotov A 2014 *Phys. Lett.* A **378** 3254–3257
[22] Bulanov S, Mur V, Narozhny N, Nees J and Popov V 2010 *Phys. Rev. Lett.* **104** 220404